A TO Z
OF
EARTH SCIENTISTS

NOTABLE SCIENTISTS

A TO Z
OF
EARTH SCIENTISTS

ALEXANDER E. GATES

Facts On File, Inc.

A TO Z OF EARTH SCIENTISTS

Notable Scientists

Facts On File, Inc.
132 West 31st Street
New York NY 10001

Library of Congress Cataloging-in-Publication Data

Gates, Alexander E., 1957–
    A to Z of earth scientists / Alexander E. Gates.
        p. cm. — (Notable scientists)
    Includes bibliographical reference and index.
    ISBN 0-8160-4580-1 (hardcover)
    1. Earth scientists—Biography. I. Title. II. Series.
QE21.G38 2002
551′.092′2′—dc21   2002014616

Facts On File books are available at special discounts when purchased in bulk quantities for businesses, associations, institutions, or sales promotions. Please call our Special Sales Department in New York at 212/967-8800 or 800/322-8755.

You can find Facts On File on the World Wide Web at http://www.factsonfile.com

Text design by Joan M. Toro
Cover design by Cathy Rincon
Chronology by Dale Williams

Printed in the United States of America

VB FOF 10 9 8 7 6 5 4 3 2 1

This book is printed on acid-free paper.

CONTENTS

LIST OF ENTRIES

PREFACE

HOW THE EARTH SCIENTISTS WERE CHOSEN

A to Z of Earth Scientists was originally intended to include only Earth scientists who have made contributions to our understanding of the Earth since World War II, with emphasis on those who are currently active or were recently active. It was intended that there should be a relatively even distribution across the subdisciplines as well as geographically, although it was realized from the outset that there would be more American scientists included. However, many of the society and government agency awards are named in honor of previously active Earth scientists so many of them are included as well.

Research for this book showed that there is very little biographical information available for currently active Earth scientists. There is basic information on employment, awards, and date of birth available for most living individuals in the volumes *American Men and Women of Science.* The American Geological Institute's Bibliography of Geology also contains their publications. The information on their contributions to the Earth sciences, however, is very difficult to obtain and commonly must come from award citations from societies, if available. In many cases, the only information available is that on web sites and even that is usually scant. As a result, information had to be solicited directly from the Earth scientists to be included in the book. In some cases, it took several solicitations to obtain the information and in others, because of a lack of response, the individual could not be included in the book. As a result of the exhaustive amount of effort required in research, unavailability of information, and a major change in the structure of the book during writing, the list of biographies changed radically during the writing of the book and is generally shorter than planned. The choices may almost seem arbitrary and capricious to some readers. The number of biographies could easily be doubled to include those who deserve recognition for their contributions to Earth science. Language problems made it especially difficult to obtain information on scientists outside of the United States. In no way should the final list of Earth scientists included in this book be construed to indicate that these are the only people who made contributions to the field or that their contributions are of greater importance than many other exceptional scientists in the profession. The hope is that this book will be popular enough to warrant a second edition in which many more deserving Earth scientists might be included.

This book is intended for people in high school, early college, and perhaps at a more advanced level of study. There are many technical terms that are briefly explained where possible.

In many cases, they are not explained at all. It is recommended that if the reader has not attended at least an introductory course in Earth science (physical geology) that an Earth science (geologic) dictionary be kept handy.

Acknowledgments

This book would not have been possible without the contributions of many people. I first and foremost wish to thank the many Earth scientists who agreed to be included in the book and who sent biographical data. Many of them reviewed early drafts of their biographies and suggested changes that improved the accuracy of the book. Many also sent photographs that are included with their biographies. A special thanks to Arthur Sylvester who voluntarily opened up his rogues' gallery of photographs to me, many of which appear in the book. James Skehan, S.J., similarly provided me with several photographs of East Coast geologists which I otherwise would not have obtained. Joseph McGregor of the U.S. Geological Survey in Denver was helpful in providing excellent photographs of geologists in a very short time.

In addition to obtaining information directly from the participants, there were several other great sources. The American Geophysical Union provided biographical information directly to me, in addition to that which is available over its web site. Much information was obtained from published material from the Geological Society of America, the Mineralogical Society of America, the Geological Society of London, the Paleontological Society, the Society for Exploration Geophysicists, and the Geochemical Society. The librarians at the Dana Library at Rutgers University in Newark, New Jersey, were extremely helpful and even more patient during the writing of this book. Veronica Calderhead and Ann Watkins were especially helpful, although most of the staff were very cooperative. The search engine google was used extensively in the locating of biographical material.

The progress in the writing of this book was memorable to say the least. Frank K. Darmstadt, senior editor at Facts On File, Inc., was extremely patient and understanding. The project could easily have collapsed if not for his willingness to adjust as unexpected situations arose. My agent, Max Gartenberg, deserves recognition for his patience, as well. If it were not for Gayle Martinko's urging to undertake the project, and agreeing to serve as coauthor at the outset, the book would not have been completed. Finally, I wish to thank my son, Colin Gates, who pitched in and compiled all of the information for the appendices when I ran into time problems. Thanks also to Maxine, Jasper, and Tom for putting up with their dad always having his nose in the computer.

INTRODUCTION

In the 19th century, Earth science was king; it was deemed the most attractive and of highest potential of all sciences by the *New York Herald* and *Knickerbocker Magazine* in the 1830s. One-fourth of all scientists in the United States between 1800 and 1860 were Earth scientists, and they charted the direction of American science. JAMES DANA from Yale University was their leader and considered on a par with Charles Darwin in terms of respect and prestige. He virtually controlled the American Association for the Advancement of Science, which was the premier scientific society of the time. Even at the end of the century the newly established U.S. Geological Survey was headed by some of the most influential scientists in the country, like John Wesley Powell, G. KARL GILBERT, and CHARLES D. WALCOTT. The Geological Society of America, the premier society of the Earth sciences, was founded in 1888.

The 20th century saw the rise of the other sciences and decline of the Earth sciences in comparison. They were influential at the beginning of the century when the quest for oil by an adventurous group called wildcatters brought great fame and fortune. The Rockefeller fortune was built in this industry with the establishment of the mammoth Standard Oil Company. Even President Herbert Hoover was an Earth scientist. In World War I, the famous Earth scientist ARTHUR L. DAY averted a major American crisis of a critical short-

age of optical glass for the war effort through an ingenious and concerted effort. This work, however, was only marginally related to Earth science research.

Earth scientists played prominent roles in the World War II effort, but generally not for their specialties. DAVID T. GRIGGS developed methods to conduct aerial bombing missions using radar guidance. Before that, bombing was done by purely visual methods. A. FRANCIS BIRCH was the lead scientist on the Hiroshima bomb (Little Boy) team and even helped load the bomb onto the *Enola Gay.* HARRY H. HESS and SIR EDWARD C. BULLARD developed methods to virtually eliminate the threat of both mines and submarines to ships. Although these and many other Earth scientists were instrumental in the war effort and many observations were made that would later help with interpretations, few direct breakthroughs in the science were realized as they were in other sciences like physics and medicine.

In the late 1950s through the 1960s, Earth sciences again drew public attention with the documentation of the plate tectonic theory. The idea that the solid Earth below our feet was actually moving boggled the imagination. Again, some of the most influential scientists in the world were Earth scientists like Harry Hess, J. TUZO WILSON, and ROGER REVELLE, among others, and they were involved in this revolution. This powerful concept has been referred to as the "glue that holds geology together" because it resulted in the rapidly radiating subdisciplines of the Earth sciences being drawn together. Plate tectonics was the mechanism by which they could be interrelated. It still drives much of the solid Earth research that is done some 40 years later.

The cold war that was in full swing at the time of the plate tectonic revolution also involved Earth scientists. In order to monitor nuclear tests, a worldwide seismic monitoring network was established. Earth scientists like FRANK PRESS, INGE LEHMANN, and LYNN R. SYKES became as much diplomats as they were scientists, serving on numerous top-level international advisory boards. They were even involved in treaty negotiations. Other Earth scientists like ARDEN L. ALBEE, DON U. WISE, CARL E. SAGAN, and WILLIAM R. MUEHLBERGER got involved in the space race. They served in high-level capacities to ensure the success of NASA's Apollo lunar missions.

Overlapping this exciting period of revolution in the Earth sciences came a societal crisis that only Earth scientists could solve. In 1973, the Arab oil embargo sent the country into a crisis that would dominate the next decade. Record numbers of students were entering the Earth sciences in colleges, and once again the Earth sciences topped the lists of the best career choices in popular magazines like *Time* and *Newsweek*. Earth scientists could write their own ticket to outstanding careers in the petroleum industry. This interest and funding led to numerous new developments and Earth scientists discovered enough new petroleum reserves to rescue the country from the crisis. Earth scientists like ALBERT W. BALLY and GERALD M. FRIEDMAN figured prominently in these efforts.

In the early 1980s, however, energy supplies became abundant and with only a few perturbations have remained so ever since. Exploration for petroleum reserves declined dramatically and the Earth sciences went into the doldrums. Although vital contributions were made in the areas of environmental science by scientists such as CRAIG M. BETHKE and SUSAN L. BRANTLEY and climate change research by WALLACE S. BROECKER, JOHN IMBRIE, and SIR NICHOLAS J. SHACKELTON, the image of Earth science declined and has remained behind the scenes. Fields such as biotechnology, particle physics, and material sciences have taken center stage. It is for this reason that this book was written at this time.

THE ENTRIES

Entries in *A to Z of Earth Scientists* are arranged alphabetically by surname, with each entry given under the name by which the Earth scientist is most commonly known. The typical entry provides the following information:

Entry Head: Name, birth/death dates, nationality, and field(s) of specialization.

Essay: ranging in length from 750 to 1,500 words, with most averaging around 1,000 words. Each contains basic biographical information—date and place of birth, family information, educational background, positions held, prizes awarded, etc.—but the greatest attention is given to the scientist's work. Names in small capital letters within the essays provide easy reference to other scientists represented in the book.

In addition to the alphabetical list of scientists, readers searching for names of individuals from specific countries can consult the Country of Birth appendix. The Country of Major Scientific Activity appendix lists scientists by the countries in which they conducted their work and research. The Field appendix cites them by the area of Earth science in which they were most notable. The Index lists page references for scientists and scientific terms used in the book. Finally, the Chronology lists entrants by their birth and death dates.

Indeed, Earth science is largely responsible for sparking the scientific revolution of the 20th century. In addition, no matter how many marvelous technological breakthroughs we have made, this Earth is still the only place we have found thus far that is able to sustain life. We must therefore keep its importance in proper perspective in this fast-paced world. Earth scientists are still making important contributions to society, far beyond what is imagined by the general public. If this book can in some small way help to bring recognition to these Earth scientists, then it will be deemed a success.

Albee, Arden L.
(1928–)
American
Geochemist, Metamorphic Petrologist

When Earth scientists write and publish a research paper, they hope it is successful. The definition of success varies by the individual, but if at least 200 people read the paper and 50 or more cite it in other research papers, then most would consider the paper to be a success. In 1968, Arden L. Albee and his student A. E. Bence published the paper, "Empirical Correction Factors for the Electron Microanalysis of Silicates and Oxides," whose methods are still employed by geologists an average of 300 times per day. In the late 1960s, new analytical procedures allowed scientists to quantitatively analyze the chemistry of individual minerals. The electron microprobe bombards individual mineral grains with a focused stream of high-energy electrons. The individual atoms in the minerals give off X rays upon impact, which are then received by detectors. The data that the detectors supply is then converted into weight percent of an oxide of the element and then into an exact mineral formula. Chemical reactions can be precisely determined with these data in contrast to the purely qualitative chemical reactions that preceded this technique. Bence and Albee devised the correction factors needed to convert counts on an X-ray detector into oxides and minerals. Those corrections are programmed into likely every single electron microprobe in the world. Electron microprobes are used on a daily (and nightly) basis at most universities that operate them. That number includes essentially all of the large universities in the world. Albee supervised the electron microprobe facility at California Institute of Technology.

Arden Albee's interest in the electron microprobe is as a tool for his research on regional metamorphism. While with the U.S. Geological Survey, Albee performed regional geologic mapping in Vermont, Colorado, and Maine. After leaving the USGS, he continued his work in northern Vermont, west Greenland and the Death Valley area of California. The goal of his research is to understand the conditions under which these metamorphic rocks formed. To accomplish this goal, he analyzed the partitioning of elements among minerals as well as with theoretical thermodynamics.

Albee has a second research career studying extraterrestrial rocks. He was an investigator of the Apollo lunar samples for many years. As a result, he became chief scientist for NASA's Jet Propulsion Laboratory from 1978 to 1984, which is operated by the California Institute of Technology. He was project scientist for the Mars Observer Mission that was launched in September, 1992, but with which contact was lost in August, 1993. He is still mission scientist for NASA's Mars Global Surveyor Mission. Albee's role in this work is not only to help plan the scientific objectives of the mission but also to design and implement instrumentation. His paper, "Development of a Miniature Scanning Electron Microscope for In-Flight Analysis of Comet Dust," in 1983, is an example of such instrumental work. He directed the design of the equipment that analyzes the rocks in situ on Mars including an onboard scanning electron microscope. He is also involved in developing the remote sensing equipment that is used to map the surface of Mars from the spacecraft. Albee is a member of the U.S.–Russian Joint Working Group on Solar System Exploration that governs the scientific cooperation on joint missions including the International Space Station.

Arden L. Albee was born in Port Huron, Michigan, on May 28, 1928. He spent his childhood in Michigan. He received his undergraduate and graduate education at Harvard University, where he earned his bachelor of arts, master of science, and doctor of philosophy degrees in geology in 1950, 1951, and 1957, respectively. Albee worked for the U.S. Geological Survey as a field geologist-petrologist during his graduate studies and until he joined the faculty at California Institute of Technology, where he remains today. He served as the dean of Graduate Studies from 1984 to 2000. Albee is married, has eight children and 13 grandchildren, and lives in Altadena, California.

Albee has been very active professionally, producing numerous papers in international journals, professional volumes, and governmental reports. He is an author of some of the most important papers in the field of metamorphic petrology, analytical techniques, and space exploration. He has also been of great service to the profession. He served on numerous advisory committees and project review boards for NASA. He also served as chair for a number of working groups on Martian missions. He is the recipient of the NASA Medal for Exceptional Scientific

Achievement for this service to space exploration. Albee has been an officer and/or editor for a number of professional societies and organizations, including the Geological Society of America, Mineralogical Society of America, and the American Geophysical Union. He has served as associate editor for the *Annual Reviews of Earth and Planetary Sciences* since 1979.

⊠ **Allègre, Claude**
(1937–)
French
Geochemist

After establishing an outstanding career in the Earth sciences, Claude Allègre became one of the few scientists to participate successfully in governmental policy. Claude Allègre is the architect of the subdiscipline of isotope geodynamics. This area involves the study of the coupled evolution of the mantle and continental crust of Earth through a multi-isotopic tracer approach. These radiogenic isotopes include such systems as strontium, neodymium (and samarium), lead, xenon, argon, helium, osmium (and rhenium), and thorium. The studies provide evidence for very early degassing of volatile elements and compounds from Earth with limited subsequent mixing between the upper mantle and the lower mantle. They also show that the atmosphere was primarily formed early in the history of the Earth with only volumetrically small additions since. Helium and neon were trapped in the Earth's interior and have been escaping at a slow rate ever since. He geochemically modeled the early solar system, the early evolution of planets and the formation of meteorites in his paper, "Cosmochemistry and the Primitive Evolution of Planets," among others. This cosmochemical research is the reason that Allègre was chosen by NASA to participate in the Apollo lunar program. In that role, he was among the first scientists to determine the age of the Moon.

With regard to the visible part of the Earth, his paper "Growth of the Continents through Time," in which he uses isotopic evidence to address the topic, exemplifies Claude Allègre's research. Once isotopes are removed from the open whole-Earth system into a closed continental system, they evolve separately. Allègre's best-known research is on the Himalayan mountains, both in terms of structural and geochemical evolution of the Asian crust. However, in considering the isotopic systematics produced by erosion of the continental crust, he also looked at Africa and South America. The other main area of research that Allègre has undertaken is to apply his considerable physics and mathematical background to scaling laws of fractures, earthquakes, geochemical distributions and energy balance which mathematically relates sizes to distributions.

Claude Allègre is probably best known by the public for his extensive governmental work. He is currently the minister for National Education, Research and Technology for France, where he has created quite a controversy by attempting to overhaul the public educational system. Certainly, it takes plenty of prior policy work to be appointed to such an important position. Allègre served as a member of the Socialist Party Executive Bureau, a National Delegate for Research, and a special adviser to the first secretary of the Socialist Party. He was a member of the European Parliament, a city councilman of Lodeve, and a member of the Languedoc Roussillion Regional Council.

Claude Allègre was born on March 5, 1937, in Paris, France. He attended the University of Paris, where he studied physics under Yves Rocard as well as geology, earning a Ph.D. in physics in 1962. He was an assistant lecturer in physics at the University of Paris from 1962 to 1968 before accepting a position as assistant physicist with the Paris Institut de Physique du Globe. He has been the director of the geochemistry and cosmochemistry program at CNRS (French National Scientific Research Center) since 1967. Allègre joined the faculty at the University of Paris VII in 1970, a position he retains. In 1971, he was appointed as director of the Department of Earth Sciences, a position he held until 1976. He was then named the director of the Paris Institut de Physique du Globe from 1976 to 1986. In 1993, he was named as a member of the Institut Universitaire de France (Denis Diderot University). Allègre was recently granted a leave from his academic position to serve as minister for National Education, Research and Technology for France. Over the years, Allègre has held several visiting scientist positions on an international basis. He was a White professor at Cornell University, New York, a Crosby Professor at Massachusetts Institute of Technology, and a Fairchild Professor at the California Institute of Technology, in addition to positions at the U.S. Geological Survey, Denver; the Carnegie Institution of Washington, D.C.; University of California at Berkeley; and at Oxford University, England. Claude Allègre is married with four children.

Claude Allègre is an author of more than 100 scientific articles in both English and French. Many of these papers are seminal studies on the evolution of the Earth, especially using isotopic evidence. They appear in respected international journals. He has also written 11 books spanning the range from widely adopted textbooks to science and policy topics, even to popular books. In recognition of his scientific contributions, Allègre has received numerous honors and awards. He is a member of the U.S. National Academy of Sciences, the American Academy of Arts and Sciences, and the French Academy of Science, as well as an officer of the Legion of Honor. He received the Crafoord Prize from the Swedish Royal Academy of Science, the Goldschmidt Medal from the Geochemical Society (U.S.), the Wollaston Medal from the Geological Society of London, the Arthur L. Day Medal from the Geological Society of America, the Gold Medal from CNRS (French National Scientific Research Center), the Arthur Holmes Medal from the European Union of Geoscience, and the Bowie Medal of the American Geophysical Union.

⊠ Alley, Richard B.
(1957–)
American
Glaciologist, Climate Modeler (Climate Change)

Because the polar ice caps are located in an area where temperatures are constantly below freezing, all precipitation that occurs on them must be frozen. Therefore, it is preserved for as long as the ice sheets remain. Because there is precipitation every year, and the precipitation traps a little bit of the atmospheric gas that it passes through, ice sheets contain a continuously preserved record of the Earth's atmosphere for up to hundreds of thousands of years. Richard Alley analyzes deep cores taken from the continental ice sheets in Greenland and Antarctica, conducting physical studies to complement isotopic, chemical, and other measurements made by collaborators. The combined data provide detailed information on climatic conditions in the past. He showed that accumulation rates for ice sheets are extremely variable depending upon whether the conditions were glacial or interglacial (between ice ages). Surprisingly, almost half of the glacial-interglacial change was achieved in a few years. This discovery means that climate changes are not slow as was previously envisioned by Earth scientists but instead can be alarmingly rapid. Some of these abrupt climate changes are linked to great surges of the ice sheets in the great ice ages. They left evidence in sedimentary deposits around the North Atlantic. Alley's findings about the mechanisms that caused these surges have led him to the idea that surges of the West Antarctica ice sheet are possible in the future. This research is summarized in the 2002 book *The Two-Mile Time Machine: Ice Cores, Abrupt Climate Change and Our Future.*

Richard B. Alley, in full field gear, standing in front of a ski-equipped LC-130 Hercules aircraft of the 109th U.S. Air National Guard in Sondrestrom, Greenland *(Courtesy of Richard Alley)*

In studying the movement of glaciers, Alley found that subglacial sediments with meltwater serve to lubricate the basal contact of the glacier with the ground, allowing it to attain relatively high velocity. This research transformed Alley into one of the foremost authorities on continental glacier mechanics and processes. He also became one of the leading proponents of the view that the radical and abrupt climate changes that occurred during transitions to and from ice ages might have implications for future climate changes. He discovered further supporting evidence for this stand using a newly devised ice-isotopic thermometer. By analyzing the stable isotopes in the ice he could determine paleotemperatures. He calibrated this thermometer using modern ice. The result of his analysis of ice cores is that there were average surface temperature changes of some 20°C (36°F) between ice ages and interglacial periods. It was not expected that these variations would be so drastic. By compiling all of these results of the ice core research with sedimentary and deformational data, Alley formed a new dynamic model for the advance and retreat of continental glaciers. They can no longer be considered as slow-moving static bodies with little to no variation but rather very active bodies with diverse processes and modes of operation.

Richard Alley was born on August 18, 1957, in Ohio. He attended Ohio State University in Columbus, and earned a bachelor of science degree in geology and mineralogy in 1980, summa cum laude and with honors and a master of science degree in geology in 1983. He earned a Ph.D. in geology with a minor in material sciences from the University of Wisconsin at Madison in 1987. He remained at University of Wisconsin for one year as an assistant scientist before joining the faculty at Pennsylvania State University at University Park, where he remains today. Alley was named the Evan Pugh professor of geosciences beginning in 2000. Alley is married and the father of two children.

Richard Alley is leading an extremely productive career. He is an author of more than 120 articles in international journals and professional books and volumes. Many of these papers are benchmark studies of ice mechanics and climate modeling and many appear in the prestigious journals *Nature* and *Science.* They have been cited an astounding 3,500 times to date. He is also an author or editor of four books. His research and teaching contributions have been well recognized by the profession in terms of honors and awards. He is the recipient of the Horton Award from the American Geophysical Union, the D. L. Packard Fellowship, and the Presidential Young Investigator Award. From Pennsylvania State University, he won the Wilson Teaching Award and the Faculty Scholar Medal. He was also invited to give testimony to then-U.S. vice president Al Gore.

Alley has also performed significant service to the profession. He serves or has served on panels and committees for the American Geophysical Union, the National Science Foundation including the Augustine Panel, the International Glaciological Society, the Polar Research Board, and the National Research Council, among others. His work has also attracted the attention of the popular media. The British Broadcasting Corporation and National Public Radio have featured him in special programs.

Alvarez, Walter
(1940–   )
American
Stratigrapher, Tectonics

Walter Alvarez has been the leader of one of the greatest revolutions in geology, extraterrestrial impacts. He decided to address one of the big questions in geology: what caused the great extinction of the dinosaurs? He chose the most complete section of rock that includes the extinction event, which occurred at the Cretaceous-Tertiary (K-T) boundary about 65 million years ago. The area chosen is in the Umbria region of the northern Appenines, Italy. The unit is a reddish limestone called Scaglia Rossa, which has its most complete section at Gubbio. There Alvarez found a 1-cm thick layer of clay right at the boundary. Across this boundary, fossils record a major extinction event of foraminifera (plankton) and other marine life. Walter Alvarez consulted with his father, Nobel Prize–winning physicist Luis Alvarez, at the University of California at Berkeley, where Walter Alvarez had just taken a position. These two collaborated with two other nuclear chemists in an attempt to determine the amount of time that it took to deposit the layer by measuring the amount of the element iridium, assuming a constant flux of this cosmic dust. Much to their surprise, the layer contained an anomalously high concentration of iridium, far greater than what could be deposited by cosmic influx. This work was released in a landmark paper entitled "Extraterrestrial Cause for the Cretaceous-Tertiary Extinction: Experimental Results and Theoretical Interpretation." The team analyzed several other sections at Stevns Klint, Denmark, and in New Zealand, and found that this iridium anomaly was worldwide. The team postulated that the collision of an asteroid or comet of 10-km diameter with the Earth could be the culprit. They proposed that the impact produced a huge crater from which an enormous mass of dust was emitted. The dust settled all over the Earth but not before blocking sunlight for a long period of time. The dust cloud inhibited photosynthesis and thus caused a collapse of the food chain from the ground up.

The evidence amassed and the theory quickly evolved after this initial discovery. Over 75 localities have been identified with the K-T iridium anomaly. Enrichment in platinum, osmium, and gold in roughly chondritic proportions was also found in the layer. At other localities, distinctive microspherules (glass melt) have been found as well as shocked quartz, a texture that can only be produced by extreme pressure. Even a potential impact crater called Chixulub has been identified off of the coast of the Yucatán Peninsula in Mexico. There was a tremendous cooling of the atmosphere as recorded in pollen samples and tremendous loss of species as a result of this event.

Walter Alvarez was born on October 3, 1940, in Berkeley, California, where he spent his youth. He attended Carleton College, Minnesota, where he majored in geology. He graduated in 1962 with a bachelor of arts degree. He attended Princeton University, New Jersey, for graduate studies, where his adviser was HARRY H. HESS. He graduated with his doctoral degree in 1967 with the thesis topic "Geology of the Simarua and Carpintero areas, Guajira Peninsula, Colombia."

He worked in the petroleum industry for several years before joining the Lamont-Doherty Geological Observatory in 1971. He started as a resident scientist (1971–1973) and became a research associate (1973–1977). In 1977, he joined the faculty at University of California at Berkeley where he currently holds the rank of professor.

Walter Alvarez has received numerous awards and honors. He was a Guggenheim Fellow in 1983–1984 and a Fellow of the California Academy of Sciences in 1984. He received the G. K. Gilbert Award of the Geological Society of America in 1985. In 1986, he became an Honorary Foreign Fellow of the European Union of Geosciences as well as a Miller Research Professor at University of California, Berkeley. In 1991, he was elected to the National Academy of Sciences as well as receiving the Rennie Taylor Award of the American Tentative Society for science journalism. He was elected as a foreign member of the Royal Danish Academy of Sciences in 1992 and as a member of the American Academy of Arts and Sciences in 1993. In 1998, he received the Journalism Award of the American Association of Petroleum Geologists.

Walter Alvarez has other interests besides pure scientific research, namely journalism. His book *T. Rex and the Crater of Doom* (Princeton University Press, 1997) is a prime example of his interest in translating the results of research to works that could be appreciated by the general population. In addition to this popular work, he is an author of numerous scientific articles in international journals and professional volumes. Many of these are seminal works of the Earth sciences.

⊠ **Anderson, Don L.**
(1933–   )
American
Geophysicist

The structure of the deeper parts of the Earth cannot be viewed from the surface and therefore

Don Anderson demonstrating geophysical relations at the California Institute of Technology in Pasadena *(Courtesy of Don Anderson)*

must be imaged using geophysical techniques. The best probes of this region are seismic waves. The seismic waves are generated at the earthquake foci, pass through the Earth, and return to the surface where they are recorded by seismographs. The greater the spacing between the earthquake and seismograph, the deeper the waves will probe the Earth. By studying minute changes in travel times of these waves and the relative travel times among waves, composition and even temperature of the deep subsurface can be determined. By performing a 3-D image analysis of these data, the structure of the deep interior of the Earth may be determined similar to how a CAT scan is used on the human body. CAT scanning of the Earth is called seismic tomography, and its main pioneer is Don L. Anderson.

Through seismic tomography, Don Anderson has shown a richly complex structure in the upper mantle and the lower crust with patterns of hot and not-so-hot areas. This work is summarized in his paper "Slabs, Hot Spots, Cratons and Mantle Convection Revealed from Residual Seismic Tomography." As a result of this seismic tomography, Anderson has also proposed a theory that contrasts with previous notions that the internal engine of the Earth is like a pot on the stove with deeply derived heat sources driving convection cells that move the plates and cause earthquakes and volcanoes. Instead, he proposes that the surface features of the Earth may exert significant control on mantle convection and related processes as described in the paper "The Inside of Earth: Deep Earth Science from the Top Down." Plate interactions and geometries may affect how the mantle moves.

These new ideas led to a reconsideration of how the Earth evolved through time both physically and chemically. Indeed, his new models have implications for how any planet evolves through time as described in the paper "A Tale of Two Planets." Anderson developed new models to explain the location of volcanoes based upon crustal stress fields. He uses types and amount of volatiles (gases) in lava to propose that the generation of virtually all magma is very shallow rather than the deep source hypothesis that still prevails with many scientists. Even the classically deep "hot spots" like Hawaii may be from a shallow source. Indeed, Anderson has nearly single-handedly redefined the function and importance of the asthenosphere, lithosphere, and his "perisphere." His top-down (rather than the classic bottom-up) approach to the Earth, both physically and chemically, is revolutionary and establishes Anderson as a true pioneer in the Earth sciences.

Even without all of the fame of his work in seismic tomography, Don Anderson has a distinguished career studying seismology of the Earth. Many of his studies establish new benchmarks in the processes by which seismic waves travel through the Earth and what information can be gleaned from their study.

Don L. Anderson was born on March 5, 1933, in Frederick, Maryland, but he was raised in Baltimore. He always enjoyed science and rock collecting, so when he went to college at Rensselaer Polytechnic Institute, New York, he majored in geology and geophysics and graduated with a bachelor of science degree in 1955. From 1955 to 1956, he worked as a geophysicist for Chevron Oil Co. From 1956 to 1958, he worked as a geophysicist in the U.S. Air Force Cambridge Research Center. In 1962, he earned a Ph.D. at the California Institute of Technology in geophysics and mathematics. From 1962 to 1963, he was a research fellow at the California Institute of Technology before becoming an assistant professor in 1963. He was promoted to associate professor in 1964 and finally to full professor in 1968. From 1967 to 1989, he directed the Seismological Laboratory at the California Institute of Technology

before becoming the Eleanor and John R. McMillan Professor of geophysics in 1989, the position he holds today. During this time, he was a Cox Visiting Scholar at Stanford University, a Green Visiting Scholar at the University of California at San Diego, an H. Burr Steinbach Visiting Scholar at Woods Hole Oceanographic Institution in Massachusetts, and a Tuve Distinguished Visitor at the Carnegie Institution, Washington, D.C.

Don Anderson has had an extremely productive career publishing more than 200 articles in international journals and professional volumes. The list of honors and awards that he has received in recognition of his research contributions is staggering. Foremost among these awards is the National Medal of Science which President Bill Clinton bestowed on him in 1999. He received an honorary doctorate from Rensselaer Polytechnic Institute. In addition, he received the James B. Macelwane Award in 1966 and the Bowie Medal in 1991 both from American Geophysical Union, the Apollo Achievement Award in 1969 and the Distinguished Scientific Achievement Award in 1977 both from NASA, the Arthur L. Day Medal from the Geological Society of America in 1987, the Newcomb-Cleveland Prize from the American Association for the Advancement of Science in 1976, the Emil Wiechert Medal from the German Geophysical Society in 1976, the Gold Medal from the Royal Astronomical Society in 1988, and the Craafoord Prize from the Royal Swedish Academy of Science in 1998. He is a Fellow at the National Academy of Sciences and the American Academy of Arts and Sciences.

Anderson has served on many important committees at the National Academy of Sciences, National Research Council, NASA, National Science Foundation, American Geophysical Union (Fellow and president (1988–1990)), Geological Society of America, American Association for the Advancement of Science, Carnegie Institution of Washington, D.C., and several others. He has served in an editorship capacity for some of the top international journals including *Journal of Geophysical Research, Tectonophysics, Geological Society of America Bulletin,* and *Journal of Geodynamics,* to name a few. He was an evaluator of some of the top geophysical programs worldwide, including Princeton University, Harvard University, University of Chicago, Stanford University, University of California at Berkeley, and University of Paris, among others. Indeed, Don Anderson participated in many of the committees, review panels, and projects that truly shaped the current state of the Earth sciences.

Ashley, Gail Mowry
(1941–)
American
Sedimentologist

Geology is an interdisciplinary science that can be closely engaged in research with other scientific fields like biology, chemistry, physics, meteorology, and oceanography. Gail Ashley has not only collaborated on projects in these fields but also in archaeology and paleoanthropology. Her specialty is modern depositional systems, which impact and interact with modern human dwellings and communities. In fact, in some cases, it is these water sediment systems that attract the human communities. This is the idea of a geological-archaeological project with which she is associated. Ashley is studying a 7-to-8m-thick section of sediments in the Olduvai Gorge, Tanzania, Africa, to determine the type of environment that existed there approximately 2 million years ago when early hominids populated the area. She is studying the ecological link between freshwater springs and these early humans with the idea that springs are more reliable sources of water than rivers and lakes as described in the paper "Archaeological Sediments in Springs and Wetlands." The work will also touch upon the paleoclimate of the region at that time. She and a team of archaeologists and anthropologists are making great new discoveries in the old stomping

grounds of Louis and Mary Leakey, as well as the australopithicine named Zinjanthropus.

The Olduvai research may be Gail Ashley's highest-profile project, but it is one of many of equal importance. She is an expert on glacial geomorphology and glacial marine sedimentation. Her work has taken her to Antarctica, the Brooks Range in Alaska, and Ireland, as well as the northeastern United States. This research has the general theme of determining the effect of glaciation on the Earth but most of her research projects deal with the effect of sediment and water flow on glacial stability. Her travels to Antarctica mark a personal triumph, as well as a triumph for women scientists. Ashley was denied a research opportunity in 1970 because there were no facilities for women. Twenty years later, times had changed. Possibly as an offshoot of her glacial work, she also conducts research on marshes,

Some of Gail Ashley's research requires her to be airlifted in by helicopter, especially that in Alaska *(Courtesy of Gail Ashley)*

rivers, and wetlands mainly in New Jersey and British Columbia, Canada. It is the glaciation that causes poor drainage and wetlands in most of the northeastern United States and other areas. Because wetlands are so environmentally sensitive, this research has also received much attention.

Gail Mowry was born on January 29, 1941, in Leominster, Massachusetts. She became interested in geology at age 14 when her next-door neighbor, a geology professor from Smith College, Marshall Schalk, introduced her to the field. She attended the University of Massachusetts at Amherst and earned a bachelor of science degree in geology in 1963. While raising a family, she returned to the University of Massachusetts and earned a master of science degree in 1972. She completed her doctoral degree at the University of British Columbia, Canada, in 1977 with a dissertation on modern sediment transport in a tidal river. She joined the faculty at Rutgers University, New Brunswick, New Jersey, in 1977 and is currently a full professor and director of the Quaternary Studies graduate program. Gail Ashley was married to Stuart Ashley for 22 years. They have two children. She is now married to Jeremy Delaney, a geochemist at Rutgers University.

Gail Ashley has produced six edited volumes, 50 professional journal and volume articles, and 21 technical reports and field guides. She served as editor for the *Journal of Sedimentary Research* from 1996 to 2000 (the first woman to hold the position) and associate editor for *Geological Society of America Bulletin* from 1989 to 1995 and for the *Journal of Sedimentary Research* from 1987 to 1990 and 1992 to 1995.

Ashley has also performed outstanding service to the professional societies in geology. She was the president of the Society of Economic and Petroleum Mineralogists (SEPM) from 1991 to 1992 and only the second-ever female president of the 15,600-member Geological Society of America from 1998–1999. She was the vice president of the International Association of Sedimentologists from

1998 to 2002 and the chairman of the Northeast Section of the Geological Society of America from 1991–1992. She has also been active in the Association of Women Geoscientists and other groups to integrate more women into the fields of science and math. Gail Ashley has taken this effort to a personal level where she is well known as an excellent mentor to her students.

⊠ Atwater, Tanya
(1942–   )
American
Tectonics, Marine Geophysicist

After the initial documentation of seafloor spreading by some of the giants of geology, there were still many details about the ocean floor to unravel. One of the main researchers in the second wave of plate tectonics is Tanya Atwater. Incredibly, of her first five professional articles, three were deemed so important that they were reprinted in textbooks and professional volumes. Two of these papers include "Changes in the Direction of Sea Floor Spreading" and "Implications of Plate Tectonics for the Cenozoic Tectonic Evolution of Western North America." She collaborated with several noted scientists including H. WILLIAM MENARD and Fred Vine. Part of her research included the defining of new processes on the ocean floor including the mechanics and topographic expression of oceanic fracture zones (transform faults), and shifting directions of seafloor spreading. She also defined new processes of mid-ocean ridges and the formation of new ocean crust. This research established Atwater as one of the leaders in tectonics of ocean basins and a pioneer for women in this field. To undertake the research, she was a member of several research cruises that previously had been restricted to male participants only. She studied the deep ocean floor at 2.5 to 3.5 km in the famous ALVIN submersible on a dozen occasions. She was on drilling expeditions worldwide and participated in collecting some of the data that provide the definitive evidence for the accepted theories now appearing in introductory textbooks worldwide.

Atwater became not only an expert on processes but also on specific geologic features. She is one of the foremost experts on the tectonics of the northeast Pacific Ocean as well as the mid-Atlantic Ridge near and in Iceland. Much of her process-oriented work was gleaned from the study of these areas. But she not only worked undersea. She was the first to determine the origin and evolution of the San Andreas Fault of California, which she did early in her career. Later in her career, she is again investigating the tectonics of southern California. This time much of her work has been geared toward geoscience education and communication. She has devoted great efforts to educating science teachers on the rich geology of California. It has been like a second career for Atwater, who now conducts workshops for science teachers and consults on the production of written media as well as with museums, television, and on video productions about geosciences. She has also continued her role as one of the principal spokespersons for the integration of women into geology and the physical sciences in general.

Tanya Atwater was born on August 27, 1942, in Los Angeles, California. She attended the Massachusetts Institute of Technology from 1960 to 1963 but transferred to the University of California at Berkeley and earned a bachelor of science degree in geology in 1965, Phi Beta Kappa. She did her graduate studies at Scripps Institution of Oceanography, University of California at San Diego, and earned a Ph.D. in 1972. She joined the faculty at Scripps Institution in 1972 but accepted a position in the joint program of Massachusetts Institute of Technology and Woods Hole Oceanographic Institution, Massachusetts, in 1974. Atwater moved back to the University of California at Santa Barbara in 1980 and remains there today. She was a USA–USSR

Tanya Atwater on a field trip to the Kelso Dunes, Central Mojave Desert in 1994 *(Courtesy of Arthur Sylvester)*

exchange scientist in 1973. Atwater was married to fellow Massachusetts Institute of Technology geologist PETER MOLNAR and together they have one son.

Tanya Atwater has had a productive career. She is an author on 50 articles in international journals, professional volumes, and major reports, as well as video presentations. Many of these are benchmark studies on marine geophysics and tectonics that were reprinted in definitive volumes on the topics. Seven of these papers appeared in the high-profile journals *Nature* and *Science*. Atwater has been recognized for her contributions to the profession through numerous honors and awards. She is a member of the National Academy of Sciences. She was declared Scientist of the Year for the 1980 *World Book Encyclopedia*. The same year she won the Newcomb Cleveland Prize from the American

Association for the Advancement of Science. In 1984, she won the Encouragement Award from the Association of Women Geoscientists and she was a Sloan Fellow in 1975–1977. She was also named to endowed distinguished lectureships at Carleton College, Minnesota, and San Antonio State College, Texas.

Atwater also performed service to the profession. She served on several national and international committees and panels including being chair of the Ocean Margin Drilling Advisory Committee and a member of the International Drilling Project. She also served on numerous committees for the American Geophysical Union.

B

⊠ **Bally, Albert W.**
(1925–   )
Dutch
*Petroleum Geologist, Structural
Geologist*

Albert Bally has had great success both in the petroleum industry and in translating those successes into scholarly works in academia. Although there are many other geologists who have made similar industry-academic connections, none have been as effective as Bally. Perhaps the best example of this duality was his 1975 paper entitled "A Geodynamic Scenario for Hydrocarbon Occurrences." It is a worldwide look at types of sedimentary basins, and explains the dynamics of the hydrocarbon-bearing basins using the theory of plate tectonics. Two updated versions of this paper were published in 1980. Also during this time, Bally was an author on the *Stratigraphic Atlas of North and Central America,* which includes a series of maps that detail the origin of sedimentary materials and the paleogeography of North and Central America.

One of his best-known areas of research was released in a now classic 1966 paper which showed that deformation in the Canadian Rocky Mountain fold and thrust belt only involved the sedimentary strata of the cover sequence over a relatively undeformed crystalline basement. This deformation is termed "thin-skinned" and it is analogous to a rug (sedimentary cover) sliding on a wood floor (crystalline basement). It quantitatively describes the great amount of thrusting that took place and its effect on lithospheric processes. Many of the important concepts he discovered in this paper were influential in future papers, such as his extraordinary work in the Melville Islands, Canada, fold and thrust belt and his development of the concept of the orogenic float. This process involves the sideways thrust faulting of rock sheets parallel to synchronous strike-slip faults.

During his career Albert Bally did extensive work on the geology of the Gulf of Mexico. He proposed that the Gulf province was the type example of a passive margin that experienced complex stratigraphic and structural deformation primarily due to gravitational instabilities. Even though common and extensive listric (shallowing dip with depth) normal faulting and its effects on sedimentation were known for a long time, the movement of salt into deformational features was not recognized until the late 1970s. Salt domes are the locations of the largest petroleum deposits in the Gulf Coast. He extended this work on halokinesis (salt tectonics) to other areas as well. Bally showed how the importance of allochthonous salt located in a fold and thrust belt can explain certain complex structural relationships in the Betic Cordillera of Spain.

Bert Bally on a field trip in California in 1984 *(Courtesy of Arthur Sylvester)*

Much of Bally's work involves the integration of seismic reflection profiling (like a sonogram of the Earth) with structural and stratigraphic observations and principles. With his access to the excellent industry seismic reflection data, he was able to document these sedimentary-deformational processes well ahead of the rest of the profession. Through an impressive feat of negotiation, Bally was able to publish some of this very expensive proprietary data in a series of atlases that are unparalleled in the field. Through this work, he established himself as one of the world's foremost experts on seismic stratigraphy and seismic interpretation. As a result of this expertise he led the way to a major national seismic profiling research project on the continental shelves called the EDGE project.

Albert Bally was born in The Hague, Netherlands, on April 21, 1925. He became interested in geology as a boy exploring the volca-

noes and foothills around Rome, Italy. He attended high school in Switzerland and upon graduation continued his education at the University of Zurich, where he earned a bachelor of science and a Ph.D. (1925), both in geology. Upon graduation, Bally accepted a post-doctoral fellowship at the Lamont-Doherty Geological Observatory of Columbia University, New York, where he remained for one year. In 1954, Bally was offered a position with Shell Oil Company where he remained until he retired in 1981. He began with Shell Canada in Alberta where he explored for prospects in the Rocky Mountain overthrust belt. He moved to Houston, Texas, in 1966 as a manager of Geological Research at the Shell Bellaire Research and Development Laboratory. In 1968, he became the chief geologist for the U.S. Division of Shell Oil and was appointed exploration consultant in 1976, and senior exploration consultant in 1980. Upon retirement from Shell, he accepted the position of Harry Carothers Weiss Professor of geology at Rice University in Houston, Texas, where he remains today. He was initially appointed department chair at Rice as well.

Albert Bally has led a dual career with impressive productivity in each regard. He produced numerous reports on his exploration and research at Shell Oil, as well as numerous scholarly publications, both at Shell and in his academic role. Many of these papers and reports are true classics in the structural, stratigraphic, and plate tectonic processes primarily as they relate to hydrocarbon accumulation. In recognition of his contributions to geology, Albert Bally has received numerous honors and awards. He was the recipient of the Career Contribution Award for the Structural Geology and Tectonics Division of the Geological Society of America for 1998, the Sidney Powers Medal from the American Association of Petroleum Geologists, the William Smith Medal from the Geological Society of London, and the Gustav Steinmann Medal from the Geologische Vereinigung of Germany.

Bally also contributed to geology in terms of service to the profession. He served as Centennial President of the Geological Society of America in 1988 in addition to councilor and numerous other roles. In his role as president, he initiated the famous Decade of North American Geology (DNAG) project. He was also very active with the American Association of Petroleum Geologists and served roles for the National Academy of Sciences as well.

⊠ Bascom, Florence
(1862–1945)
American
Field Geologist, Petrologist

Florence Bascom is considered the "Grand Dame of American Geology." She was truly a pioneer who enabled women to make a name for themselves in the traditionally male-dominated field of geology. Even though she was the second woman in the country to earn a Ph.D. in geology (following Mary Holmes, who earned a Ph.D. in geology from the University of Michigan in 1888), Bascom was a first for women in geology in almost every aspect of her geological career. The U.S. Geological Survey hired her as their first woman geologist; she was the first woman to present a paper before the Geological Society of Washington; she was the first woman admitted to the Geological Society of America in 1924; and the first woman officer of Geological Society of America (vice president, 1930). She was an associate editor of the *American Geologist* from 1896 to 1905. In the first edition of *American Men of Science* published in 1906, she was regarded as a four-star geologist. This meant that her peers and colleagues regarded her as one of the top 100 geologists in the United States.

Florence Bascom became an expert in mineralogy, petrology, and crystallography. Her dissertation was her earliest and one of her most important contributions to geology. Using petro-graphic (optical microscope for rocks) methods, she showed that rocks that were previously considered sedimentary were actually metamorphosed lava flows. She continued research in this area and established herself as one of the foremost experts on crystalline rocks in the central Appalachian Piedmont. This research included mapping vast areas in Pennsylvania and Maryland but also topical studies on metamorphic processes. Her tight coordination of petrographic work with the fieldwork was considered cutting-edge methodology at the time. Her contributions to Piedmont geology are still valued and used by geologists working in that area today. Later in her career, she expanded this interest to include the development of mountain belts in general, especially with regard to crystalline rocks.

Florence Bascom was born in Williamstown, Massachusetts, on July 14, 1862, the youngest of six children. Her father, John Bascom, was a pro-

Portrait of Florence Bascom *(Courtesy of the U.S. Geological Survey)*

fessor of oratory and rhetoric at Williams College, Massachusetts. He was a strong supporter of women's rights and spoke publicly regarding the importance of a college education for women. In 1874, he became president of the University of Wisconsin at Madison and in 1875 the university began admitting women. Florence Bascom enrolled in the fall of 1877. Even with limited access to the library and the gym and not being allowed to enter classrooms already filled with men, she earned a bachelor of arts and letters degree in 1882 and a bachelor of science degree in 1884. She then became interested in geology and went on to earn a master of science degree in 1887 also from the University of Wisconsin. She continued her graduate studies at Johns Hopkins University in Baltimore, Maryland, but was required to sit behind a screen during classes so she did not distract the male students. She graduated with a Ph.D. in geology in 1893.

Florence Bascom began her college teaching career as soon as she completed her undergraduate degree. She taught at the newly founded Hampton School of Negroes and American Indians (currently Hampton University) in Ohio, 1884–1885, Rockford College, Ohio, from 1887 to 1889, and Ohio State University from 1893 to 1895. After completing her Ph.D., she began her famous teaching career at Bryn Mawr College, Pennsylvania, in 1895 where she single-handedly developed the Department of Geology. In 1896, she was also hired as an assistant geologist by the U.S. Geological Survey to map the geology of Pennsylvania, Maryland, and New Jersey during the summer months. In 1909, she was promoted to geologist. Florence Bascom retired to professor emeritus from Bryn Mawr in 1928, but remained active until her death on June 18, 1945, as the result of a stroke.

Florence Bascom was an author on research publications that total nearly 40, including USGS bulletins and portfolios, as well as journal articles. Most of these papers are on the crystalline rocks of the central Appalachian Piedmont. She also

published the extensive research she conducted on Piedmont geomorphology (provenance of surficial deposits). Her accomplishments must also include the training of some of the most prominent women geologists of the time. Louise Kingsley, Katherine Fowler Billings, petrologists ANNA I. JONAS STOSE and Eleanora Bliss Knopf, crystallographer Mary Porter, paleontologist Julia Gardner, all of whom went on to have careers with the USGS, were among Bascom's protégées. Also included among her students were petroleum geologist Maria Stadonichenko, Barnard's glacial geomorphologist Ida Ogilvie, Isabel Fothergille Smith of Scripps College, Bryn Mawr's Dorothy Wyckoff, and Anna Heitonen. Florence Bascom firmly established the first gateway at Bryn Mawr College for women to enter the field of geology. It would be many decades before any other school even approached her success.

⊠   **Berner, Robert**
(1935–   )
American
Geochemist, Sedimentologist

Robert Berner has an interesting approach to scientific research. His work is a prime example of how small-scale, curiosity-driven science can produce big-scale scientific results. He works with modest funding, relatively simple equipment, and small groups of highly motivated scientists. Collaboration with biologist Alfred Redfield in his early career began this approach for him as well as convincing him of the advantages of a holistic approach to science. Berner attacks his research problems using all available resources and methods regardless of the subdiscipline or even the field, be it geology, biology, chemistry, physics, meteorology, or oceanography, or what techniques it may entail. He can be considered the "father of Earth system science," the newest and among the most popular directions in Earth science. Earth system science involves the collapse of walls between disciplines of the Earth and related sciences, as well as those in ecology and related biosciences. The interactions of processes take precedence over their individuality. There are now many books on the subject and even research projects must have their interactions demonstrated in order to receive federal funding in many programs.

Robert Berner has been among the leading innovators of scientific thought in the field of sedimentary geochemistry. His research interests are in geochemical cycles of carbon, phosphorus, and sulfur within these sediments largely using stable isotopes (nonradioactive) as tracers. Related areas that he researches include biogeochemistry, diagenesis, mathematical modeling of Earth's surface geochemistry, chemical oceanography, and chemical weathering. His research on the early stages of diagenesis of sediments revealed the complexity of interrelationships among physical, chemical, and biological processes occurring near the sediment-water interface. This research inspired his mathematical models for diagenesis, which were the first of their kind. His work on the physical chemistry of carbonate minerals in seawater set the mark for chemical oceanography as well as current work on climate modeling. His research on the surface chemistry of silicate minerals undergoing weathering set the standard for much of the research that was to follow. He has modeled the global carbon cycle and the role it plays in controlling atmospheric oxygen and carbon dioxide, and global climate over Phanerozoic time (the past 535 million years). He is particularly interested in how the evolution of land plants may have influenced global weathering rates and the carbon cycle. This work provides the basis for the climate change analysis that is currently being conducted at a remarkable pace. Robert Berner is a true pioneer in this, the most vigorous field in Earth science today.

Robert Berner was born on November 25, 1935, in Erie, Pennsylvania, where he spent his childhood. He attended the University of Michigan, Ann Arbor, and earned bachelor of science and master of science degrees in geology in 1957

and 1958, respectively. He earned his Ph.D. at Harvard University, Massachusetts, in 1962. He joined Scripps Institution of Oceanography, University of California at San Diego, as a Sverdrup Postdoctoral Fellow in 1962 to 1963. He then joined the faculty of the University of Chicago, Illinois, in 1963. In 1965, he moved to Yale University, Connecticut, where he remains today.

Robert Berner has had a very productive career. His accomplishments are reflected in the fact that he is among the most frequently cited earth scientists in scientific literature. In addition to having published more than 200 articles in international journals, he wrote four successful books including *Principles of Chemical Sedimentology* in 1971, *Early Diagenesis: A Theoretical Approach* in 1980, *The Global Water Cycle,* which he wrote with his wife, E. K. Berner, in 1987, and *Global Environment* (also with E. K. Berner) in 1996.

Berner has been recognized with numerous honors and awards for his groundbreaking research. He was elected to the U.S. National Academy of Sciences at a young age and he is a Fellow of the American Academy of Arts and Sciences. He was awarded the Huntsman Medal in Oceanography from the Geological Society of Canada in 1993 and the V. M. Goldschmidt Medal from the Geochemical Society in 1995. He was awarded the Murchison Medal from the Geological Society of London in 1996 and the Arthur L. Day Medal from the Geological Society of America in 1996. He was awarded the Bownocker Medal from Ohio State University in 2001 and an honorary doctoral degree, Doctor Honoris Causa, Université Aix-Marseille III, France, in 1991.

Berry, William B. N.
(1931–)
American
Invertebrate Paleontologist

Just as John McPhee described in his popular books on geology, there is a sharp contrast between East Coast geology and West Coast geology. Many of the notable Earth scientists of the East Coast are "grand old geologists of the Appalachians." Even though they are basically doing the same sort of research with the same cutting edge, West Coast geologists are regarded as young mavericks. This impression may reflect the age of the rocks (the East Coast is much older) or the age of the schools or perhaps an historical migration of many of the new doctorates from the East Coast to the West Coast in the 1950s and 1960s or some combination thereof. William Berry has managed to be from both coasts. Early in his career, Berry established himself as one of the true leaders in Appalachian paleontology among a very talented group and he has never really abandoned that position through all of his other work. He is a specialist in graptolites and especially Ordovician graptolites, mainly from the Taconic Mountains of New York, although his earliest reports were from Texas. Graptolites are small enigmatic sawlike fossils that are mainly found in deepwater shales. His work expanded into Silurian graptolites from the Maine slate belt and nearby Canada and later to Devonian graptolites, among the others, and his area of interest spread to the entire United States and even western Ireland.

Berry's interest in graptolites has never really faded but he became more interested in biostratigraphy, especially with regard to regional correlations. With his colleague Arthur Boucot, Berry began a mammoth project of correlation of Silurian rocks worldwide including North America, South America, Southeast Asia and the Near East, Africa, Australia, New Zealand, New Guinea, and China. All the while, he kept expanding his graptolite studies to northern Canada and Greenland but periodically he returned to his roots in the Appalachians. Eventually his research further evolved into paleoenvironmental and paleoclimate analysis of these ancient settings. Berry studied the evolution of the platforms and basins and considered the stimuli that caused animals to

evolve. He considered several mechanisms of change including ocean venting, destabilization of ocean density gradients, and even meteorite impacts. Most of this research was conducted on black shales where the graptolite fossils are found. These studies drew Berry into the modern group of environmental geologist and climate change modelers. His administrative work and service to the profession also moved in this direction concurrently with his research.

William Berry was born on September 1, 1931, in Boston, Massachusetts. He attended Harvard University and earned a bachelor of arts degree in 1953 in geology and a master of arts in geology in 1955. He completed his graduate studies at Yale University, Connecticut, where he earned a Ph.D. in 1957. Upon graduation, he accepted a position at the University of Houston, Texas, but moved to the University of California at Berkeley the next year (1958) and remains there today. While a faculty member, Berry has also held numerous positions with the Museum of Paleontology at the University of California at Berkeley including the curator of Paleozoic and Mesozoic invertebrate fossils (1960–present), associate director (1962–1966), acting director (1966, 1972–1976), and director (1976–1987). He served as chair of the Department of Paleontology (1975–1987) and the director of the environmental sciences program (1979–1993). He has also been a marine scientist in the Lawrence Berkeley National Laboratory since 1989. William Berry has been married to Suzanne Spaulding since 1961; they have one child.

William Berry has led a very productive career. He is an author on some 165 articles and reports in international journals, professional books and volumes, governmental reports, and conference proceedings. Many of these papers are benchmark studies on graptolites, paleo-oceanography, and biostratigraphy that appear in journals such as *Nature* and *Science*. He is an author or editor of 12 books and volumes. Two of these books, *Principles of Stratigraphic Analysis* and *Growth of a*

Prehistoric Time Scale Based on Organismal Evolution, are a widely adopted textbook and a more popular scientific book that have been reprinted several times. Berry was a Guggenheim Fellow in 1967.

Berry has performed significant service to the profession and the public. He has served in numerous roles for the International Stratigraphic Commission, the National Research Council, the Geological Society of America, the American Geological Institute, and the American Association for the Advancement of Science. He was also the director of the Environmental Sciences Curriculum Development Program for the San Francisco, California, Unified School District. He was similarly an adviser for the Catalan (Spain) Ministry for the Environment to develop an environmental health and safety program. Berry served in numerous editorial capacities including as associate editor of *Paleoceanography* (1986–1992) and a member of the board of editors for the University of California Publications in the geological sciences.

⊠   Bethke, Craig M.
(1957–   )
American
Hydrogeologist

One of the most important applied aspects of geology today is the study of how fluid flows through rock and soil. It not only dictates our ability to find clean sources of groundwater for drinking and industrial uses, but because there is such a close interaction between ground and surface water, it also affects our surface water quality. In addition, oil and gas flow through rock as they migrate into a reservoir where they can be drilled and produced much in the same manner as groundwater flow. Craig Bethke has quickly established himself as one of the leading hydrogeologists in the field. He mathematically models fluid flow and chemical interactions, both at the surface and in

Craig Bethke demonstrates the models produced by computer software that he developed *(Courtesy of Lillian Morales)*

the subsurface, using sophisticated computer programs. Even the more standard computer techniques for such analyses, taken from engineering applications using procedures called finite elements and finite differences modeling, can have many tens of thousands of lines of code. Bethke's work goes beyond the standard applications and many of his projects require the use of a supercomputer. He is in an elite class in the whole geological community to be able to so quantify such complex phenomena.

Bethke's main interest has been the fluid migration history in the evolution of sedimentary basins. When sediments are deposited, they are saturated with fluids that typically contain minimal amounts of dissolved solids. As those sediments are progressively buried beneath additional strata, the pressure and temperature build. The pore fluids dissolve material from the sediments in which they are contained or, depending upon conditions, precipitate chemicals in which they are saturated. By this process, they take on the chemical signature of their host sediment. They are then forced to migrate from this building pressure and will chemically interact with any other sediment that they pass through. Bethke models these very complex multicomponent chemical interactions between fluids and sediments during migration. Their study may require him to study detailed clay mineralogy, petrology of the sediments, isotopic analysis, and detailed geochemistry. These studies have great application to petroleum exploration. They have been done on the Denver Basin, Colorado, the Los Angeles Basin, California, and the Illinois Basin, among others.

In addition to these paleohydrogeologic studies which concentrate on flow in ancient systems, Bethke also studies the environmental aspects of aqueous geochemistry (water chemistry) which involve currently active systems. He uses the distribution of isotopic tracers to map out the flow patterns in active settings. This work has taken him to the Western Canada sedimentary basin and the Great Artesian Basin of Australia. A new aspect of his research is to add the interaction of microbiology with geologic processes. This multidisciplinary study of complex geologic processes is opening up, with great success, an aspect of aqueous geochemistry that has traditionally been overlooked.

Craig Bethke was born on June 6, 1957, in Rolla, Missouri. He attended Dartmouth College, New Hampshire, where he earned a bachelor of arts degree in Earth sciences with distinction in 1979. He did his graduate studies at the Pennsylvania State University at University Park in geosciences and at the University of Illinois at Urbana-Champaign where he earned a Ph.D. in 1985. During his graduate study, he worked as an exploration geologist at ARCO Oil and Gas Co. and at Exxon Production Research Co. and Exxon Minerals Co. Bethke joined the faculty at University of Illinois in 1985, and he remains there today. Bethke has been a visiting professor at both the Académie des Sciences in Paris, France, and École Nationale Supérieure des Mines de Paris in Fontainebleau, France. Craig Bethke is married to Abigail Bethke; they have three children.

Craig Bethke is in the early stages of what promises to be a very productive career. He has been an author of 34 articles in international journals and professional volumes. He has also written one advanced textbook and four pieces of software documentation. Several of these are seminal papers on fluid migration in sedimentary basins, including one paper in the prestigious journal *Science*. He is also the primary author of several widely used software packages including The Geochemist's Workbench and Basin2. Considering the relatively early point in his career, Craig Bethke has received an astounding number of honors and awards for his research contributions to the science as well as his teaching. He received the Meinzer Award from the Geological Society of America, the Lindgren Award from the Society of Economic Geologists, a Presidential Young Investigator Award from the National Science Foundation, and he was chosen as a Shell Faculty Career Fellow. As a student he was a National Science Foundation Graduate Fellow; he received the Best Student Paper Award from the Clay Minerals Society and the Upham Prize at Dartmouth College. At University of Illinois, he was named a Beckman Associate and a Fellow at the Center for Advanced Study, as well as having been cited for excellence in teaching numerous times.

⊠ Billings, Marland P.
(1902–1996)
American
Structural Geologist

When John McPhee contrasted mountain building events from the East Coast of the United States with those from the West Coast in his book *In Suspect Terrain,* he did so metaphorically by describing geologists. In contrast to the glitzy modern image portrayed for the geologically young mountains of the West Coast, the Appalachians were portrayed as a New England geologist. This gritty, aging New Englander, tough as nails and self-sacrificing yet with a wry sense of humor, is personified by Marland Billings. But this image should not convey the idea of a geologist who is hopelessly rooted in archaic theories and methods. Rather, the New England geologists began unraveling the histories of unbelievably complex geologic terranes well before the West Coast geologists. When Marland Billings began his assault on the New England Appalachians, they were considered to be largely Precambrian crystalline rocks that were so complex that they would never be understood. Billings was undaunted. Using the most modern of petrologic and structural techniques at the time and devising new ones as he went, Billings and a group of some of the top geologists in the world, many of whom he trained, put New England into context. Three distinct Paleozoic (535 million–245 million years ago) orogenies (mountain building events) emerged (Taconian, Acadian, and Alleghenian) and the rock units were assigned ages, many by continuous long-distance correlations with rocks that contain fossils. New England should really be viewed not as a stodgy old regional study but as an early cutting-edge and continuing regional study.

With some background experience in the Alps and the Rocky Mountains, Marland Billings began his work in central New Hampshire on Paleozoic rocks where fossils had been discovered. From there, he worked his way into the regionally metamorphosed rocks that comprise most of New England. These studies stretched into Vermont and throughout Massachusetts and even into southern Maine. At the time, having Marland Billings work on an area meant that it would soon be brought into a modern context. He is probably best known for his research on the White Mountain magma series and surrounding rocks where he and his wife, Kay, trudged through the Presidential Range, the most rugged terrain in New England. These observations set the stage for a reinterpretation of the geologic style of New England.

Marland Billings was born on March 11, 1902, in Boston, Massachusetts. He received his precollegiate education at Roxbury Latin School, Massachusetts, before enrolling at Harvard University in Cambridge, Massachusetts. He earned a bachelor of arts (magna cum laude), master of arts, and Ph.D. in geology in 1923, 1925, and 1927, respectively, and was an instructor during his last year. He accepted a position at Bryn Mawr College in Pennsylvania upon graduation but returned to Harvard University as a faculty member in 1931. He remained at Harvard University throughout his career, serving as department chair from 1946–1951 as well as curator of the Geological Museum. He retired to professor emeritus in 1972. During World War II, Billings served with the U.S. Office of Field Service in the South Pacific in 1944 where he evaluated strategic nickel deposits in New Caledonia. Marland Billings married geologist and former student Katherine Stevens Fowler in 1938. They had one son. Marland Billings died on October 9, 1996, in Peterborough, New Hampshire.

Marland Billings was an author of numerous scientific articles in international journals and professional volumes. He is also the author of the widely adopted textbook *Structural Geology* that was first published in 1942 but still used at colleges into the 1970s. He also was an author on the book *Bedrock Geology of New Hampshire* and of the state geological map of New Hampshire (1955). Many remember him best for his avid participation in the annual field trip of the New England Intercollegiate Geological Conference. In recognition of his research contributions to geology, Marland Billings received numerous honors and awards. He was a member of the National Academy of Sciences. He received honorary doctorates from Washington University in Saint Louis, Missouri, and the University of New Hampshire. In 1987, he was presented with the Penrose Medal, the top award from the Geological Society of America.

Billings's service to the profession was exceptional. Among numerous panels and committees, he served as the president of the Geological Society of America (1959) and the vice president of the American Association for the Advancement of Science in 1946–1947. He was also president of the Boston Geological Society in 1940. He was a member of the Mineral Resources Committee of New Hampshire from 1935 on and at that time he was the de facto state geologist. From 1958, he consulted for the Metropolitan District Commission for Boston, Massachusetts, and evaluated the bedrock for virtually all of the water supply and other tunnels around Boston at that time.

⊠ **Birch, A. Francis**
(1903–1992)
American
Geophysicist

Francis Birch is famous not only for his contributions to geophysics and geology but also to the World War II effort in his role in development of the atomic bomb. He is considered one of a few founders of the science of solid Earth geophysics. His most famous research was to determine the basic architecture of the deep Earth. In a 1952 paper entitled "Elasticity and the Constitution of the Earth's Interior," he conclusively showed that the mantle of the Earth is mainly composed of silicate minerals and that the upper mantle and lower mantle regions are each basically homogenous but of different composition. The two regions are separated by a thin transition zone associated with silicate phase transitions from open structured minerals in the upper mantle to denser, closed structured minerals in the lower mantle. Birch also showed that the inner and outer cores are alloys of crystalline and molten iron respectively. This breakthrough remains a benchmark in Earth science that appears in every textbook in physical geology.

Francis Birch combined theory with experimental practices in his research. His geological research was combined with the disciplines of physics and electrical engineering. By combining these three disciplines, Birch was able to successfully solve many virtually otherwise unaddressable geologic problems. He had the uncanny ability to recognize a geologic problem, decide an approach to the problem, and use that approach to find a result to the problem. Birch's research dealt with elasticity, phase relations, thermal properties, and the composition of the Earth's interior as summarized in his paper "Elasticity and the Earth's Interior." He knew that there was a limited amount of data regarding high-pressure physical properties of rocks and minerals. These had to be addressed in order to better interpret measurements made by seismological and gravity techniques. To these ends, Birch's experimental research concentrated on elasticity, phase relations, thermal properties and heat flow, and the composition of the Earth's interior. His laboratory studies of seismic wave velocities in rocks and their variation with pressure and temperature discovered the first approximations of density-pressure relationships at high compressions. Birch used these data that he collected to interpret global seismic data in regard to composition and structure of the interior.

Another major contribution Birch and his research team made was to our knowledge of terrestrial heat flow. By combining experimental data on thermal conductivities of rocks with temperature gradient measurements from boreholes and tunnels, they helped distinguish heat flow as one of the most important conditions of continental geophysics. This work is presented in several major papers, including "Heat from Radioactivity" and "Heat Flow in the United States."

Albert Francis Birch was born in Washington, D.C., on August 22, 1903, where he spent his youth. He graduated from Western High School in 1920. He attended Harvard University and participated in the ROTC program. He graduated magna cum laude in 1924 with a bachelor of science degree in electrical engineering. He worked for two years for the New York Telephone Company in engineering when he decided to change his course of study to physics. Birch received an American Field Service Fellowship that led to two years of study (1926–1928) at the Institut de Physique, University of Strasbourg, France. Birch studied under Pierre Weiss who was one of the founders of modern magnetism. As a result, Birch decided to return to Harvard in 1928 as a graduate student in physics. He worked in the high-pressure laboratory of Percy W. Bridgman who received the Nobel Prize in physics in 1946. Birch was an instructor and tutor in physics from 1930 to 1932. He received his master of science degree in 1929 and his Ph.D. in 1932.

Just prior to graduation, Harvard University offered Birch the opportunity to work in the newly established high-pressure research program as the first research associate in geophysics. The next year, he became director of the program. It was at this time that Francis Birch married Barbara Channing; they had three children. Francis Birch took a leave of absence in 1942 to participate in the World War II effort. He began at the Radiation Laboratory at Massachusetts Institute of Technology, but in 1943 he accepted a commission as a second lieutenant in the U.S. Navy at the Bureau of Ships in Washington, D.C. This assignment was short-lived because he was quickly chosen by Robert Oppenheimer to participate in the Manhattan Project. He moved to Los Alamos, New Mexico, where he was soon promoted to commander and the head of the Uranium-235 fission bomb project (code name Little Boy). Birch personally supervised the assembly and loading of "Little Boy" onto the B-29 *Enola Gay* prior to the bombing of Hiroshima, Japan. Francis Birch was awarded the Legion of Merit by the U.S. Navy for these outstanding efforts.

Francis Birch returned to Harvard University in 1945 to resume his academic career. He quickly advanced to become the prestigious Sturgis

Hooper professor of geology in 1949 and later he would be the chairman of the Geological Sciences Department. He retired to professor emeritus in 1974, but continued his research at a bit slower pace until his death on January 30, 1992, at 88 years old.

Francis Birch led an extremely productive career serving as author of numerous scientific articles in international journals and professional volumes. Many of these papers are benchmark studies in mantle structure and processes, heat flow, and the propagation of seismic waves through the Earth. In recognition of his numerous contributions to Earth sciences, Francis Birch received numerous prestigious honors and awards. He was a member of the National Academy of Sciences. He received honorary doctoral degrees from the University of Chicago and Harvard University. He also received the National Medal of Science from President Johnson in 1968. He was recipient of the Gold Medal from the Royal Astronomical Society of London in 1973, the Vetlesen Medal for 1960, both the Arthur L. Day Medal in 1950 and the Penrose Medal in 1969 from the Geological Society of America, the William Bowie Medal from the American Geophysical Union in 1960, and the Bridgman Medal from the International Association for the Advancement of High Pressure Research in 1983.

Bloss, F. Donald
(1920–   )
American
Mineralogist

For many years, the analysis of minerals was done using a microscope and then wet chemical methods for further resolution if necessary. With the advent of X-ray analysis and spectroscopy, these old optical methods, although still used to give general results and to guide the choice of further analysis, were considered archaic for detailed analysis. Donald Bloss almost single-handedly kept these optical methods alive through all of these years and he did so using innovative new approaches. Probably the most impressive of these new techniques is the Bloss Automated Refractometer, or BAR, which was patented in 1987. One of the most distinctive optical properties of minerals is the index of refraction, basically the speed at which light travels through the mineral. Just as objects appear bent as they pass from air to water, so does light bend or refract through minerals. The BAR shoots a laser beam through a mineral and then precisely measures the angle at which it bends. This angle will not only uniquely define certain minerals but can even give accurate chemistry of minerals with simple two-component solid solutions like olivine (Mg-Fe) or plagioclase (Na-Ca). The ultimate version of this device is to be called the automated petrographer in which a thin section (microscope slide of rock) is mapped out, including all of the minerals and

Don Bloss working in his office at Virginia Tech in 1988 *(Courtesy of Don Bloss)*

their orientations. It therefore can determine composition and fabric of a rock.

This is not the only device that Bloss invented. He also designed and produced the "spindle stage," a simple microscope attachment that allows single minerals to be observed in all directions for more accurate analysis. The name Bloss is known by virtually all students of geology not for these inventions but for his popular textbooks, *Crystallography and Crystal Chemistry* and *Introduction to the Methods of Optical Crystallography.* These two books have been the standard bearers for their subjects for 40 and 30 years, respectively. Their success is based largely on his ability to inject his own teaching philosophy into the writing.

Bloss's scientific contributions are mainly involved with defining the optical properties of a variety of minerals under a variety of conditions. Much of the information that we have on the more recent quantitative optical properties of minerals, especially concerning solid solutions, resulted from this research. He also looked at the physics of light as it passes through minerals.

Don Bloss was born on May 30, 1920, in Chicago, Illinois, where he spent his youth. He enrolled at the University of Chicago, Illinois, where he earned a bachelor of science degree in geology in 1947, Phi Beta Kappa. He remained at the University of Chicago for graduate studies and earned a master of science degree in geology in 1949 and a Ph.D. in mineralogy in 1951. He joined the faculty at the University of Tennessee at Knoxville in 1951. He moved to the University of Southern Illinois at Carbondale in 1957. In 1967, he accepted a position at Virginia Polytechnic Institute and State University (Virginia Tech) and remained there for the rest of his career. He was named the first-ever alumni distinguished professor at Virginia Tech in 1972. He also served as department chair from 1988 to 1990. He retired to professor emeritus in 1991. Bloss was a National Science Foundation Senior Postdoctoral Fellow at Cambridge University, England, and the Swiss

Federal Institute, Zurich, in 1962–1963. He was also the first-ever Caswell Silver Distinguished Visiting Professor at the University of New Mexico in 1981–1982. Bloss is a chess enthusiast and has written four books on the subject, including one with his grandson, Andrew Kensler, entitled *Sammy Seahorse Teaches Chess.*

Don Bloss has led a very productive career. He is an author of some 70 articles in international journals as well as six geology books, one chapter in a book, and numerous entries in encyclopedias, as well as papers in collected volumes. Many of these studies define the state of the science for optical properties of minerals among others. These research contributions have been recognized in terms of honors and awards. Besides those already mentioned, Bloss received the Award of Merit and Honor from the State Microscopical Society of Illinois, the Ernst Abbe Award from the New York State Microscopical Society, and he was inducted into the Hall of Fame at the Carl Schurz High School, Chicago, Illinois. He also had the mineral "blossite" named after him.

Bloss has also performed service to the profession. He was president (1977) and vice president (1976) of the Mineralogical Society of America, among numerous other positions, panels, and committees. He was also of service to the Geological Society of America and the National Science Foundation. Bloss served as chief editor for *American Mineralogist* in 1972–1975.

Bodnar, Robert J.
(1949–   )
American
Geochemist

When a mineral crystallizes, it can trap a minute bubble of fluid, melt, and/or vapor that is present during the crystallization process, whether igneous, metamorphic, or sedimentary during diagenesis (lithification). This encapsulated bubble is called a fluid inclusion. The fluid within it tells geologists something about the composition of the fluid that accompanied the crystallization of a pluton or the metamorphism of a terrane, or the mineralization of a vein, among other things. Heating or cooling the inclusion until all of the liquids and gases combine on a stage attached to a microscope may use an experimentally determined "isochore" to determine the pressure and temperature of formation as well. There are many Earth scientists who study fluid inclusions. Robert Bodnar is a pioneer in the production of synthetic fluid inclusions to model those formed in nature. By experimentally reproducing fluid inclusions, Bodnar determines the conditions of their formation. This experimental process allows him to understand and develop models for fluid/rock and fluid/magma interactions at crustal and upper mantle conditions. Bodnar has written numerous papers on the uses and relations of synthetic fluid inclusions, including "Synthetic Fluid Inclusions in Natural Quartz II," and "Applications to Pressure-Volume-Temperature Studies."

To conduct the analyses on the fluid inclusions, Robert Bodnar established the Fluids Research Laboratory at Virginia Tech. In addition to the experimental apparatus and standard cooling-heating stages, there is a Laser Raman Microprobe and a Fourier Transform Infrared Microprobe to determine the composition of the inclusions. Bodnar has worked on a variety of projects from around the world and beyond. He studied fluid inclusions in the proposed Martian meteorites. Terrestrial projects include fluids from the Somma-Vesuvius volcanic system, melt inclusions from Ischia in Naples, Italy, the volcanic system of White Island of New Zealand, among others. Closer to home he worked on problems in the southern Appalachians and in Massachusetts. He also works extensively on economic deposits. He studied the genesis of Egyptian gold deposits as well as sulfide deposits in Ducktown, Tennessee, and porphyry copper and precious metal deposits in Wyoming; Arizona; New South Wales, Australia; Gyeongsang Basin, South Korea; and the Morning Star deposit in California, among others. He also studied inclusions in the Marmarosh Diamonds and mantle xenoliths from the folded Carpathian Mountains of eastern Europe. In terms of petroleum exploration, he has worked on problems in the North Sea oil province as well as the Monterey Formation in California, among others. Papers on economic deposits include "Hydrothermal Fluids and Hydrothermal Alteration in Porphyry Copper Deposits" and "Fluid Inclusion Studies in Hydrothermal Ore Deposits."

Robert Bodnar was born on August 25, 1949, in McKeesport, Pennsylvania, where he grew up. He attended the University of Pittsburgh, Pennsylvania, where he earned a bachelor of science degree in geology in 1979. He undertook graduate studies at the University of Arizona and earned a master of science degree in 1978. Upon graduation, he obtained a position as research assistant with the U.S. Geological Survey in Reston, Virginia, in the experimental geochemistry and mineralogy section. Robert Bodnar got married in 1979; he and his wife have two children. By 1980, he heeded the advice of his colleagues and returned to graduate school at the Pennsylvania State University in University Park. He earned a Ph.D. in geochemistry in 1985 as an advisee of EDWIN ROEDDER from the U.S. Geological Survey. Bodnar obtained a position as research scientist at the Chevron Oil Field Research Company of Chevron, U.S.A., in La Habra, California, in 1984. In 1985, he was offered and accepted a faculty position at Virginia Polytechnic Institute and State University in Blacksburg where he remains today. He has been named to a prestigious C. C. Garvin endowed chair in 1997 and later as a university distinguished professor.

Robert Bodnar is an author of some 120 scientific articles in international journals, professional volumes, and major governmental and industry reports. Included in these well-cited and seminal papers on all aspects of fluids and melts in geology are the benchmark papers on synthetic fluid inclusions. In recognition of his many contributions to Earth sciences, Robert Bodnar has received several honors and awards in addition to those already mentioned. He received a Presidential Young Investigator Award from the National Science Foundation, the Lindgren Award from the Society of Exploration Geologists, and the Alumni Award for Research Excellence from Virginia Tech. Pennsylvania State University named him a Centennial Fellow, and the Society of Exploration Geologists named him a Thayer Lindsley Lecturer.

Bodnar has served on committees and panels for the Geochemical Society, the Mineralogical Society of America and the National Science Foundation. Among his editorial work was the position of associate editor of *Geology*.

⊠ Bouma, Arnold H.
(1932–)
Dutch
Sedimentologist

Unstable accumulations of sediments at the shelf edge and especially in submarine canyons can slide down the continental slope in essentially an underwater avalanche. This flow of unconsolidated debris is called a turbidity current or turbidite and it involves no movement of water, just material (sediments). During the Grand Banks Earthquake of 1929, such a turbidite was determined to have sped down the slope at 30 miles per hour as documented by sequential breaking of underwater telephone and telegraph cables. When these turbidites come to rest, they form a very characteristic deposit known as a Bouma sequence, named after its discoverer, sedimentologist Arnold Bouma.

Arnold Bouma is undoubtedly the world's foremost expert on turbidite deposits and the submarine fans that they commonly form. He produced two volumes, *Turbidites* and *Submarine Fans and Related Turbidite Systems*, among many other papers that summarize this work. The Bouma sequence divides turbidite deposits into A-D intervals, based upon grain size and sedimentary structures, and as a reflection of proximity to channels in submarine fans. Full sequences only occur in or near channels, whereas distal areas of a fan contain only partial sequences of finer grained material. The dominant types and sequences of sediments can then subdivide the fan. Bouma documented these deposits and the processes that produce them, both in modern settings using sediment cores taken from research vessels, as well as in ancient deposits on land all over the world. Every sedimentology class at every college in the world studies the groundbreaking work of Arnold Bouma. Oil companies also took an interest in his work because these deposits can contain petroleum reserves. Bouma spent many years applying his research to petroleum exploration. He even applied his work to environmental issues like coastal protection and dredging.

Arnold Bouma was born on September 5, 1932, in Groningen, Netherlands. He attended R.H.B.S. (high school and junior college) in Groningen, Netherlands, from 1944 to 1951. He attended the State University at Groningen from 1951 to 1956 and earned a bachelor of science degree in general geology. He earned a master of science degree in geology, sedimentology, and paleontology and a doctorate in sedimentary geology from the State University at Utrecht, Netherlands, in 1959 and 1961, respectively. Bouma won a Fulbright Post-doctoral Fellowship to Scripps Institution of Oceanography, La Jolla, California, in 1962–1963. From 1963 to 1966, he was an instructor at the Geological Institute at Utrecht, Netherlands, and a member of the faculty of oceanography at Texas A & M University from 1966 to 1975. From 1975 to 1981, he was a research marine geologist with the U.S. Geological Survey, first in the Pacific-Arctic branch and later in the Atlantic-Gulf of Mexico branch. He held positions of senior scientist, manager, chief scientist, and acting vice president for Gulf

Portrait of Arnold Bouma *(Courtesy of Arnold Bouma)*

Research and Development Company in Harmarville, Pennsylvania, and Houston, Texas, between 1981 and 1985. In 1985, Gulf Oil Company was bought by Chevron USA, Inc. and Bouma became a senior research associate at Chevron Oil Field Research Company (research and development branch) in Houston, Texas, and La Habra, California. He left Chevron in 1988 to become the Charles T. McCord chaired professor of petroleum-related geology at Louisiana State University in Baton Rouge, where he remains today. He also served as director of the Basin Research Institute and head of the School of Geosciences at Louisiana State University in 1989–1990 and 1990–1992, respectively. Arnold Bouma married Mechilina Kampers in 1961; they have three children.

Arnold Bouma has been phenomenally productive in his career. He has written or edited 11 books and volumes and authored or coauthored 119 articles in professional journals and volumes. Many honors and awards have been bestowed upon him throughout his career, including being a distinguished lecturer for the American Association of Petroleum Geologists in 1982, Francis P. Shepard Award from the Society of Economic Paleontologists and Mineralogists in 1982, Best Paper Award at the American Association of Petroleum Geologists Annual Meeting in 1984, Outstanding Education Award from the Gulf Coast Association of Geological Societies in 1992, and keynote speaker at the International Geological Congress in Rio de Janeiro, Brazil, 2000, and at the GEOSCIENCE 98 Conference at University of Keele, United Kingdom, in 1998, among many others.

His service to the profession is perhaps even more impressive than his papers and awards. Arnold Bouma was editor in chief for *Geo-Marine Letters* from 1980 to 2000, editor in chief of *Marine Geology* from 1963 to 1966 (and he is still on the editorial board), and series book editor for *Frontiers in Sedimentary Geology* from 1985 to present, among several other editorial positions and president of the Society for Sedimentary Geology (SEPM) from 2000 to 2001. He organized Leg 96 (a deep-ocean expedition) of the Deep Sea Drilling Project in 1980–1985. He organized the first COMFAM (Committee on Submarine Fans Meeting). He has served as a member and chair of international professional and government panels and committees too numerous to list here. He organized and convened multiple international conferences, short courses and field trips both topical (on turbidites) and general. He even helped to produce a BBC-AAPG film, *Deep Water Sands,* in 1985–1986.

⊠ **Bowen, Norman L.**
(1887–1956)
Canadian
Petrologist, Geochemist

Norman L. Bowen was the greatest petrologist of the 20th century and one of the most influential geologists of all time. His name is known by anyone who has attended a college course in physical geology by virtue of the famous Bowen's Reaction Series which appears in every physical geology and petrology textbook in the world. This diagram and concept shows the crystallization sequence of common minerals in igneous rocks of "average" compositions. Plagioclase forms the continuous reaction series because it continuously changes composition with temperature from calcium-rich at high temperature to sodium-rich at low temperature. The discontinuous reaction series shows the crystallization of a sequence of iron-magnesium-rich minerals during cooling of magma or lava from about 1,400 to 750 degrees centigrade. The continuous reaction series crystallizes at the same time as the discontinuous series to form all of the common igneous rocks. Conversely, the diagram and concept shows how minerals melt if rocks are heated to their melting point. It neatly explains assemblages of minerals in igneous rocks, their temperatures of formation and many igneous textures. Although a simplification of a very complex series of processes, the Bowen's Reaction Series concept is surprisingly applicable in most rocks.

This widely applicable concept was derived through years of research. Norman Bowen solved many of the basic petrologic (study of rocks) field problems by defining laws and principles derived from experimentally determined chemical relationships (phase diagrams) of common minerals. As a result of this groundbreaking research, petrologists were able to approach igneous rocks quantitatively, whereas previously the main focus was only on description and classification. His experimental work involved the melting and quenching of rocks at a series of temperatures to determine their relations of crystallization. From these data he would construct a "phase diagram" from which melt percentages, melt compositions, types, and percentages of minerals crystallized could be determined at any given temperature. The nepheline-anorthite diagram was the first completed very efficiently using 17 different mixtures and 55 quenching experiments. This system was the first example found in silicates of solid solution. Bowen then studied the two-component system of plagioclase, albite-anorthite. These results helped determine the basis for Bowen's views on magma differentiation and crystal fractionation. Both of these theories had not been demonstrated experimentally prior to the research Bowen and his colleagues had accomplished. Bowen subsequently experimented with many other systems.

Bowen published numerous papers but probably his most famous work was his 1928 book, *The Evolution of Igneous Rocks.* In this book, he explains phase diagrams for common rock systems. Although still a simplification, the results apply so well to field and petrographic observations of igneous rocks that it became an instant handbook for igneous petrologists. It still remains one of the most important books in geology.

Norman L. Bowen was born in Kingston, Ontario, on June 21, 1887. He completed his elementary and high school education in Kingston public schools, and entered Queen's University, Canada. Bowen had his sights set on becoming a teacher but after one year decided to join an Ontario Bureau of Mines geological mapping party to Larder Lake with the allure of money and travel. It was a revelation for him and he enrolled in the School of Mining upon his return, registering in mineralogy and petrology. He graduated with a bachelor of science degree in chemistry and geology in 1909. He received medals in chemistry and mineralogy and was named the 1851 Exhibition Scholar. Bowen continued with his graduate studies at Massachusetts Institute of Technology in Cambridge.

In 1910, Bowen applied to the Geophysical Laboratory at the Carnegie Institution of Washington, D.C., to complete an experimental study related to a geological field problem as part of the requirement for his Ph.D. During this time, Bowen married his college sweetheart, Mary Lamont, on October 3, 1911. The following spring (1912) Bowen graduated with a Ph.D. in geology and was busy fielding job offers. Bowen accepted the position as assistant petrologist at the Geophysical Laboratory. Besides a 10-year period of teaching at the University of Chicago, Illinois (1937 to 1946), including two years as department chair, Bowen remained at the Geophysical Lab for his entire career and directed it for most of the time. He embodied the Geophysical Laboratory. Bowen officially retired in 1952 and the next year he moved to Clearwater, Florida, to enjoy his golden years. However, he grew restless after only a few months and returned to Washington, D.C., and was appointed research associate at the Geophysical Laboratory. Norman L. Bowen died on September 11, 1956.

Norman Bowen led a phenomenally productive career not only in terms of total publications but also in terms of impact on the field. For example, between 1945 and 1954, five of the 20 most often cited articles in all of geology were written by Bowen and his associates. There are no truer classics in petrology than those written by Bowen. As recognition for these outstanding contributions, he received numerous honors and awards. He was a member of the U.S. National Academy of Sciences, the American Academy of Arts and Sciences, the Indian Academy of Sciences, and the Finland Academy of Sciences. He received honorary degrees from Harvard University, Yale University, and his alma mater, Queen's University. He also received the Bigsby Medal and the Wollaston Medal from the Geological Society of London, the Penrose Medal from the Geological Society of America, the Roebling Medal from the Mineralogical Society of America, the Miller Medal from the Royal Society of Canada, the Hayden Medal from the Academy of Natural Sciences of Philadelphia, and the Bakhuis Roozeboom Medal from the Royal Netherlands Academy. The American Geophysical Union named a medal in his honor.

Bowen was also very active in service to the profession. In addition to serving as president of both the Geological Society of America (1946) and the Mineralogical Society of America (1937) he was a member and chair of numerous committees and panels for both societies and the government.

Bowring, Samuel A.
(1953–   )
American
Isotope Geochemist

The Cambrian-Precambrian boundary is the most profound transition in the geologic record in terms of life. This boundary marks the demise of a rich diversity of invertebrate fauna that lack shells, including some jellyfish and worms, but also some complex forms. They were replaced by a whole new group of shelled invertebrate fauna in the Cambrian, many of which are the ancestors of our modern marine invertebrates. Classically, this boundary was considered to have occurred approximately 600 million years ago but was later revised to 570 million years ago. Recently, however, Sam Bowring, working in conjunction with sedimentologist John Grotzinger and paleontologist ED LANDING, among others, has revised that age to 534 million years using new high-precision geochronology. This work is a

major contribution to the science. He further determined the ages for the appearance and changes of certain animals in the Cambrian age, thus determining rates of evolution during this period of rapid diversification. This detailed geochronology of individual volcanic layers coupled with detailed stratigraphy and paleontology on a layer-by-layer basis as shown in his paper "A New Look at Evolutionary Rates in Deep Time: Uniting Paleontology and High Precision Geochronology," sets a new precedent in evolutionary analysis. It has already led to new insights and will likely lead to more in the future. During this project he performed research on rocks from the Avalon Terrane in Nova Scotia, Canada, as well as those in Namibia and Madagascar, Africa, and the White Sea in Russia.

The other major area of research for Sam Bowring is the development of the continental crust. Ocean crust is created at the mid-ocean ridges and destroyed at the subduction zones within about 200 million years. Continental crust, on the other hand, has been built throughout the history of the Earth. Because there are numerous and complex processes in the assembly of a continent, all of which overprint and modify each other in complex ways, deciphering the geology of continents becomes a monumental task. Bowring collaborates with tectonic and regional geologists to provide the geochronologic (age) constraints on some of these events. He has done research on 3.96-billion-year-old gneiss in the Slave Province of the Northwest Territories of Canada, which are among the the oldest rocks on Earth and therefore among the earliest continental crust. He defined 2- to 2.4-billion-year-old crust in the western United States in Arizona and New Mexico. He also performed research on Precambrian rocks from the Natal Province of South Africa, as well as those from Namibia, Botswana, and Zimbabwe, Africa. It is clear that Bowring will travel to the ends of the Earth to find the best location to research the particular process that he is studying at the time. This care in the details of

Sam Bowring (left) on a field trip with a student
(Courtesy of Sam Bowring)

his isotopic analysis, his effort in forming solid collaborations with geologists with complimentary expertise, and his care in choosing only the best examples on which to perform research has propelled Bowring to a position of one of the premier scientists on Precambrian research.

Samuel Bowring was born on September 27, 1953, in Portsmouth, New Hampshire, where he spent his youth. He enrolled at the University of New Hampshire in Durham and earned a bachelor of science in geology, cum laude, in 1976. He completed a master of science degree at New Mexico Institute of Mining and Technology in geology in 1980. He earned a Ph.D. in geology from the University of Kansas at Lawrence in 1985. Bowring joined the faculty at Washington University in Saint Louis, Missouri, in 1984. He moved to Massachusetts Institute of Technology in Cambridge in 1991, and he remains there as professor of geology today.

Sam Bowring is leading a very productive career. He is an author of 82 articles in international journals and professional volumes. He also has several other publications in field trip guides and governmental reports. Many of these papers are seminal works on the early history of the continental crust or the defining works on the Precambrian-Cambrian boundary. Many are published in high-profile journals like *Science*. Bowring has received several honors and awards for his contributions to the science. He is a member of the American Association for the Advancement of Science. He was named the Louis Murray Fellow at the University of Cape Town, South Africa, in 1995. As a graduate student, he received the Dean A. McGee and McCollum Burton Scholarships and the Erasmus Haworth Honors in geology. He has also been invited to present several important keynote addresses worldwide.

Bowring has been involved in significant service to the profession. He served as associate editor for *Geology Magazine,* and *Journal of Geophysical Research*. He was also on the editorial board for *Precambrian Research.*

Bragg, Sir (William) Lawrence
(1890–1971)
English
Mineralogist

Although Sir Lawrence Bragg was trained as a physicist and employed as a physicist or chemist throughout his career, he was tremendously influential in Earth sciences, as he was in metallurgy and medicine as well. After Röntgen discovered X rays in 1895, von Laue demonstrated that the X rays were diffracted in a three-dimensional scattering if passed through the mineral zincblende in 1912. With his father, physicist Sir William Bragg, Lawrence Bragg showed that this complex scattering could be perfectly explained by reflections of the X rays from successive planes of atoms in the mineral structure. A paper on this work is entitled, "The Analysis of Crystals by X-Ray Spectrography." He determined the mathematical conditions of this diffraction in an equation that has been named Bragg's Law. He and his father then developed an X-ray spectrometer which was used to determine the atomic structure of rock salt, diamond, flourspar, pyrite, calcite, cuprite, corundum, and metallic copper. For this breakthrough, the father-and-son team were jointly awarded the Nobel Prize for physics in 1915. Lawrence Bragg was 25 years old at the time.

It is almost a curse to begin a career with such success because everything else tends to pale in comparison. This, however, was not the case with Lawrence Bragg. He went on to apply his X-ray diffraction techniques to minerals, to metallurgy, and finally to medical problems. He slowly worked his way through minerals of increasing structural complexity, finally addressing the silicates. His 1934 book, *Atomic Structure of Minerals* (rewritten as *Crystal Structure of Minerals* in 1965) is a summary of these findings. Amazingly, scientists went from having no idea how the atoms are arranged in minerals to a general understanding of crystal chemistry through this single development. This development came from a man who really knew nothing else about minerals, as he admitted in a famous speech to the American Mineralogical Society.

Bragg would later show how X-ray patterns and thus atomic structure of deformed metal differs from undeformed metals. He devised an X-ray microscope and a "fly's eye" apparatus to provide the basis for high-magnification optical methods that would be devised later by others. He also proved the crystal chemistry of hemoglobin and of proteins still later, which had profound implications for the medical field. By this time, Bragg's notoriety had reached throughout Great Britain. He was asked to give the vacation lectures by the Royal Institution and even a highly successful series of television broadcasts on the properties of matter, which further increased his fame. He devised a series of simple but elegant experiments that became classics of British television.

William Lawrence Bragg was born on March 31, 1890, in Adelaide, South Australia, the son of Sir William Bragg, a professor of physics at the University of Leeds, England. He attended Cambridge University, England, where he earned several degrees, including a Ph.D. in physics in 1913. He joined the cavalry in 1915 to serve in World War I. He devised a method for location of enemy artillery using sound and was decorated with the Military Cross in 1918 as a result. In 1919, he joined the faculty at the University of Manchester, England, as a professor of physics. He was a visiting professor at Cornell University, New York, in 1934. Lawrence Bragg was married in 1921; he and his wife, Lady Alice, had four children and numerous grandchildren. In 1937–1938, Bragg served as the director of the National Physical Laboratory of Great Britain before accepting the position of Cavendish Professor of experimental physics at Cambridge University, England, in 1939. His final move came in 1953 when he accepted the position of Fullerian Professor of chemistry at the Royal Institution in London, England. He assumed the role as director in 1954 and remained as such until 1966, when he retired.

He lived the rest of his life enjoying his family and occasionally giving public lectures. Sir Lawrence Bragg died on July 1, 1971.

The career of Sir Lawrence Bragg can be described as nothing less than distinguished. He is an author of more than 180 international publications ranging from cutting-edge scientific to general interest. The impact of many of these publications on geology as well as physics, metallurgy, and medicine cannot be overstated. In recognition of this illustrious career, Bragg received numerous prestigious honors and awards. In addition to the Nobel Prize in 1915 and a knighthood, which he received in 1941, Bragg was named as a Fellow of the Royal Society in 1921. He received the Hughes Medal, the Copley Medal, and the Royal Medal from the Royal Society of London, the Roebling Medal from the Mineralogical Society of America, and the Companion of Honor, a rare distinction for a scientist, among many others.

⊠ Brantley, Susan L.
(1958–   )
American
Aqueous Geochemist

Environmental geology commands the most interest and concern of all of the fields in Earth science today. A large part of this field is the chemical system formed by the interaction of rocks, fluids, and gases. It not only affects issues like water quality and pollution control, but air quality and the greenhouse effect. These complex interactions are the mainstay of the field of aqueous geochemistry, of which Susan Brantley is one of the premier experts. She studies the chemical processes and compositional control of natural waters both at the surface of the Earth as well as deeper in the crust. This research is conducted both with laboratory experimentation as well as in field areas from the deserts of Peru to the glaciers of Iceland. Experimental work involves dissolution studies of certain minerals under a variety of conditions to simulate weathering as well as subsurface process. Brantley has produced two important volumes on this work, including *Geochemical Kinetics of Mineral-Water Reactions in the Field and in the Lab* and *Chemical Weathering Rates of Silicate Minerals.* She is especially interested in feldspars, including the effect of coatings, and the release of trace components. However, she has also conducted dissolution studies on many other minerals including olivine, anthophyllite, and pyroxene among others. Another interesting aspect of this research is the study of the effect of bacteria on weathering. Bacteria and other microbes are situated right at the interface between the fluid and the mineral surfaces where all weathering and other fluid-rock interactions take place. She has documented the removal of iron from the silicate mineral surfaces by long-lived bacteria colonies. This field of biogeochemistry is one of the most promising directions in geology. It not only has implications for environmental sciences but for material sciences and possibly even for the medical field. Bacteria can enhance dissolution or stabilize surfaces as well as provide a buffering control on the composition of fluids.

In addition to this experimental and detailed mineral research, Brantley and her students have conducted some interesting field projects. The field projects are typically designed to complement the experimental research. She does research on the hydrogeochemistry of active volcanoes. Several of these volcanoes have included Volcan Poas in Costa Rica, Grimsvotn in Iceland, and Ol'Doinyo Lengai in Tanzania. The Costa Rica research was completed on the chemistry of Laguna Caliente, one of the most acidic natural waters in the world with pH consistently below zero. Brantley is also studying the carbon dioxide flux from the geysers and other geothermal features of Yellowstone National Park, Wyoming. A similar study documents the degassing of the volcanic field in Campi Flegrei in Italy.

Susan Brantley was born on August 11, 1958, in Winter Park, Florida. She attended Princeton University, New Jersey, where she earned a bachelor of arts degree in chemistry (magna cum laude) in 1980. In 1980–81, Brantley was a Fulbright Scholar in Peru. She also completed her graduate studies at Princeton University, where she earned a master of arts and a Ph.D. degree in geological and geophysical sciences in 1983 and 1987, respectively. Brantley was a National Science Foundation Graduate Student Fellow and an IBM Student Fellow for much of her graduate career. In 1986, she joined the faculty at Pennsylvania State University at University Park, where she remains as of 2002. She became the director of the Center for Environmental Chemistry and Geochemistry in 1998 and the director for the Biogeochemical Research Initiative for Education in 1999. In 1995, Brantley was a visiting scientist at both the U.S. Geological Survey in Menlo Park, California, and Stanford University, California.

Susan Brantley is still in the early stages of a productive career. She is an author of some 78 scientific publications in international journals, professional volumes, and conference proceedings. Many of these papers are seminal studies on geomicrobiology and processes of aqueous geochemistry and appear in the best of journals including the high-profile journal *Nature.* She is also an editor of two professional volumes. Brantley has received several honors and awards in recognition of her research contributions to the geologic profession. She received the Presidential Young Investigator Award through the National Science Foundation (1987–1992), the David and Lucile Packard Fellowship (1988–1993), and the Wilson Research Award from Pennsylvania State University (1996).

Brantley has performed outstanding service to the profession at this point in her career. She served as councilor for the Geochemical Society as well as a member of several committees. She also served on committees and panels for the National Research Council, the National Academy of Sciences, and the National Science Foundation. She served in several editorial capacities including editor for both *Chemical Geology* and *Geofluids* as well as assistant editor for *Chemical Geology.*

⊠ **Bredehoeft, John D.**
(1933–)
American
Hydrogeologist

With the mounting pressure to find clean sources of groundwater as the population of the world increases, hydrogeology has emerged from the shadows of the Earth sciences to become perhaps its most important discipline. Indeed, protection of water resources is one of the most pressing needs of society today. One of the true pioneers in shepherding this emergence is John Bredehoeft. He was among the first to apply quantitative methods to modeling the underground flow of water. He developed numerical models to predict the direction and speed of this flow as well as the transport of contaminants and wrote them into widely adopted computer programs. These models were not only applied to contaminated sites like the San Francisco International Airport, but also for economic analysis for optimal groundwater development in a modified version. His expertise in groundwater flow and environmental impact was also applied to the disposal of high-level nuclear waste. He developed his own plan for the burial of waste in crystalline rocks beneath a cover of sediment that contradicted accepted practices. In this role, Bredehoeft evaluated and advised on the Waste Isolation Pilot Plant in New Mexico and the Yucca Mountain Repository in Nevada.

In more of a pure research role, Bredehoeft conducted several investigations into the hydrodynamics of fluid flow in the deep subsurface. These studies are regional in nature to explain large-scale movement. Among these investigations are a model of the Dakota Sandstone and associated aquifers (water-bearing rock units) and artesian (pressurized) systems in South Dakota and similar studies of the Denver Basin, Colorado, the Big Horn Basin of Wyoming, the Uinta Basin of Utah, and the Illinois Basin. He also produced an analytical flow model for the Caspian Basin of Russia. Many of these studies provided new and innovative explanations for the patterns including the role of geological membranes and partitioning of aquifers with shale layers. He was even involved with the high-pressure injection of fluid into deep wells to produce earthquakes in Rangely, Colorado. He attempted to use the information gleaned in this project to help predict earthquakes in California using data from water wells near active faults.

John Bredehoeft was born on February 28, 1933, in Saint Louis, Missouri. He attended Princeton University, New Jersey, where he earned a bachelor of science degree in geological engineering with honors in 1955. He completed his graduate studies at the University of Illinois at Urbana-Champaign, earning a master of science degree in geology in 1957 and a Ph.D. in geology with a minor in civil engineering in 1962. Between his graduate degrees, from 1957 to 1959, he worked as an exploration geologist for Humble Oil in Vernal, Utah. John Bredehoeft married in 1958; he and his wife, Nancy, have three children. During the later stages of his doctoral work, Bredehoeft also worked as a groundwater hydrologist for the Nevada Department of Conservation and Natural Resources in Reno in 1961–1962. He joined the U.S. Geological Survey in 1962 as a research geologist in Arlington, Virginia, and remained until his retirement in 1994. During that time, he held positions as deputy chief hydrogeologist for research (1970–1979), regional hydrogeologist in Menlo Park, California (1980–1984), and research geologist supergrade, also in Menlo Park (1984–1994), among others. He was also a visiting professor at the University of Illinois (1967–1968) and a consulting professor at Stanford University, California (1989–1991). In 1995, Bredehoeft established his own environmental consulting company, The Hydrodynamics Group, in La Honda, California, where he is still the principal today.

Productivity in John Bredehoeft's career can be measured in a variety of ways, including governmental and industrial reports in addition to some 100 research papers in scientific literature. Many of these papers contain widely adopted methods to model groundwater flow and contaminant transport in addition to site specific studies. In recognition of his research contributions to hydrogeology, Bredehoeft has received several honors and awards. He was named to the U.S. National Academy of Engineering and the Russian Academy of Natural Sciences. He received the Horton Award from the American Geophysical Union, the Penrose Medal, and the O. E. Meinzer Award from the Geological Society of America, the Meritorious Service Award and the Distinguished Service Award from the U.S. Department of the Interior, the Boggess Award from the American Water Resources Association, the M. King Hubbert Award from the National Ground Water Association, and the Alumni Achievement Award from the University of Illinois, among others.

Bredehoeft performed extensive service to the profession and the public through his governmental position. He served on the board of directors for the National Ground Water Association, the Council for the International Exchange of Scholars, and on numerous committees for nuclear waste for the National Research Council. He also served on advisory committees for the U.S. Department of Energy, the National Science Foundation, UNESCO, American Geophysical Union, and the Geological Society of America, among others. Bredehoeft has also served as the editor for the journal *Ground Water* for many years.

Broecker, Wallace S.
(1931–)
American
Chemical Oceanographer

Wallace Broecker has said that his entire research career is simply an elaboration on several chapters of his Ph.D. dissertation in 1958. Considering the magnitude of accomplishments in his career, this statement is analogous to the saying, "All I ever needed to know I learned in kindergarten." It is true that two themes, evidence for an abrupt change in climate 11,000 years ago and the distribution of radiocarbon around the Atlantic Ocean generalize the basic ideas of his research, but they hardly summarize his career.

There are several groundbreaking contributions in Wallace Broecker's career. His early dating of ocean sediments using carbon-13 methods set the stage for him to use decay products of uranium to obtain an older range. Instead of the 25,000-year limit of dating using radiocarbon, he extended his dating to 320,000 years. He studied the impact of glaciation on sedimentation in relationship to astronomical cycles. His work on reefs and carbonate banks led him to investigate the carbon and oxygen cycles with regard to these deposits and the role of atmospheric gases. He used stable isotopes to define these relationships and mapped the relative CO_2 concentrations in the oceans.

In his most famous work, Broecker defined the interaction of the sediments, oceans, and atmosphere using evidence from carbon dioxide flux. He studied CO_2 content of ice cores from Greenland in addition to stable isotopes. He found that the CO_2 content showed abrupt changes between two basic levels, glacial and interglacial. It appears that the content of this greenhouse gas could be abruptly changed based upon ocean circulation in the North Atlantic. This research means that excessive production of greenhouse gases could cause a sharp rather than gradual change in climates, which would be dev-

Wallace Broecker outside of the Lamont-Doherty Earth Observatory in 1990 *(Courtesy of Susan Rogers)*

astating. Broecker studied the downslope limits of alpine glaciers as a reflection of temperatures and found the same abruptness of change. Deep-sea coring of sediments in the North Atlantic supported periodic catastrophic bursts of ice into the ocean which completely disrupted the chemical systems.

This studying of the surface of the Earth in terms of a single large chemical system in various stages of equilibrium and disequilibrium in response to flux of various gases defines virtually all of Broecker's research. He studied the flux of gas by measuring radioactive radon in the water. Because it has such a short life, concentration gradients show flux direction and rate. Using all of these chemical systems, he wrote his acclaimed textbook *Tracers in the Sea*. He also studied flux in terrestrial environments in lakes eutrophied with phosphate in Canada. Considering the vast and ever-growing eutrophication of our lakes in devel-

oped areas, this work has great applied importance. He is even involved in carrying out experiments in the famous Biosphere 2. Because it is a closed chemical system, experiments may be carried out on mass balances of CO_2 and other gases. As evident from this review of his work, Wallace Broecker has produced research that is not only groundbreaking within the profession, it also has profound implications for our continued existence on the planet. In a field crowded with giants, Wallace Broecker has established himself as a leader.

Wallace Broecker was born on November 29, 1931, in Chicago, Illinois. He grew up in Oak Park, Illinois, during the Great Depression and graduated from Oak Park-River Forest High School in 1949, the alma mater of Ernest Hemingway. He attended Wheaton College, Illinois, but transferred to Columbia College, New York, where he earned a bachelor of arts degree in 1953. He married his wife, Grace, in 1952. He remained at the Lamont-Doherty Geological Observatory of Columbia University for his graduate studies and earned a master of arts and a Ph.D. in 1956 and 1958, respectively. His dissertation adviser was J. Laurence Kulp, but several prominent classmates and colleagues, both internal and external, appear to have had as much, if not more, influence on his work. Broecker never left Lamont-Doherty Observatory. He started as an instructor in 1956 and remains there today as the Newberry Professor of Earth and environmental science since 1977. He served as chair of the department from 1977 to 1980. He was also a visiting professor several times to places like California Institute of Technology and Heidelberg, Germany, where he was a von Humboldt Fellow.

The productivity that Wallace Broecker has demonstrated is simply astounding. He is an author of 385 articles in international journals, professional volumes, and technical reports. One year he published 23 scientific articles. These studies are some of the best recognized of their kind and appear in the most prestigious journals.

He has also written six books, two of which are highly regarded textbooks. His achievements have been well recognized by the profession in terms of honors and awards. He is a member of the National Academy of Sciences and the American Academy of Sciences. He was awarded a Sloan Fellowship in 1964, the Arthur L. Day Medal by the Geological Society of America (1984), the Huntsman Award by the Bedford Institute of Canada (1985), the Vetlessen Award from the Vetlessen Foundation (1987), and the Goldschmidt Award from the Geochemical Society (1988).

Broecker has performed service to the profession too extensive to list here. He was president of the Geochemical Society in 1981. He has served on numerous panels and committees for the National Science Foundation, National Research Council, Joint Oceanographic Institute for Deep Earth Sampling, several national and international oceanic boards, the Geochemical Society, and the Geological Society of America, among others.

⊠ **Bromery, Randolph W. (Bill)**
(1926–)
American
Geophysicist

Randolph Bromery pioneered the integration of African Americans into the world of Earth scientists and has achieved the greatest success in this regard. Not only has he been successful in contributing to the science, but also he has assumed a number of high-profile positions in government, academia, and industry, all with equally impressive results. He serves on the board of directors for such companies as Exxon, Chemical Bank, NYNEX, John Hancock Insurance, Singer, Southern New England Telephone, and Northwestern Life. He helped found the Weston Geophysical International Corporation in 1981, and served as manager from 1981 to 1986. He also founded the Geoscience Engineering Corporation in 1983.

Utilizing his pilot training, Bromery's first research efforts were to become involved in pioneering efforts of airborne geophysical surveying. This work involved not only the development and testing of new equipment but also the interpretation of the data obtained. The research began with airborne magnetic surveying of the United States by flying numerous parallel straight paths at one-mile spacing or less and taking regular individual readings until the whole target area was covered. The next airborne geophysical method to be investigated was radioactivity, which was also done for the contemporaneous United States, Alaska and Hawaii. This research was extended to West Africa, where he planned and executed a survey of Liberia in a U.S. State Department–sponsored program. The main goal of the program was to search for economic mineral deposits and was extended to other countries as well. Bromery also conducted land-based gravity and other geophysical investigations.

Randolph (Bill) Bromery was born on January 18, 1926, in Cumberland, Maryland, where he grew up. The Great Depression made his youth financially difficult. He attended the segregated and poorly funded Frederick Street School, but he was fortunately able to attend the new George Washington Carver High School, where he was a member of the first graduating class in 1942. He supplemented his formal education with tutoring to make up for the deficiencies of the school. Because he had advanced machine shop training in an after-school program through President Roosevelt's National Youth Administration, Bromery was able to obtain a machinist job in Detroit, Michigan. It was short-lived, however, because he enlisted in the U.S. Army Air Corps and was called to active duty in 1943. He was trained as a pilot and assigned to the 99th Air Squadron as part of the famous Tuskegee (Alabama) Airmen. He was stationed in southern Italy, where he flew fighter escort missions.

After his discharge in 1945, Bromery took a correspondence course in mathematics from the University of Utah to achieve admission to the University of Michigan at Ann Arbor. Unfortunately, his mother became ill and he transferred to Howard University in Washington, D.C., to be near her after only one year. He left Howard University in 1948 before graduating to take a position with the Airborne Geophysics Group at the U.S. Geological Survey in Cabin John, Maryland. Bill Bromery married Cecile Trescott that year; they have five children. Bromery finally returned to Howard University to complete his bachelor of science degree in mathematics in 1956. He entered the graduate program in geology at the American University in Washington, D.C., as a part-time student, and was awarded a master of science degree in 1962. He attended the Johns Hopkins University in Baltimore, Maryland, for the remainder of his graduate studies and earned a Ph.D. in geology on a Gilman Fellowship in 1968. His adviser was ERNST CLOOS.

In 1967, Bromery accepted a faculty position at the University of Massachusetts at Amherst. He became chair of the department in 1969, but moved to vice chancellor for student affairs in 1970. In 1971, however, he was made acting chancellor and finally chancellor for the University of Massachusetts. In 1977, he was also named executive vice president and then senior vice president as well. By 1979, Bromery was tired of administrative work and returned to the faculty as Commonwealth Professor of geophysics. Soon administrative work beckoned again and he served as president of Westfield State College in Massachusetts from 1988 to 1990. He moved directly to interim chancellor of the board of regents of higher education of Massachusetts from 1990 to 1991. In 1992, he was named president of Springfield College, Massachusetts, where he remained until 1999. He then moved back to the position of full chancellor of the board of regents where he remains today.

Among all of his numerous administrative and industrial positions, Randolph Bromery managed to lead a productive scientific career. He is an author of some 100 scientific publications in international journals, professional volumes, and governmental reports. Many of these form the basic groundwork for airborne geophysical surveying both in the United States and overseas. In recognition of his contributions to geology and pioneering efforts for minority participation in science, Randolph Bromery has received several honors and awards. He has received honorary doctoral degrees from Western New England College, Frostburg State College, Westfield State College, Hokkaido University in Japan, and North Adams State College. He was named Outstanding Black Scientist by the National Academy of Sciences and he received the Distinguished Service Award from the Geological Society of America, where he has served on numerous committees in leadership roles.

⊠ **Brown, Michael**
(1947–)
British
Metamorphic Petrologist

One segment of the rock cycle contains the transformation of metamorphic rocks into igneous rocks by heating and melting. However, it is not quite so simple. Because minerals melt at different temperatures and pressures, there is a point where rocks are part newly melted material and part old metamorphic rock. Once crystallized, these rocks are called migmatites (mixed rocks) and composed of minimal melts of granitic composition called leucosome and leftover metamorphic rock called melanosome. Migmatites are important in the genesis of granites but also extremely complex both chemically and in the way they deform. Michael Brown has taken the challenge to research these complex rocks and has firmly established himself as one of the foremost experts. This work began with his doctoral dissertation on the St. Malo Migmatite Belt in northeastern Brittany, France. He extended this work to southern Brittany but also to basement rocks in Timor and Proterozoic rocks in peninsular India and more recently in the northern Appalachians and in Proterozoic rocks from Brazil. In the Appalachians, the studies include work on contact metamorphism around plutons in Maine, whereas the other studies are largely on regionally metamorphosed rocks. The chemical complexity of these rocks involves the melting reactions and concentrations of incompatible elements which do not fit easily into common rock forming minerals. The structural complexities result from the deformation of liquid rock sandwiched between layers of gumlike "solid" rock. Common techniques that are used on solid rocks do not work well on migmatites.

Michael Brown has a research interest in the genesis and emplacement of granitoid plutons. He conducted research on plutons in the Channel Islands between England and France, the Qoqut granite complex in Greenland with the Geological Survey of Greenland, plutonic rocks of the Atacama Region of Chile with the Servicio Nacional de Geologia y Mineria de Chile, and on granite plutons in Maine with support from the National Science Foundation. The investigations into the genesis of these plutons is largely an extension of the work on migmatites in that they can act as the source of magma. This research involves detailed whole rock and mineral geochemistry, as well as isotope geochemistry. He also studies the mechanics of the emplacement of the granite plutons and the contact metamorphism they impose upon the country rocks. Many of these plutons were emplaced into regions undergoing strike-slip deformation producing intriguing geometries.

Michael Brown was born on March 19, 1947, in Hayes, Middlesex, England. He attended the University of Keele, United Kingdom, where he graduated with a bachelor of arts degree with double honors in geography and geology and mi-

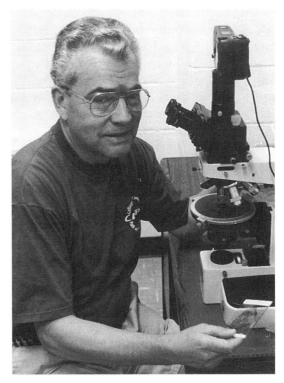

Michael Brown works at the petrographic microscope at the University of Maryland *(Courtesy of Michael Brown)*

nors in chemistry and politics in 1969. He remained at University of Keele for his graduate studies and earned a Ph.D. in geology in 1974, supported by a National Environment Research Council Studentship. Brown accepted a position as lecturer at Oxford Brookes University in 1972. He became department head in 1982. In 1984, he moved to Kingston University, United Kingdom, as head of the School of Geological Sciences. He also served as assistant dean of academic affairs from 1986 to 1989. Brown moved to the United States in 1990 to become the chair of the department at the University of Maryland at College Park where he remains today. Brown was a visiting professor at University of Kyoto, Japan, in 1993 and at Kingston University in 1990–1992. Michael Brown has three children.

Michael Brown is in the middle of a productive career. He is an author of some 100 scientific articles in international journals, professional volumes, and governmental reports. Many of these are seminal works on migmatites and other highly metamorphosed rocks as well as granites. Several of these studies appear in high-profile journals like *Nature.* Brown has performed significant service to the geologic profession. He has served in several capacities for the Geological Society of America, the Geological Society of London, where he served as member of the council, the Mineralogical Society of Great Britain, the American Geophysical Union, and the Mineralogical Society of America. His editorial roles are also numerous. He served as founding editor of the *Journal of Metamorphic Geology,* subject editor for the *Journal of the Geological Society of London,* and coleader of the International Geological Correlation Program Project 235 "Metamorphism and Geodynamics."

⊠ **Buddington, Arthur F.**
(1890–1980)
American
Petrologist

Arthur Buddington is a geologist famous for his work with the geochemistry and classification of rocks and minerals, as well as ore deposits. His research was based on his perceptive observations during the field mapping of geologic terranes that vary considerably. Buddington may be best known for classification of anorthosites, which are igneous rocks composed mainly of the mineral plagioclase feldspar. He defined a Grenville type (approximately 2 billion years of age) that can be formed in one of two ways. Anorthosites can intrude up into the ground in huge crystalline masses or by the formation of crystals which settle out within layers of gabbro-based complexes. These two types were based on his fieldwork in the Grenville terrain of the Adirondacks and his

observations on the Stillwater Complex of Montana. This work is summarized in a 1970 volume entitled *The Origin of Anorthosites and Related Rocks,* published by the New York State Museum. In a 1959 paper, Buddington proposed a system for the origin of various igneous intrusive rocks based upon the depth at which they were formed. He devised this system which would have to wait for the advent of plate tectonics to be fully appreciated based upon observations in Newfoundland, the Alaska Coast Ranges, the Adirondack Mountains, and the Stillwater Complex of Montana.

Arthur Buddington was also involved in economic geology. During his field studies of shallow intrusive igneous rocks of the Oregon Cascades and their related iron ores he defined a class of iron ore deposits he termed "xenothermal" which means shallow depth and high temperature. It was originally believed that the temperature during the ore formation and the depth at which they were formed correspondingly affected one another. Buddington's research on this class of iron ores showed that this was not the case. He also conducted research on iron ore deposits in New York and New Jersey. With DONALD H. LINDSLEY, Buddington completed research that contributed to the development of advanced geothermometers and oxygen activity-meters through large amounts of data he collected during many studies on magnetite-hematite-ilmenite ore deposits of the Adirondack region.

Perhaps the main contribution of Buddington to geology was his administrative and organizational skills. He enticed NORMAN L. BOWEN to give a series of lectures at Princeton University in the mid-1920s and then to write them up in the famous *Evolution of the Igneous Rocks.* He also was a major positive influence on his many students. Two of these students included HARRY H. HESS and J. TUZO WILSON, who contributed greatly to the theory of plate tectonics. It was mainly through Buddington's influence that Hess would lead his illustrious career at Princeton University.

Arthur Buddington was born in Wilmington, Delaware, on November 29, 1890, to parents who operated a small poultry and produce farm. He attended elementary and junior high school in Wilmington and in Mystic, Connecticut, and graduated from Westerly High School, Rhode Island, in 1908. Buddington attended Brown University, Rhode Island, and graduated with a bachelor of arts degree in geology second in his class in 1912. He continued his graduate studies at Brown University and earned a master of science degree in 1913. He switched to Princeton University, New Jersey, for the rest of his graduate studies and received his Ph.D. in geology in 1916. He remained at Princeton University on a postdoctoral fellowship until he accepted a teaching position at Brown in 1917. In 1918, he enlisted in the army and worked in the Chemical Warfare Service during World War I. After the war ended, Buddington returned to Brown University, but quickly accepted an appointment to the Geophysical Laboratory at the Carnegie Institution of Washington, D.C., in 1919. He finally joined the faculty at his alma mater of Princeton University in 1920, and remained for the rest of his career. Arthur Buddington married Jene Elizabeth Muntz of David City, Nebraska, in 1924. They had one child. During this time he worked with the U.S. Geological Survey in 1930 and from 1943 on and he served as department chairman from 1936 to 1950. He retired to professor emeritus in 1959 but remained active for many years. Arthur Buddington died on December 25, 1980; his wife had predeceased him five years earlier.

Arthur Buddington was an author of some 70 scientific articles in international journals, professional volumes, and governmental reports. The most impressive aspect of them is the tremendous range of subject matter and extraordinary quality. In recognition of these contributions to the science, Buddington received several honors and awards. He was a member of the National Academy of Sciences and the American Academy of Arts and Sciences. He received honorary degrees from Brown University, Franklin and Marshall College, and the University of Liège. He also received the Penrose Medal from the Geological Society of America, the Andre Dumont Medal from the Geological Society of Belgium, and the Distinguished Service Award from the U.S. Department of the Interior. The mineral buddingtonite was named in his honor.

Buddington was also active in service to the profession. He was president of the Mineralogical Society of America, vice president of the Geological Society of America, and section president of the American Geophysical Union, among service on many other committees and panels. He was also chair of the geology section for the National Academy of Sciences.

⊠ **Bullard, Sir Edward C.**
(1907–1980)
British
Geophysicist

Sir Edward Bullard was one of the giants of geophysics. Even though he started out as a physicist, carefully experimenting on electron scattering in groundbreaking research, he quickly realized that his true calling was in geophysics. His work set the stage for modern methods in geophysics in four areas: heat flow, generation of Earth's magnetic field, and gravity and seismic refraction methods. Bullard devised the methods that are currently used for heat flow surveying shortly after World War II. He devised methods to measure thermal conductivity, which unless incorporated into heat flow analysis leads to erroneous results. He used thermal gradients in South African gold mines, as well as British coal mines, in addition to numerous borehole measurements to show that thermal conductivity greatly affects the actual temperatures that are measured at various depths. He established the first reliable average heat flux value for the continents. Perhaps even more impressive were his methods to measure heat flow on the ocean floor. He invented a device that was driven into the ocean floor to obtain temperatures of the underlying rock. He then tested it on deep-sea cruises with ROGER REVELLE and others at the Scripps Institution of Oceanography, California. The result was the identification of elevated heat flow at the mid-ocean ridges, which later served as an important piece of evidence for plate tectonics.

His interest in designing new practical, portable equipment to measure geophysical quantities also spread to gravity and seismic refraction techniques. His earliest geophysical work was to redesign a pendulum apparatus for measuring gravity and then to conduct a field study with it over the East African Rift System, which earned him a Smithson Fellowship from the Royal Society. He designed a portable short period seismograph (to take a sonogram-like picture of the Earth) which he used to measure the depth to basement in a field survey in southeast England. Although these pieces of equipment were later redesigned, his success in carrying out field geophysical surveys set the stage for modern geophysical work.

After his extensive work in magnetics during World War II, Sir Edward Bullard investigated how motions in the Earth's core might induce the main magnetic field. He measured secular (short-term) changes in the magnetic field, like the rate of westward drift of the poles, and then used mathematical models to obtain numerical solutions for fluid mechanical problems in the Earth's core. He was the first to model the core as a self-exciting dynamo using primitive computers, thus setting the stage for later more appropriate studies. Bullard even got involved with the plate tectonic revolution. He was the first to apply Euler's theorem to determine poles of rotation to better fit the dispersed continents back together into the supercontinent of Pangea. Clearly, Sir Edward Bullard was one of the most influential Earth scientists of the 20th century.

Edward Crisp Bullard was born on September 21, 1907, in Norwich, England. His family produced Bullard's Ales. He attended Cambridge University to study natural sciences but switched to physics and graduated with his doctoral degree in 1932. In 1931, he married Margaret Ellen Thomas. They had four daughters and the marriage ended in divorce in 1974. That same year, he accepted a position as demonstrator in the Department of Geodesy and Geophysics, where he was the second member. Later Sir Harold Jeffreys would join the department. When World War II broke out, Bullard became an experimental officer attached to the HMS *Vernon* in 1939. The Germans had developed a very effective magnetic mine that could be dropped from airplanes. They sank 60 ships in three months. With his knowledge of magnetics, Bullard developed mine sweepers that drastically reduced the threat. He even anticipated German advances in trigger mechanisms for mines, developing sweepers before the mines even came into use. He became assistant director of Naval Operational Research and oversaw projects on mine development and submarine warfare. He returned to Cambridge University after the war, but in 1948 he accepted the position of chair of the Department of Physics at the University of Toronto, Canada. After spending several months at Scripps Oceanographic Institute in La Jolla, California, in 1949, Bullard returned to England as the director of the National Physical Laboratory, where he remained until 1955. He returned to Cambridge University as the chair of the Department of Geodesy and Geophysics where he remained for the rest of his preretirement career. He retired to professor emeritus in 1974 with his health beginning to fail. He married Ursula Cooke Curnow that year and returned to Scripps Oceanographic Institute, where he remained for the rest of his life. Sir Edward Bullard died in his sleep on April 3, 1980.

It is difficult to overestimate the impact of Sir Edward Bullard's career on the profession. Many of his papers are true milestones in geophysics (magnetics, seismic refraction, and heat flow) and especially in marine geophysics. In recognition of his contributions to the profession, Bullard received numerous honors and awards. In addition to having been knighted for his work at the National Physical Laboratory he was a Fellow of the Royal Society. He was a member of the U.S. National Academy of Sciences and the American Academy of Arts and Sciences. He received the Hughes Medal and the Royal Medal from the Royal Society (England), the Bowie Medal and the Maurice Ewing Medal from the American Geophysical Union, the Chree Medal from the Physics Society (England), the Arthur L. Day Medal from the Geological Society of America, the Gold Medal from the Royal Astronomical Society (England), the Wollaston Medal from the Geological Society of London, and the Vetlesen Prize from the Vetlesen Foundation of Columbia University.

Bullard also performed great service to the public and the profession. Among others, he was an adviser to the British government on nuclear disarmament, chair of the Anglo-American Ballistic Missile Commission and the chair of the Committee for British Space Research. He was active in industry as well, including as the director of IBM, U.K., and various positions in the family brewing business.

⊠ **Burchfiel, B. Clark**
(1934–   )
American
Structural Geologist, Tectonics

With the vast number of highly talented geologists doing research in the frontier field of tectonics, progress has become one of inches rather than the leaps and bounds that it started out with. For that reason, it is always astonishing when some geologists seem to be able to make those impressive advances in spite of the usual slow pace. Clark Burchfiel is the epitome of that geologist. Although Burchfiel has concentrated the bulk of his

Clark Burchfiel leads a field trip to the Clark Mountains, Basin and Range Province, in California with Greg Davis (background) in 1971 *(Courtesy of Arthur Sylvester)*

efforts on the southwestern United States and Tibet, one of the reasons for his great success is that he will travel to the ends of the Earth to address a problem in the ideal example. His travels have additionally taken him to the Alps of eastern Europe, the Scandinavian Caledonides, and the Peruvian, Bolivian, and Colombian Andes. Typically, his approach to these problems is with detailed field geology coupled with geophysical data where available.

Some of the topics that Clark Burchfiel is best known for include a pull-apart (extensional) origin for Death Valley, California, many studies on the plate tectonic evolution of the western Cordilleras, and studies contrasting plate subduction and convergence in the Carpathians, the

Caledonides, and the Cordilleras. The studies he is most famous for in recent years are the defining of different types of extension as plates are broken apart during the formation of an ocean basin. This work with student Brian Wernicke established a whole new vigorous direction in geology. He and student Leigh Royden found concurrent normal faulting with parallel reverse faulting in the Himalayas. Prior to this the generally accepted idea was that all faults in a compressional area should be reverse or if there are normal faults, they lie perpendicular to the reverse faults. This work showed that mountains can reach only a certain height and thereafter the crust collapses any way it can. Many tectonic geologists had to rethink their models and start reexamining their field areas as a result of this work. This response of the profession is common after Burchfiel publishes a paper.

Clark Burchfiel was born on March 21, 1934, in Stockton, California. He attended Stanford University, California, where he earned a bachelor of science degree and a master of science degree in geology in 1957 and 1958, respectively, in addition to playing varsity football. He earned his Ph.D. from Yale University, Connecticut, in structural geology and tectonics in 1961 under advisement of JOHN RODGERS. He earned the Silliman Prize for his work. Burchfiel worked as a geologist with the U.S. Geological Survey during his final year at Yale University before accepting a position at Rice University, Texas, in 1961. He was named a Carey Croneis Professor of geology from 1974 to 1976. In 1977, he joined the faculty at Massachusetts Institute of Technology and was named Schlumberger Professor of geology in 1984, a chair that he still holds today. He held several exchange and visiting professor positions during his career at the Geological Institute of Belgrade, Yugoslavia (1968), the Geological Institute of Bucharest, Romania (1970), the Australian National University (1976), the University of Athens, Greece (1986), and University of Lund, Sweden (1992). Burchfiel married fellow geologist and

Massachusetts Institute of Technology professor Leigh Royden in 1984; they have four children.

Clark Burchfiel has had an extremely productive career publishing well over 100 articles in international journals and professional volumes. Many of these papers are benchmark studies that are repeatedly cited in other scientific articles. In addition to his own superb research, he has mentored many of the top geologists in the field of tectonics today. Thus his influence has even a more far-reaching effect. Burchfiel has been recognized for his contribution to the field with numerous honors and awards. He has been a member of the National Academy of Sciences since 1985 and was a Guggenheim Fellow from 1985 to 1986. He is a Fellow of the American Academy of Arts and Sciences and an honorary Fellow of the European Union of Geologists as well as a member of the Chinese Academy of Sciences. He received many awards from professional societies including the prestigious Career Contribution Award from the Structural Geology and Tectonics Division of the Geological Society of America in 1996. Burchfiel served as editor for the journal *Tectonics* and is on the editorial board for *Tectonophysics* as well as prominent journals in Norway, Switzerland, Germany, and China.

⊠ Burke, Kevin C. A.
(1929–)
British
Tectonic Geologist

After the pioneers of the plate tectonic theory proved that there was such a process, the monumental task of placing the rest of the features of the Earth into that framework remained. The first group of giants included the likes of ALLAN V. COX, HARRY H. HESS, W. MAURICE EWING, and J. TUZO WILSON. The second group is no less daunting, and prominent among them is Kevin Burke. When Kevin Burke began his research, the geology of the continents was an open book for study, just waiting to be interpreted in this new context. Kevin Burke and his comrades formed a kind of "tectonic cavalry" attacking mountain belts and ocean basins alike with voracity. Burke's area of interest is primarily stratigraphy and the development of sedimentary basins but he teamed with others to address virtually every tectonic problem. His primary research collaborators include JOHN F. DEWEY, A. M. ÇELAL SENGOR, and Bill Kidd. They applied the theory of triple junctions in continental extension not only in the type example of Saudi Arabia where the Gulf of Aden and the Red Sea form active arms and the East African Rift System forms the failed arm, but worldwide where the relations are not as clear. They were especially interested in the failed arm, which forms a large sedimentary basin with potential petroleum reserves called an aulocogen as in the 1977 paper "Aulocogens and Continental Breakup." He compared aulocogens to basins created by continental collisions which he named impactogens in the 1978 paper "Rifts at High Angles to Orogenic Belts: Tests for their Origin and the Upper Rhine Graben as an Example." He took the then-nascent idea of escape or extrusion tectonics as defined in the Himalayas and while most geologists were still trying to understand the process, he identified every place it was currently happening on Earth. This 1984 paper is entitled, "Tectonic Escape in the Evolution of the Continental Crust," with A. M. Çelal Sengor. Escape tectonics is the lateral movement of continental mass out of the way of a progressing continental collision. It is analogous to how clay squirts sideways when impacted with a fist or other object. He also addressed the processes of sedimentation in island arc settings with the Caribbean Sea as his area of choice.

However, Burke was not satisfied with working only on currently active orogens and basins, he also addressed ancient examples. Everything from small regional studies to large-scale sedimentary basin analysis were placed into plate tectonic context. Because plate tectonics control the distri-bution of hydrocarbons, much of Burke's work was of great interest to oil companies as exemplified by his 1975 paper, "Petroleum and Global Tectonics." He consulted with such companies as Exxon USA for a decade during the oil crisis. He also worked extensively with NASA on basalt volcanism as well as planetary evolution. His latest endeavor is the study of phenomenally large sand oceans that formed after the breakup of the supercontinent of Rodinia in the Late Proterozoic time (about 600 to 700 million years ago). They had an order of magnitude more sand than any other observed deposits ever. The largest of these is in North Africa and underlies the current Sahara Desert. Africa is one of Burke's favorite areas, but he is also well known for his work in the Caribbean Sea and in Asia.

Kevin Burke was born on November 13, 1929, in London, England. He attended University College, London, where he was a Goldsmid Scholar from 1948 to 1951 and the recipient of a DSIR Research Studentship from 1951 to 1953. He earned a bachelor of science degree in geology in 1951 and a Ph.D. in geology in 1953. He worked as a lecturer in geology at the University of Gold Coast, Africa, from 1953 to 1956, before working in exploration for raw materials for atomic energy both for the Geological Survey of Great Britain and as an international atomic energy adviser for the Republic of Korea. He was a member of the faculty and department head at the University of West Indies, Jamaica, from 1961 to 1965 before accepting a professorship at the University of Ibadan, Nigeria, where he remained until 1971. He was a visiting professor at University of Toronto, Canada, in 1972 and 1973 and then moved to the State University of New York at Albany as professor and chair, where he remained until 1982. During that time, Burke was a visiting professor at California Institute of Technology (1976), University of Minnesota (1978), and University of Calgary (1979). He moved to Houston in 1982 where he was the deputy director and later the director of the Lunar and Planetary Institute of NASA as well as professor of geology at the University of Houston, where he continues today. From 1989 to 1992, Burke left the university to work at the National Research Council in Washington, D.C., on a major project to decide the future of solid Earth sciences. Kevin Burke married Angela Phipps in 1961; they have three children.

Kevin Burke has contributed some 140 scientific articles in international journals, professional volumes, and governmental reports. Many of these papers are true classics of plate tectonics and appear in the most prestigious journals, like *Nature*. He was named a Du Toit Memorial Lecturer by the Geological Society of South Africa in recognition of these achievements.

Kevin Burke served as editor or associate editor for some of the most prestigious journals in the world, including *Tectonics, Tectonophysics, Geological Society of America Bulletin, Geology, Journal of Geology,* and *Journal of Geophysical Research,* among others. He has served on some of the most prestigious boards and panels for agencies and societies such as NASA, National Academy of the Sciences, International Geological Congress, CO-CORP (continental seismic program), NATO Advanced Study Institute, Ocean Drilling Program, and NFR Sweden.

Carmichael, Ian S.
(1930–)
British
Igneous Petrologist

The introduction of high-energy analytical techniques in the mid-1960s revolutionized the field of geochemistry and igneous petrology. Rather than performing time-consuming and tedious wet chemical analyses, scientists could generate much more data quickly and accurately. Data collection was no longer an end unto itself and scientists could devote an order of magnitude more of their time and energy to addressing the processes and relations involved. This breakthrough rapidly advanced the science to its modern pace and depth of inquiry. One of the true leaders of this revolution is Ian Carmichael. The freedom to address this new depth of inquiry led Carmichael to the most basic and yet the most difficult problems in igneous petrology, namely the origin and evolution of magmas. His research was accomplished through detailed analytical work coupled with comprehensive field studies of classic areas. The results were explained in terms of classical thermodynamics. In the early days, that approach was often hampered by the lack of data to constrain the theories. This need led Carmichael to embark on an experimental program to measure the thermodynamic properties of minerals and melts. He pioneered a new kind of experimental petrology in which he determined volumes, compressibilities, heat capacities, viscosities and other properties (e.g., oxidation state) of these magmas and minerals in very complex systems. At the time, the custom was to consider relatively simple systems. These experimental results were then used to model phase equilibria instead of the other way around, as was also common at the time.

In addition to all of this theoretical and experimental research, Carmichael has always maintained a strong field program. He has studied volcanic systems from Iceland to New Guinea and Alaska to Mexico, where he has concentrated for the past 25 years. In addition to his boundless energy and productivity, Carmichael is known for mentoring some of the premier petrologist-geochemists in the profession. He is well known for his prowess as a mentor.

Ian Carmichael was born on March 29, 1930, in London, England. He left England at 17 years of age, first going to a school in Connecticut, then to Cuba, and finally to the Colorado School of Mines in Golden before returning to England after 18 months. He then served as a second lieutenant in the paratroopers of the British Armed Forces for two years, where he was stationed in the deserts of the Sinai and the Sudan. He attended Cambridge University, England,

Portrait of Ian Carmichael in 2001 *(Courtesy of Ian Carmichael)*

where he earned a bachelor of arts degree in geology in 1954. Carmichael then accepted a position as a geologist, prospecting in northern Ontario and the Canadian Arctic coast (in winter!) before returning to the Imperial College of Science at the University of London, where he earned a Ph.D. in 1960. He married in 1955.

Carmichael's first academic position was as a lecturer at Imperial College in London, which he held from 1958 to 1963. He took a sabbatical leave as a National Science Foundation senior scientist at the University of Chicago, Illinois, to use the new electron microprobe and when he was denied an extension of his leave, he quit. In 1965, he joined the faculty at the University of California at Berkeley where he remained for the rest of his career. While at Berkeley, he served as chairman of the department from 1972–1976 and 1980–1982 and associate dean for the graduate school (1976–1978), for graduate academic affairs (1985) and associate provost for research (1986–2000). He was also an acting provost for research (1989) and served as the acting director of the botanical garden (1996–1998) and the director of the Lawrence Hall of Science (1996–present). Ian Carmichael was married again in 1970 and again in 1986 to Kathleen O'Brien.

Ian Carmichael has led an extremely productive career. He is an author of numerous articles in international journals and professional volumes. Many of these papers are benchmark studies on igneous petrology and igneous processes. He is also an author of a classic textbook *Igneous Petrology* and an editor of a classic volume *Thermodynamic Modeling of Geological Materials: Minerals, Fluids and Melts.* For his research contributions to the profession, Carmichael received a number of honors and awards. He was named a Fellow of the Royal Society of London. He received the Roebling Medal from the Mineralogical Society of America, the Murchison Medal from the Geological Society of London, the Schlumberger Medal from the Mineralogical Society of Great Britain, the Arthur L. Day Medal from the Geological Society of America, and the Bowen Award from the American Geophysical Union. Carmichael was a Guggenheim Fellow and also named a Miller Research Professor at the Miller Institute for Scientific Research.

Carmichael has performed significant service to the profession at the National Science Foundation, National Research Council, Geological Society of America, Mineralogical Society of America, and the American Geophysical Union. He is probably most noted for his editorial work. He was editor in chief and executive editor to the prestigious journal *Contributions to Mineralogy and Petrology* from 1973 to 1990. Since 1990 he has been an associate editor.

⊠ Cashman, Katharine V.
(1954–)
American
Petrologist (Volcanology)

When a major volcanic eruption occurs, a huge amount of gas is released into the atmosphere, an explosion may occur, ash and other particles may be ejected into the atmosphere, and lava may flow down the volcano slopes. Depending upon the location of the volcano, it may impact human habitations in one way or another. If the volcano exists in a populated area, the intensity and extent of those components of an eruption will determine whether the eruption can be used as a tourist attraction or if it will cause significant death and destruction. A major control on this intensity and extent is the mechanisms, rates, and timing of the release of gas from the magma. Katharine Cashman has developed intricate observational methods of textures in volcanic rocks to predict and constrain some of these destructive events. She looks at the size and density of holes (vesicles) left by the passage of gas through the magma and lava. If the volcanic gases are released quickly, the eruption will be explosive, just like the carbon dioxide being released quickly from soda. Several papers document this work including "Vesiculation of Basalt Magma During Eruption." She also measures crystal sizes and crystallization rates in the same rocks. If the crystallization rates are quick, the lava flow will not travel as far. She measures crystal density and crystal sizes and the distribution of them to study rates and relative viscosity. She also uses Ar/Ar thermochronology (age dating of the individual minerals) to put absolute timing on these rates. By comparing these natural results with physical and thermodynamic models currently being developed for processes occurring in magma reservoirs a predictive capability may be provided for the hazardous aspects of volcanic eruptions.

Cashman is also studying the formation of pumice in volcanic systems especially with regard to submarine eruptions. She has done extensive work on Mount Saint Helens both regarding the 1980 eruption and the minor eruptions that have occurred since. This research is the most exhaustive of any on this famous eruption and is leading to new insights on the processes of the eruption. This work is establishing new protocols on how to study volcanic eruptions that will be applied to other eruptions worldwide. Some of the other projects that Cashman is involved with include the crystallization, degassing and physical properties of lava from the Mauna Loa volcano, Hawaii, the growth of crystals in lava from Mount Erebus, Antarctica, flow patterns and cooling histories of volcanic feeder pipes in Brazil, and studies of fallout from pyroclastic volcanism from the Shirahama Formation, Japan.

Katharine Cashman was born on July 19, 1954, in Providence, Rhode Island. She attended Middlebury College, Vermont, where she earned a bachelor of arts degree in geology and biology with honors in 1976. She earned a master of science degree in geology with first-class honors from Victoria University, New Zealand, in 1979. She returned to the United States to accept a position as research scientist at the U.S. Geological Survey at Woods Hole, Massachusetts, from 1979 to 1981. She earned a doctorate in geology from the Johns Hopkins University, Maryland, in 1986. Cashman joined the faculty at Princeton University, New Jersey, in 1986, but moved to the University of Oregon in Eugene, where she remains today. During her tenure, she was a visiting professor at Queen's University, Kingston, Ontario, Canada, in 1991, at the Woods Hole Oceanographic Institute of the Massachusetts Institute of Technology in 1993, and at GEOMAR in Kiel, Germany, in 1999.

Katharine Cashman is leading a productive career. She has published some 50 articles in international journals and professional volumes. Several of these studies establish new processes in volcanism and the crystallization of volcanic rocks. She has received several professional awards in recognition of her achievements. She is a member of the American Association for the Advancement of Science. She received several honors as a student, including being inducted into Phi Beta Kappa, and receiving the Charles B. Allen Award and a Fulbright scholarship. She received the Group Cash Award in 1982 for her work on Mount Saint Helens.

Cashman's service to the geologic profession is extensive. She served on numerous panels and committees, both national and international, on volcanic processes and volcanic hazards. She is a section president for the American Geophysical Union, for which she has served on several committees. She has also served for the Geological Society of America and the Geochemical Society. She has held a number of editorial positions for *Geology, Earth in Space,* and *Journal of Geophysical Research.*

⊠   Chan, Marjorie A.
(1955–   )
American
Sedimentologist

Many scientists travel the globe to study geology but Marjorie Chan has found that some of the type examples of her research interests are in her own backyard in Utah. The University of Utah geologist has research projects that span some 800 million years of geologic time from the Precambrian to the Pleistocene.

The development of the Basin and Range Province in the southwestern United States provides one of the great examples of combined topographic-structural features of the Earth. The sedimentation that results from this extreme geometry is also noteworthy. Marjorie Chan is one of the primary geologists studying the processes in the Great Salt Lake, Utah, and its precursor, Lake Bonneville. During the ice ages of the Pleistocene epoch, the huge glacial Lake Bonneville filled the Great Salt Lake basin. Chan studies the

shoreline deposits of the lake along the Wasatch front. This research documents the processes of one of the premier closed basin, intermontane lakes that formed in response to global glacial conditions. Chan's studies also stress that the shoreline records of Lake Bonneville comprise important geoantiquities, records of recent Earth history. Many of these geoantiquities hold valuable scientific information and are important assets to society and should be preserved. This idea is emphasized in her paper, "Geoantiquities in the Urban Landscape."

Both surface and ground waters from the bounding Wasatch Range and related mountains fed the Lake Bonneville basin under tremendous pressure that still affects the area to a lesser extent today. Chan's research tests the validity of climate modeling utilizing her sedimentologic and geomorphic field data superimposed onto a database that utilizes geographic information systems. These studies show the control of depositional units on modern groundwater flow. With the abundant hypersaline nonpotable waters in the basin and the lack of abundant rainfall, the area is extremely environmentally sensitive. Contamination of the groundwater, which is driven deeply into the basin from the slopes of the bounding mountain ranges, could be potentially devastating to a very large area.

Marjorie Chan has a second area of research on sedimentary rocks of much greater age but in the same area. The Proterozoic Cottonwood Formation of central Utah contains the oldest known evidence of lunar-solar tidal cyclicity (in the paper *Oldest Direct Evidence of Lunar-Solar Tidal Forcing Encoded in Sedimentary Rhythmites: Proterozoic Big Cottonwood Formation, Central Utah*). The goal of this research is to detect evidence of ancient climate change, of tidal friction and to document the lunar retreat rate over time. Chan has also investigated evidence of climate and global change reflected in ancient aeolian deposits of the Permian Cedar Mesa Sandstone in southeastern Utah.

Marjorie Chan was born in 1955 in a car on the Ford U.S. Army base in California (her parents did not get to the hospital on time!). Her father was a marine biologist and field trips with him kindled her interest in science. She attended the College of Martin in Kentfield, California, where she earned an associate of arts degree in mathematics and an associate of science degree in physical science in 1975. She continued her undergraduate studies at the University of California at Davis, where she earned a bachelor of science degree in geology in 1977. Chan completed her graduate studies at the University of Wisconsin at Madison in geology in 1982 with a Kohler Fellowship and as an advisee of ROBERT H. DOTT JR. She joined the faculty at the University of Utah in Salt Lake City upon graduation and has remained ever since. Chan has also worked as an intern and consultant for numerous companies and organizations over the years, including the California Division of Oil and Gas, Lawrence Livermore National Laboratory (California), Marathon Oil Company (Colorado), ARCO Oil company (Texas), and the Utah Geological Survey. Marjorie Chan is married and has two sons.

Marjorie Chan is amid a productive career. She is an author of 43 publications in international journals, professional volumes, and governmental reports. Many of these papers are seminal reading about the tidal, fluvial, lacustrine, glacial, and aeolian depositional systems in Utah, both recent and ancient. She is also an author of one book and one videotape. Chan has been an investigator on numerous external grants totaling some $3 million. She has received several honors and awards in recognition of both her research and teaching contributions to geology. She received the Outstanding Young Women of America Award (1983), two Excellence of Presentation Awards from the Society of Economic Paleontologists and Mineralogists, the Telly Award for videos, and numerous awards from the University of Utah, including the Distinguished Faculty Teaching Award and the Distinguished Faculty Research Award. The Asso-

Marjorie Chan on a field trip with her two sons in Utah in 2001 *(Courtesy of Marjorie A. Chan)*

ciation for Women Geoscientists, among others, also named her a distinguished lecturer.

Chan performed significant service to the profession. She has served on numerous committees and functions for the Geological Society of America, the American Association of Petroleum Geologists, the Society of Economic Paleontologists and Mineralogists, and the Drilling, Observation and Sampling of the Earth's Continental Crust, Inc.

Cherry, John A.
(1941–   )
Canadian
Hydrogeologist

The transport of contaminants and pollutants from landfills, underground storage tanks, and other point sources into the groundwater system is the greatest threat facing our ability to obtain clean and safe drinking water. Horrifying stories like Love Canal in New York are commonly reported in newspapers and on television and radio and are even made into popular movies. The study and remediation of such problems has become one of the primary legislative goals of many governments. Many millions of dollars are devoted to this work, including the naming of numerous U.S. Environmental Protection Agency Superfund sites and their equivalents in other countries. John Cherry is one of the true leaders in the development and application of the science of contaminant hydrogeology and remediation. His record of innovative research and publication is unrivaled. He has the rare combination of the understanding and prediction of problems, the

Portrait of John Cherry *(Courtesy of John Cherry)*

technical expertise to pursue them, and the intellectual rigor to properly analyze and interpret the results. This work is carried out at the world's leading center for hydrogeological research at the University of Waterloo, Ontario, Canada, a reputation for which Cherry is largely responsible. He also established the Borden Research Site, which is among the first field testing hydrogeology sites in the world where the behavior of industrial contaminants introduced purposefully in a natural aquifer has been intensely studied over long periods of time.

John Cherry uses all remedial, chemical, and physical analytical and modeling techniques available to him including isotopic and chemical studies of water, flow and fluid transport process modeling, water resource evaluation, and aquifer restoration, among others. His first project was the evaluation of a low-level radioactive waste facility in western Canada, which he had to plod his way through because there was no accepted approach. It raised many questions that he has been addressing ever since. Some of his major areas of interest include the fate and transport of dense chlorinated solvents (DNAPLs) through both non-indurated (porous) soil and rock, as well as fracture systems in both clay and indurated (solid) rock. Remediation of such hazardous substances could be to isolate them from the rest of the groundwater system, the methods for which are evaluated by Cherry. He also studies the properties of clay-rich materials that can control the molecular diffusion of such pollutants. These studies have implications for the natural attenuation of pollutants versus those situations which require human intervention.

John Cherry was born on July 4, 1941, in Regina, Saskatchewan, Canada. His family moved several times in his youth and he attended high schools in Ottawa and Regina, Canada. He attended the University of Saskatchewan in Saskatoon and earned a bachelor of science degree in geological engineering in 1962. He began his graduate studies at the University of California at

Berkeley and earned a master of science degree in geology in 1964. He finished his graduate studies at the University of Illinois at Urbana-Champaign, receiving a Ph.D. in 1966. Upon graduation, Cherry became a postdoctoral fellow at the University of Bordeaux, France, but returned to Winnipeg the next year (1967) to join the faculty at the University of Manitoba, Canada. In 1971, he moved to the University of Waterloo in Ontario, and he remains there today. He was the director of the Institute for Groundwater Research from 1982–1987 but switched to the director of the University Consortium for Solvents-in-Groundwater Research in 1988. He also holds an NSERC (Environmental Research Council) chair in contaminant hydrogeology. John Cherry is married to the former Joan Getty; they have two children.

John Cherry is leading a productive career. He is an author of numerous scientific articles in international journals, professional volumes, and governmental reports. Many of these publications are often-cited benchmark studies on contaminant hydrogeology and remediation. He is also the first author of perhaps the historically most widely used 1979 textbook in the field, *Groundwater* with R.A. Freeze. In recognition of his research contributions to hydrogeology, Cherry has received numerous honors and awards. Among these are the O.E. Meinzer Award from the Geological Society of America and the William Smith Medal from the Geological Society of London. He also received several Outstanding Teacher awards from the University of Waterloo.

Cherry has performed significant service to the profession. He has served on several panels and committees for the National Research Council, NATO, the Geological Society of America, the Association of Groundwater Scientists and Engineers, and the International Association of Hydrogeologists, among others. He has also served in several editorial roles for the *Journal of Hydrology* among others.

⊠ **Clark, Thomas H.**
(1893–1996)
British
Paleontologist, Regional Geologist

Thomas Clark gained fame for his early (1924) visit to the famous Burgess Shale where he met its recent discoverer, CHARLES D. WALCOTT of the Smithsonian Institution. Clark was really only there to add fossils to his museum's collection, but the chance meeting spread his name around the geologic circles. For this reason, when he launched his major 1926 program to map the Quebec Appalachians, it met with much enthusiasm. Over the next decade, Clark mapped the geology and paleontology along the United States border in the Eastern Townships from the Sutton-Dunham area toward Phillipsburg and Lake Memphremagog. This painstakingly detailed work on complexly deformed and largely metamorphosed rocks established him as one of the leading geologists in Canada. It also resulted in the first detailed regional map of the area that was internally consistent. By 1937, Clark shifted his research area to the relatively weakly deformed rocks of the St. Lawrence Lowlands from the Ontario border to Quebec City. He produced new maps in some areas and the first geologic maps ever in others. In any event, it once again resulted in an internally consistent map of this important area of southern Canada. The Saint Lawrence Valley is likely the most seismically active area of eastern North America.

During the time of his research in the Saint Lawrence lowlands, Thomas Clark performed a reconnaissance study on the rocks of the Laval area near Montreal. He found that the existing maps were incorrect and proposed a major mapping initiative of the entire Montreal area to the Ministère des Mines. That project began in 1938 and continued for the next decade along with that of the Saint Lawrence Valley. This work formed the basis for development and industrialization of the area over the next two decades. By the late

1960s, however, the development of Montreal and the drilling for the Saint Lawrence Seaway in addition to other construction projects yielded far more geologic information than had been previously available. Using the new exposures and extensive drill cores and well log data from both construction projects and oil and gas exploration, Clark revised his geologic maps in a major research project.

Thomas Clark was born on December 3, 1893, in London, England, where he spent his youth. He immigrated to the United States to attend Harvard University, Massachusetts, where he earned a bachelor of arts degree in geology in 1917. He enlisted in the U.S. Army upon graduation, where he served in the Medical Corps during World War I. After the war, Clark returned to Harvard University, where he earned a Ph.D. in geology in 1924. Upon graduation, he accepted a teaching post at McGill University, Canada, and remained there for the next 69 years. Thomas Clark met and married Olive Marguerite Melvenia Pritchard in 1927; they had one daughter. Clark eventually became the director of the Redpath Museum at McGill University, where he began as the curator of the paleontology collection. He was also chair of the department for many years. Thomas Clark died in 1996 at the age of 103. His wife had predeceased him several years earlier. Clark's longevity is in stark contrast to most of the famous Earth scientists, who have a tendency to die young.

Thomas Clark led a productive career, especially considering that his most productive years were during a time when publication was at a far slower pace than it is today. Nonetheless, he was an author of more than 100 scientific articles, reports, and geologic maps in international journals, professional volumes, and governmental reports. He was also an author of *The Geological Evolution of North America,* which was widely regarded as a standard textbook for university-level geology. In recognition of his many contributions to geology, Thomas Clark received many honors and awards. He was elected a Fellow of the Royal Society of Canada in 1936. Among his many awards are the Harvard Centennial Medal from Harvard University, the Logan Gold Medal and the Centennial Award from the Royal Society of Canada, and the Prix Grand Mérité Geoscientifique. His service to the profession included numerous roles with the Geological Society of Canada, the Royal Society of Canada, and the National Environmental Research Council of Canada.

Cloos, Ernst
(1898–1974)
German
Structural Geologist

Normally, world travel to the prime examples of geological features is required to establish an international reputation in geology based upon field research. Ernst Cloos established a stellar reputation in geology based upon his research on the central Appalachians, essentially in his backyard. The reason for his success was his meticulous attention to detail and concentration on the processes of formation rather than just the regional relations. He felt that a careful and detailed study of a small area yields results more useful to others than even the most brilliant generalizations based upon questionable facts. At the time of Cloos's main activity, there were a large number of reputable geologists applying new ideas to the geology of New England. Cloos virtually single-handedly analyzed the central Appalachians at the same level of sophistication. His most famous paper was on South Mountain, Maryland, where he developed new methods to evaluate strain in the rocks using oolites that is now widely applied and unsurpassed for accuracy ("Oolite Deformation in the South Mountain Fold, Maryland"). He studied boudinage in a paper of the same name with the same methodical manner using examples from the central Appalachians as well as cleavage, lineation, and slicken side (fault striations) development.

Cloos also performed studies on thickness variations in folded strata as the result of the flow of rocks away from the limbs and into the hinge. This began another research direction in developing analog models to replicate structural processes. He used wet modeling clay following the example of his brother, Hans Cloos, another famous structural geologist. He modeled joint systems as well in several papers, including "Experimental Analysis of Fracture Patterns." With all of this research on the central Appalachians, by necessity he also became an expert on regional geology. Her performed many studies on the Piedmont and Blue Ridge Provinces on a variety of areas. As a second area of expertise, Cloos also studied several granite and granodiorite plutons in Maryland, California, and Ontario, Canada. His area of concern was emplacement mechanics rather than geochemical aspects.

Ernst Cloos was born on May 17, 1898, in Saarbrücken, Germany, but the family soon moved to Cologne. When his father died in 1904, the family moved to a place in the Black Forest near Freiburg. Cloos attended several boarding schools in his youth but was not much of a scholar. He enlisted in the armed forces at age 17, and wound up as the pilot of an observation plane during World War I. His plane was shot down over Switzerland and he was interned for the remainder of the war. After the war, he entered the University of Freiburg, Germany, intent on biology, but soon switched to geology. He transferred to the University of Breslau, Germany, but he also spent a short period at the University of Göttingen. In 1923, Cloos graduated with a Ph.D. in geology and married the daughter of his former instructor, Professor Spemann. They had two daughters. Upon graduation he was offered a position at the University of Göttingen, but the offer was rescinded and he wound up cataloging fossils for Seismos G.m.b.H. in Hannover, Germany. This company

was the first to conduct seismic reflection profiling in Texas and Louisiana for oil exploration and Cloos was included. The company then moved to the Middle East to continue oil exploration there. In 1930, Cloos returned to the United States to study the Sierra Nevada batholith in California with a grant from the German government. In 1931, he was offered a lectureship at the Johns Hopkins University in Baltimore, Maryland, and he remained there for the rest of his career. He served as department chair from 1952 to 1963 and single-handedly rebuilt the department. He also served as director of the Maryland Geological Survey in 1962 and 1963. Ernst Cloos retired in 1968 to professor emeritus, but remained active until his death on May 28, 1974.

Ernst Cloos led a very productive career. He is an author of at least 65 scientific publications in international journals, professional volumes, and governmental reports. Many of these papers are seminal works on the structure and tectonics of the central Appalachians as well as experimental structural investigations. In recognition of his research contributions to geology, Ernst Cloos received numerous honors and awards. He was a member of the U.S. National Academy of Sciences and the Finnish Academy of Science and Letters. He was awarded an honorary doctor of laws degree from Johns Hopkins University. He was awarded the Gustav Steinman Medal from the German Geological Society, the President's Award from the American Association of Petroleum Geologists, as well as a Guggenheim Fellowship.

Cloos was also of great service to the profession. He served on numerous committees and in several positions for the Geological Society of America, including president (1954) and vice president (1953). He also served as chairman of the Division of Geology and Geography for the National Research Council from 1950 to 1953. He was appointed to numerous positions for the Geological Survey for the State of Maryland.

⊠ Cloud, Preston E., Jr.
(1912–1991)
American
Paleontologist

Preston Cloud was an outstanding evolutionary paleontologist, biogeologist, and humanist who is best known for several popular science books, his stance against Creationism in public schools, and his dire warnings about overpopulation. Cloud was one of the founders and leaders in the burgeoning field of Precambrian Earth history. He employed a holistic approach to understanding the first 85 percent of Earth history. He developed the idea that periodic radiative explosions of life and especially metazoans were the result of evolutionary opportunism rather than an incomplete record. They reflected the availability of new ecological niches and conditions. Metazoans evolved as the result of the availability of free oxygen which resulted in complex cells. Cloud emphasized complex interrelations through the whole 4.5 billion years. His depth of investigation gave him a special appreciation of the place of humankind within this evolving environment. This deep understanding led him to write and assemble several popular science books on the evolving Earth, including *Adventures in Earth Science, Cosmos, Earth and Man* and *Oasis in Space: Earth History from the Beginning*. From this unique vantage, Preston Cloud was able to judge the condition of the Earth within an historical perspective of planetary dimensions. He was successful in warning the public of the dangers of the rapidly increasing population and the steadily decreasing availability of natural resources, a revelation that Cloud made. This warning has been taken up by many environmental groups worldwide as a result and is recognized as our most pressing problem. He was also instrumental in bringing this potentially calamitous situation to the attention of the National Research Council and the National Academy of Sciences, which also took up the cause.

Other areas that Preston Cloud mastered included the origin of carbonate atolls in the Pacific Ocean and the interplay of biological and geological processes in their formation. This theme of biogeology would recur throughout his research efforts whether regarding the evolution of brachiopods or the biostratigraphy of carbonate platforms. He was equally skilled in the biostratigraphy of clastic environments as well. Cloud was well known for his ability to adapt his research to take advantage of research opportunities and to seek out projects with the greatest potential.

Preston Cloud was born on September 26, 1912, in West Upton, Massachusetts. His father was an engineer draftsman who traveled constantly. The family moved to Waynesboro, Pennsylvania, in the late 1920s. In 1929, he graduated from Waynesboro High School. The United States was amid the Great Depression in 1930, so Cloud decided to join in the U.S. Navy. When he was discharged in 1933 in California, he hiked his way back to the East Coast. Money was still tight and he was fortunate enough to find enough money to pay for his first semester at George Washington University in Washington, D.C. A mentor, Ray Bassler, was a curator of paleontology at the National Museum and employed Cloud to assist him. Cloud completed his bachelor of science degree in 1938. He enrolled at Yale University, Connecticut, and earned a Ph.D. in paleontology/geology in 1942. He worked with CARL O. DUNBAR.

He was an instructor at the Missouri School of Mines in 1940, and then returned to Yale University as a postdoctoral fellow. In 1941, Cloud joined the U.S. Geological Survey to study manganese deposits in Maine for the World War II mineral exploration program. The next year, he was director of the Alabama Bauxite Project. In 1943, Cloud joined the Texas Bureau of Economic Geology in the Ellenburger Project to study the stratigraphy and sedimentology of the Ellenburger carbonate complex. In 1946, Cloud accepted a position as a professor of invertebrate paleontology at Harvard University, Massachusetts. However, due to a lack of support to expand the teaching and research areas, he returned to the U.S. Geological Survey in 1949. Shortly thereafter he became chief of the Branch of Paleontology and Stratigraphy. He increased the staff from 15 to almost 60 geologists and organized the first two marine biology programs.

In 1961, Preston Cloud accepted a faculty position at the University of Minnesota as a professor of geology, chairman of the Department of Geology and Geophysics, and head of the School of Earth Sciences. In 1965, he decided to take a professorship at the University of California, Los Angeles, but finally settled down permanently at the University of California campus at Santa Barbara in 1968. In 1974, he convinced the head of the U.S. Geological Survey, Vince McKelvey, to build a laboratory at Santa Barbara for the study of early organisms and to rehire Cloud to run it. He officially retired in 1979, but remained active throughout the rest of his life. He was a Luce Professor of Cosmology at Mount Holyoke College, Massachusetts, and a Queen Elizabeth II Senior Fellow at Canberra University, Australia. Preston Cloud died on January 16, 1991, of Lou Gehrig's disease. Preston Cloud had three marriages during his life, first to Mildred Porter, from 1940 to 1949; second to Francis Webster from 1951 to 1965; and finally to Janice Gibson in 1972. Cloud and Francis Webster had three children.

Preston Cloud was the author of some 200 publications ranging from scientific papers in international volumes and journals to governmental reports to popular science writing. In recognition of his many contributions, Preston Cloud received numerous honors and awards. He was a member of the National Academy of Sciences, the American Academy of Arts and Sciences, and the Polish Academy of Sciences. He was awarded the Walcott Medal from the National Academy of Sciences, the Penrose Medal from the Geological Society of America, the Paleontological Society

Preston Cloud leads a field trip in California in 1982 *(Courtesy of Arthur Sylvester)*

Medal (United States), the L.W. Cross Medal from the American Philosophical Society, the A. C. Morrison Award from the New York Academy of Sciences, the Rockefeller Public Service Award, and the Distinguished Service Award and Gold Medal from the U.S. Department of the Interior.

⊠ Conway Morris, Simon
(1951–)
British
Invertebrate Paleontologist

Simon Conway Morris gained his initial fame as one of HARRY B. WHITTINGTON's two graduate students portrayed in STEPHEN JAY GOULD's book *Wonderful Life: The Burgess Shale and the Nature of History* and his career has skyrocketed from there. Beginning with this famous work he began in graduate school on fossils in the Burgess Shale of Canada, he established himself as perhaps the foremost authority on metazoan evolution. These metazoans are arthropods that underwent tremendous evolutionary changes in the late Proterozoic through the early to middle Paleozoic (ca. 700 million to 400 million years ago), which set the stage for the metazoans we see today. His fame continued as he was asked to appear in numerous radio, television, and newspaper interviews. This early media work culminated in his giving the Royal Institution Christmas Lecture in 1996, which was broadcast by the BBC to an audience of about 1 million people in Great Britain. He has appeared on several BBC and NOVA science documentaries, and even written magazine and newspaper articles. His now famous book, *Crucible of Creation: The Burgess Shale and the Rise of Animals,* in its seventh printing, began rather discreetly. It was first published in Japan in Chinese. Part of its success is that Conway Morris disputes the opinion of Stephen Jay Gould, who maintained that each step in the history of life is necessary to end up with the life of today, otherwise everything would be different. Conway Morris maintains that within limits, the evolutionary process is more predictable and forms will evolve in certain directions regardless of each step. He is a proponent of widespread convergent evolution, in which animals will tend to evolve toward the most efficient form for a particular environmental niche and therefore look and behave the same regardless of what their ancestors started out as. Sharks and dolphins look very similar but had completely different ancestors; one evolved from fish and the other from terrestrial mammals.

Simon Conway Morris did not come by this respect purely by chance. After he began with the Burgess Shale, he expanded his investigation of early animals and the Cambrian explosion of life worldwide. He studied preskeletal and early skeletal fossils in China, Sweden, Mongolia, Greenland, southern Australia, south Oman, Alberta and Newfoundland, Canada, several places in the United States, and closer to home in Oxfordshire,

England. He took this vast paleontological experience and successfully interfaced it with molecular biology, especially in terms of phylogeny and molecular clocks. Many of his publications are in biological journals. The discovery of some unique fossil embryos and separately the oldest fossil fish (in China) ever found (by some 50 million years!) further spread his notoriety. The radically earlier development of fish than was previously thought further documents the astounding evolution that took place in the Cambrian. As an outgrowth of the theoretical aspects of the evolution of animals, Conway Morris has even become involved in the Search for Extraterrestrial Intelligence Institute (SETI). He argues that organic evolution is strongly constrained so that from DNA to any organ, the same evolutionary sequence will occur anywhere in the universe that it is able to do so.

Simon Conway Morris was born on November 6, 1951. He attended the University of Bristol, England, where he earned a bachelor of science degree with honors in geology in 1972. He attended Cambridge University (Churchill College), England, where he earned a Ph.D. in 1975 as an advisee of Harry Whittington. Conway Morris was appointed to a research fellowship at St. John's College of Cambridge University from 1975 to 1979. He was appointed to lecturer and lecturer of paleontology positions in the Open University at Cambridge University from 1979 to 1991, at which point he became a reader in evolutionary paleobiology. He was also named a Fellow at St. John's College in 1987. Since 1995, Conway Morris has been a professor of evolutionary paleobiology at Cambridge University. He was a Gallagher Visiting Scientist at the University of Calgary in 1981, a Merrill W. Haas Visiting Distinguished Professor at the University of Kansas in 1988, and a Selby Fellow at the Australian Academy of Sciences in 1992. Simon Conway Morris has been married to Zoe Helen James since 1975; they have two sons.

Simon Conway Morris is an author of more than 112 articles in international journals, professional volumes, and governmental reports. Many of these papers are cutting-edge studies that establish a new benchmark in metazoan evolution, the Cambrian-Precambrian boundary, and theoretical evolution. A number of these papers appear in high-profile journals like *Science* and *Nature.* He is also an author of some 100 book reviews, numerous encyclopedia entries, and the one book mentioned, in addition to editing five professional volumes. In recognition of his research contributions to paleontology and evolution, Conway Morris has received numerous honors and awards. He was awarded an honorary doctoral degree from the University of Uppsala in Sweden. He also received the Charles D. Walcott Medal from the U.S. National Academy of Sciences, the Charles Schuchert Award from the Paleontological Society of the United States, the George Gaylord Simpson Prize from Yale University, and the Lyell Medal from the Geological Society of London.

Conway Morris has also performed significant service to the profession. He has served numerous functions for the Geological Society of London, the Royal Society (England), the U.K. National Environment Research Council, the International Trust for Zoological Nomenclature, and the Systematics Association among others.

⊠ **Cox, Allan V.**
(1926–1987)
American
Geophysicist (Plate Tectonics)

Evidence for the existence for plate tectonics had been mounting for many years, and especially in the 1950s and early 1960s. The real clincher for the theory, however, was the documentation of seafloor spreading. In the 1950s, magnetic surveys were conducted over the mid-ocean ridges in the Atlantic and Pacific Oceans. A series of odd stripes parallel to the ridges and symmetric across the ridges were discovered but were considered enigmatic. A British geologist, Fred Vine, and his associates hypothesized that the striping was produced by repeated reversals of the Earth's magnetic field coupled with a dual conveyor belt model from the mid-ocean ridge. Allan Cox and associates including Brent Dalrymple, among others, took the next step by determining the timing of the magnetic polar reversals over the past 4 million years. They determined the age of largely volcanic rocks from locations worldwide with known polarity. Cox developed and modified the methodology for determining paleomagnetism in rock samples that is still used today. The research team constrained the points of reversal by more detailed sampling, paleomagnetic and geochronologic work. This research established the first paleomagnetic time scale, which could be used for magnetostratigraphy of sedimentary rocks with magnetic signatures, as well as volcanic rocks.

Cox and associates then applied their paleomagnetic time scale to the magnetic stripes on the ocean floor. There was already a large body of data on the magnetism of the ocean floor which could finally be made comprehensible. With this evaluation, they proved unequivocally that new ocean crust was being created at the mid-ocean ridges and moving symmetrically away in both directions. By comparing the width of the various stripes with their ages, for the first time the velocity of plate movement could be determined. This work triggered the main revelation about plate tectonics that revolutionized the science of geology. After this work, in addition to continuing his paleomagnetic research, Cox applied his findings to the mechanics of plate movements and interactions on a spherical body. He described these movements in terms of poles of rotation about which plates move and developed quantitative methods to determine these poles as well as the relative velocities.

Allan Cox was born on December 17, 1926, in Santa Ana, California. He attended the University of California at Berkeley initially as a chemistry major, but a summer job with the U.S.

Geological Survey in Alaska convinced him that his true calling was in geophysics. He earned a bachelor of arts degree in 1955, a master of arts degree in 1957, and a Ph.D. in 1959, all in geology and geophysics and all from the University of California at Berkeley. Upon graduation, Cox was employed as a geophysicist with the U.S. Geological Survey in Menlo Park, California. In 1967, he joined the faculty at Stanford University, California, as the Cecil and Ida Green Professor of Geophysics. Cox was so recognized for his innovative teaching that a faculty teaching award was named in his honor. He became dean of the School of Earth Sciences at Stanford University in 1979. Allan Cox died on January 27, 1987, as the result of a bicycle accident in the hills behind Stanford University.

Allan Cox led a productive career with authorship on more than 100 scientific articles in international journals, professional volumes, and governmental reports, as well as two popular textbooks on plate tectonics. Several of his papers are milestones in the development of the seafloor spreading model. In recognition of his research contributions to plate tectonics and geophysics, the professional community bestowed several honors and awards upon him. Allan Cox was a member of both the National Academy of Sciences and the American Academy of Arts and Sciences. He was awarded the John Adams Fleming Medal from the American Geophysical Union, the Arthur L. Day Medal from the Geological Society of America, the Day Award from the National Academy of Sciences, and the Vetlesen Medal from the Vetlesen Foundation at Columbia University.

In addition to serving on numerous panels and committees for the National Research Council, the National Academy of Sciences, the U.S. Geological Survey, and the Geological Society of America, Allan Cox served as president of both the Section of Geomagnetism and Paleomagnetism and the entire American Geophysical Union (1978–1980), in addition to serving on numerous panels and committees.

⌧ Craig, Harmon
(1926–   )
American
Geochemist, Oceanographer

If there were an Indiana Jones of the Earth sciences, it would be Harmon Craig. Not only does he work on some of the most important problems in Earth science, he does it while having the most daring of adventures. If he is not sailing over the top of an erupting submarine volcano or descending into the crater of an active underwater volcano, he might be captured by Zairian gunboats at gunpoint on Lake Tanganyika or robbed by Masai warriors at spear point.

Harmon Craig is an isotope geochemist who specializes in using helium isotope ratios as a tracer to track the release of gases from the deep interior of the Earth into the oceans and atmosphere. He discovered that the rare isotope helium 3 was trapped in the Earth at the time of formation 4.5 billion years ago and it is being continuously released by degassing from the Earth's mantle through mid-ocean ridge volcanoes and seafloor vents. This discovery has led to a greater understanding of how the oceans and atmosphere were formed in the primordial Earth. It also led Craig to search worldwide for mantle plumes or hot spots that tap the deep mantle near the Earth's core and release this helium 3 in higher quantity than in the mid-ocean ridges. He sampled volcanic rocks and gases in the East African Rift Valley from northern Ethiopia to Lake Nyasa, in the Dead Sea, in Tibet and Yunnan, China, and in all of the volcanic chains in the Pacific and Indian Oceans. He has identified 16 such volcanic hot spots with deep mantle signatures, 14 in ocean islands and two on continents in the Afar Depression of Ethiopia and Yellowstone Park, Wyoming.

In this research, while using the Scripps Institution of Oceanography Deep-Tow vehicle, Craig discovered hydrothermal sea vents in the Galapagos Islands seafloor-spreading center.

Using the submersible vehicle ALVIN he discovered similar vents on the Loihi seamount to the east of the island of Hawaii and the next Hawaiian island when it reaches the surface. He also sampled gas and rocks from the Macdonald Seamount in the Tubuai Island chain. He found more vents in ALVIN in back arc basins of the Mariana Trough some 12,000 feet below sea level.

Through this worldwide tracking of helium 3 within ocean water, he discovered that the deep water of the south Pacific Ocean circulates in the opposite direction than had previously been described and a symmetrical circulation cell exists in the north Pacific as well. Through similar deepwater circulation studies, Craig found that the element lead is rapidly scavenged by particulate material, the method by which many trace metals are removed from the ocean.

Craig was also involved with the analysis of gas trapped in Greenland ice cores. He discovered that the methane content of the atmosphere has doubled over the past 300 years. This finding has important implications for climate change modeling. He is currently measuring temperatures of past ice ages using his discovery that noble gases are gravitationally enriched in polar ice as a function of temperature.

Craig and his wife, Valerie, have an ongoing project to show the source of marble in ancient Greek sculptures and temples using carbon and oxygen isotopes.

Harmon Craig was born on March 15, 1926, in New York City. He enlisted in the U.S. Navy in 1944 during the later stages of World War II and served as a communications and radar officer on the USS *E-LSM* until 1946. He attended college at the University of Chicago, Illinois, and completed his doctorate in geology in 1951. He was an advisee of Nobel Prize laureate Harold Urey in the chemistry department. Harmon Craig married his wife, Valerie, in 1947 and they had three children. He was a research associate in geochemistry at the Enrico Fermi Labora-

Harmon Craig aboard a Scripps Institution research vessel shows a fish that was accidentally speared by an arm of the ALVIN submersible during a dive *(Courtesy of Harmon Craig)*

tory at the University of Chicago, Illinois, from 1951 to 1955. He joined the faculty at the Scripps Institution of Oceanography in 1955 and remained there for the rest of his career. He was a member of numerous oceanographic research expeditions including Monsoon in 1961, Zephyrus in 1962, Carrousel in 1964, Nova in 1967, Scan in 1970, and Antipode in 1971. He was also a member of several expeditions for Geochemical Oceanic Section Study (GEOSECS) in 1972–1977, 1982, 1983, 1985, and several expeditions to the eastern Pacific, Tibet, and China (between 1973 and 1993). He was the director of GEOSECS in 1970.

Harmon Craig has written 182 articles in international journals, professional volumes, and major oceanographic reports. Many of them are seminal studies on isotopic geochemistry of oceans, ice cores, and deep-sea hydrothermal vents.

The research contributions that Craig has made to the profession have been well recognized in terms of honors and awards. He is a member of the National Academy of Sciences. He received honorary doctorates from the Université de Pierre et Marie Curie, Paris, France, in 1983 and the University of Chicago, Illinois, in 1992. He received the V. M. Goldschmidt Medal from the Geochemical Society, a Special Creativity Award from the National Science Foundation, the Arthur L. Day Medal from the Geological Society of America, the Vetlesen Prize from the Vetlesen Foundation, the Arthur L. Day Prize from the National Academy of Sciences, and the Balzan Prize from the Balzan Foundation. He was named a Columbus Iselin Lecturer at Harvard University, Massachusetts, as well as a Guggenheim Fellow.

His service to the profession is equally impressive, including serving on numerous committees and panels for the Ocean Drilling Project, National Science Foundation, and the Geological Society of America, among others and as associate editor for the *Journal of Volcanology and Geothermal Research.*

⊠   Crawford, Maria Luisa (Weecha)
(1939–   )
American
Petrologist

Maria Luisa Crawford has about five major interests in which she has established herself as one of the foremost authorities in the field. These interests seem to evolve with time and she makes a marked shift in her research always to emerge in a leading role. Early in her career, Crawford was one of the first scientists to utilize the newly invented electron microprobe on metamorphic rocks. However, she soon switched directions and became interested in lunar petrology and geochemistry. She conducted research on mare basalts and determined the crystallization history and the origin of these flood lavas that filled craters early in the history of the Moon. The flooding of these lavas came as the result of impacts from large meteorites. The studies involved analysis of the rocks that were returned to Earth in the Apollo lunar missions. She was especially interested in basalt called KREEP, a name based on elemental enrichment (potassium: K, rare earth elements: REE, phosphorus: P).

Weecha Crawford is especially known for her research in Alaska and British Columbia, Canada. With colleagues, she studied the processes in intense continental collision at the peak of tectonism in "Crustal Formation at Depth During Continental Collision," among others. They proposed magma genesis and emplacement into rapidly developing major folds and faults that they named the "tectonic surge." These plutons are rapidly uplifted and exhumed. They have distinct large scale and contact metamorphic relations around them. These structural, metamorphic, and plutonic relations are all part of a major study of the accretion of continental fragments in the growth of continents. By adding small pieces of continent together in a series of collisions, full-sized continents may be built. The southern Alaska/British Columbia, Canada, area may be the best place on Earth to conduct such a study. This project is named ACCRETE and has been well received by the geologic community.

The other area in which Dr. Crawford has established herself as an expert is on the regional geology of the Pennsylvania Piedmont. She conducts studies on the metamorphic petrology, geochemistry, and geochronology on these complex rocks. Based upon these studies, plate tectonic reconstructions are proposed. Much of this work is done with graduate students. Crawford has been an outstanding mentor to her students as well as a teacher. She has been especially encouraging to women seeking to establish careers in geology. In this vein, Dr. Crawford has been an active member of the Association of Women Geoscientists. Through her efforts in publication, mentoring, and public relations, she has been a real factor in making pathways for women to enter the male-dominated field of geology.

Maria Luisa Crawford was born on July 18, 1939, in Beverly, Massachusetts. She attended Bryn Mawr College in Pennsylvania where she earned a bachelor of arts degree in geology in 1960. Before she began her graduate studies, she visited the University of Oslo in Norway in 1960 and 1961 on a Fulbright Fellowship. Upon her return, she enrolled at the University of California at Berkeley, where she earned her doctorate in 1965. Upon graduation, she joined the faculty at her alma mater of Bryn Mawr College, where she has remained for her entire career. Crawford was named the William R. Kenan Jr. professor of geology from 1985 to 1992. She also served as department chair from 1976 to 1988 and again from 1998 to present. She was a visiting professor at several colleges including the University of Wisconsin. She is married to William Crawford, a fellow professor of geology at Bryn Mawr.

Crawford's very productive career has included some 68 articles in international journals, professional volumes, and governmental reports. She also edited one book and wrote 18 entries for encyclopedias and similar publications. Many of these papers are required reading for lunar studies as well as metamorphic petrology of the Pennsylvania Piedmont and British Columbia, Canada. Crawford has received several awards in recognition of her research and teaching including a MacArthur Fellowship and the Outstanding Educator Award from the Association for Women Geoscientists Foundation. Crawford has also been very successful in obtaining grant funding from the National Science Foundation to support her research.

Weecha Crawford doing field research in British Columbia, Canada *(Courtesy of Maria Louisa Crawford)*

Crawford has performed significant service to the geologic community. She has served on numerous committees for the National Science Foundation, the Geological Society of America, the Mineralogical Society of America, and the National Academy of Sciences. She was a member of the advisory board for both Princeton University, New Jersey, and Stanford University, California. She was on the evaluating committee for some 11 departments nationwide, including such notable schools as Dartmouth College, New Hampshire; University of Toronto, Canada; and Bates College, Maine, among others. Crawford has also served in editorial positions including associate editor for *Geological Society of America Bulletin* and an editorial board member for *Geology* and *Computers and Geosciences.*

D

Dana, James D.
(1813–1895)
American
Mineralogist

James Dwight Dana was one of the most influential scientists of the 19th century. He was certainly the leader of American Earth sciences which at that time were the most important and popular of all sciences by virtue of their economic potential. Geology accounted for one-fourth of all scientists between 1800 and 1860 and was deemed the most attractive and of highest potential of all sciences by the *New York Herald* and *Knickerbocker Magazine.* By 1847, Louis Agassiz, another famous geologist, declared that James Dana was "at the head" of American geology. He had assumed the chief editor position of the *American Journal of Science* in 1846, which was the leading scientific journal in the United States at the time. This position gave him immense power. He was of the opinion that North American geology was the most straightforward in the world and that it should serve as the type example against which all other examples should be compared. He cited the beautifully continuous flat-lying strata that cover the entire midsection of the country with neatly arranged mountain belts at its edges as his proof. This idea found its way into his journal, which he edited through most of the second half of the cen-

tury, but also into the American Association for the Advancement of Sciences where he was the most influential member. At that time, it was the leading scientific society in America (and one of the few). Dana served as president in 1855 and using biblical references furthered his cause at every opportunity. This attitude found its way into American geology for many years to come and may still exist to some degree.

James Dana is probably best known for one of his earliest works, the 1837 *System of Mineralogy.* In this work, he first systematized minerals into groups based upon their form and chemistry. As a result, mineralogy textbooks bearing his name were still popular into the late 20th century, nearly 150 years later. He also wrote the *Manual of Geology,* which was a comprehensive treatment of geology and a major influence on geologic thought through the 19th century. Dana was the main proponent, if not the originator, of the geosyncline-contraction hypothesis for mountain building, the accepted mechanism prior to plate tectonics. He imagined a thin crust over a viscous region with cooling cracks and lateral sliding to produce valleys and mountains. He was a strong proponent of uniformitarianism in that book. He also wrote a book on *Corals and Coral Islands* in which he defined atolls, barrier reefs, and fringing reefs based upon his observations while part of the Wilke's Expedition. In this work he supported many of the observations that his compatriot Charles Darwin had made during his historic cruise aboard the HMS *Beagle,* but disagreed with the mechanisms for their formation. He identified the temperature requirements for the formation of reefs and studied the role of uplift and subsidence in their development. He produced an exhaustive work in which he classified corals and other coelenterates, entitled *Zoophytes and Geology.* This work also resulted from his South Pacific Ocean cruise and established him as one of the premier paleontologists of the time, as well. In this regard, he was at odds with Charles Darwin on occasion. In terms of the evolution of life, Dana was a proponent of catastrophism. He walked a narrow line between his strong religious beliefs and the new idea of evolution with strange results at times.

James Dwight Dana was born on February 12, 1813, in Utica, New York, where he grew up. He attended Yale University, Connecticut, and followed the fixed curriculum, but excelled in mathematics and science. Even before his graduation in 1833 with a bachelor of arts degree, Dana took a position as instructor of midshipmen on a U.S. Navy cruise aboard the USS *Delaware* to the Mediterranean Sea where he was able to visit Mount Vesuvius in Italy. He returned to the United States in 1834, but did not hold a consistent job until 1836 when he returned to Yale University as an assistant to Benjamin Silliman, where he was also able to continue his education. Dana joined the United States South Seas Exploring (Wilkes) Expedition in 1838 in the position of geologist and did not return until 1842. As a scientist traveling around the world to make scientific observations of the natural world, only Charles Darwin matched Dana. He invested the money he earned on the expedition in a store owned by his brother in Utica, New York, and lived off of the earnings as well as continuing income from the expedition for several years while writing reports on the results. James Dana married Harriet Frances Silliman, the daughter of his adviser, in 1844. In 1849, Dana was appointed to a Silliman

Professorship of Natural History at Yale University, but was not required to teach until 1856. He became the main force of the geology department, if not the United States, for many decades to come. Although he suffered from poor health throughout this time, he did not retire until 1890. In his later years, Dana spent increasing time writing hymns and love songs for the guitar. James Dwight Dana died on April 14, 1895.

⊠ Dawson, Sir (John) William
(1820–1899)
Canadian
Paleontologist

Sir (John) William Dawson was one of the most influential geologists ever in Canada. He is known not only for his detailed research in paleontology, but also for his associations with such geological dignitaries as Charles Lyell and for his work in professional service. He brought Canada from a relative backwoods reputation to one of respect within the profession on an international basis almost single-handedly. Dawson's fieldwork in Nova Scotia and Quebec yielded more than 200 new post-Pliocene fossil discoveries. He also discovered a method for perfecting the examination of thin fossil slices using a microscope. This technique aided Dawson in identifying more than 125 new Paleozoic Canadian plant fossils reaching from the coastal areas of Nova Scotia to midwestern Canada.

Sir William Dawson's relationship with Charles Lyell was legendary. While exploring Joggins, Canada, together, they discovered bones of small amphibians in fossil trunks while examining a thick section of Carboniferous strata. These fossils were the first of their kind in this area and the earliest North American Carboniferous reptile, *Dendrerpeton acadianum,* ever to be collected. Dawson also discovered the earliest land snail, Pupavetusta, and remains of Devonian plants. With all of this extensive research in eastern Canada,

Dawson wrote his opus, *Acadian Geology*, which was first published in 1855. It is the most complete volume of the geology of coastal towns in Canada ever published.

In the early 1860s, Dawson began working on SIR WILLIAM EDMOND LOGAN's Laurentian fossils. He discovered colonial forms of strange jellylike structures that contained limestone filled walls. These strange new discoveries were named *Eozoon canadense* which means "dawn animal of Canada." Dawson strongly believed that his new discoveries were foraminifers (single-celled creatures usually classified in the protist kingdom). Dawson was met with resistance from several geologists and paleontologists from Ireland and Germany who believed that his discoveries were not organic in nature at all. Dawson and his supporters continued to steadfastly maintain their position and prevailed.

John William Dawson was born October 13, 1820, in the small coastal fishing village of Pictou, Nova Scotia. He was given the opportunity to earn a reputable education through Pictou Academy, a public school in town that concentrated on teaching the natural sciences. He was also fortunate enough to be surrounded by sandstone and shale formations in his hometown of Pictou that contained Carboniferous plant fossils. This environment allowed the already science-oriented youth to investigate the disciplines of both geology and paleontology. In the fall of 1840, Dawson's father sent him to the University of Edinburgh in Scotland. There were only a few colleges and universities in the world at that time that included a natural science major that concentrated in geology and botany in their curriculum. Due to financial difficulties, however, Dawson was forced to return to Nova Scotia to help his family business for one year. During his trip back to Canada, Dawson met geologist William Logan, who was about to become director of the Geological Survey of Canada.

He returned to the University of Edinburgh in 1841 to complete his studies. It was then that he met his future wife, Margaret Ann Young Mercer. Upon completion of his second period of study in 1947, Dawson again returned to Canada and joined the General Mining Association of London. He completed a geological survey of Cape Breton and investigated coal and other mineral deposits for the government of the province and for several small mining companies. In 1849, he gave a series of lectures on several disciplines of natural history including geology to his former grade school, the Pictou Academy, the Halifax Mechanic's Institute, and at Dalhousie College in Halifax, Nova Scotia.

From 1850 to 1853 Dawson held the position of the first-ever superintendent of education for the province of Nova Scotia. The public school system was poorly managed and unorganized when Dawson took it over in 1850. During his tenure with the Nova Scotia school system he worked so tenaciously that he reformed the entire school system in less than three years. He was also able to continue his scientific research during this time. The job required extensive traveling and Dawson used this time to gather data and investigate several paleontological inquires. Some of Dawson's greatest paleontological discoveries were made during his extensive travel for the school system. Dawson had the opportunity to work with the great Charles Lyell in 1853 during one of his trips to North America to continue his research there. In 1855, Dawson became principal of McGill University in Montreal. Dawson's hard work is the reason that the university became one of Canada's most best-known and most reputable colleges. It was a huge accomplishment for a school that was attended by the non-English minority. Sir William Dawson retired from McGill University in 1893, but remained active until his death on November 19, 1899, in Montreal, Quebec, Canada.

Sir William Dawson led an extremely productive career, especially for those years. He was an author of some 200 publications ranging from popular essays on both scientific and religious topics. His technical research papers ranged from earthquake accounts to descriptions of fossil amphibians and mollusks and include such titles as the "Geological History of Plants," among others. His first popular book, *Arcadia, or Studies of the Narrative of Creation in the Hebrew Scriptures* did not raise public interest until it was reissued as *Origin of the World* after Darwin stirred up the controversy between Christian theology and science. Dawson attempted to quell that controversy in his next book, *Nature and the Bible and in Facts and Fancies of Natural Sciences.* He also wrote *Links in the Chain of Life,* which illustrated several plants and animals through geologic time.

Sir William Dawson received many scientific awards and honors in recognition of his many contributions. In 1854, he became a Fellow of the Royal Society of London. He also formed the Royal Society of Canada in 1882, serving as its first president. He was also the fifth president of the Geological Society of America. In 1884, he was knighted by Queen Victoria for his contributions to geology and was given the title of Sir William Dawson. He also received the Lyell Medal from the Geological Society of London. His son, George Mercer Dawson, also went on to become the director of the Geological Survey of Canada and the 12th president of the Geological Society of America.

Day, Arthur L.
(1869–1960)
American
Geochemist, Geophysicist

The name of Arthur L. Day is still well known for the awards in his honor as well as his extensive shaping of the profession. Although he referred to himself as a physicist, his major contributions were in the Earth sciences. He applied physics and chemistry to the solution of geological problems long before it was fashionable. He was also distinctly interested in practical ap-

plications of his high temperature experimental research while director of the Geophysical Laboratory at the Carnegie Institution of Washington, D.C. The lab was considered of little use by many until Day used it to save the day in the World War I effort. Quality optical glass for things like gun sights, periscopes, rangefinders, binoculars, and the like had come exclusively from Germany prior to 1917. When America entered the war, they found themselves in a critical shortage with military needs of 2,000 pounds of optical quality glass per day while the capacity of the country was 2,000 pounds per month. Arthur Day was appointed to the General Munitions Board (War Industries Board) in charge of optical glass production. He designed a plan to upgrade several commercial facilities and streamline production in existing facilities. Day supervised the production of more than 90 percent of the optical glass produced in the United States and the crisis was averted.

Arthur Day was also the main force in establishing the Carnegie Institution Seismological Observatory in Pasadena, California. He organized numerous agencies and institutions to design it. This facility was the most advanced of its kind at the time and the first of its kind in the United States. It would set the pace for earthquake studies including prediction and prevention. The facility would later become part of the California Institute of Technology when Day retired and boast the likes of BENO GUTENBERG and CHARLES F. RICHTER.

Arthur Day also made significant contributions with his research as well. He first extended the capabilities of standard gas thermometers to very high temperatures. He used this as a practical temperature scale for melting points. This research involved the determination of the physical properties and phase relations of solids and liquids. The first system he investigated was plagioclase feldspar, but he also looked at sulfur, platinum, graphite, and quartz glass. He turned his attention to the geophysics and geochemistry of volcanoes. He devised new gas sampling equipment and sampled exhalations of the Hawaiian volcanoes, Yellowstone National Park, Lassen Peak, and Geyserville, California. This research would help determine the composition and phase relations of these gases. He would later work on the volcanic areas of New Zealand. He was also interested in radioactivity and devised new deep-sea coring tools to investigate the radioactive contents of marine sediments.

Arthur L. Day was born on October 30, 1869, in Brookfield, Massachusetts, where he grew up. He attended the Sheffield Scientific School of Yale University, Connecticut, where he earned a bachelor of science degree and a Ph.D. in physics in 1894. He taught physics at Yale University upon graduation, but decided that he needed postdoctoral experience. In 1897, he went to the Physicalisch-Technische Reichsanstalt in Charlottenburg-Berlin, Germany, as a volunteer assistant but he was soon offered a paid position. He was offered a one-year position as a physical geologist at the newly established high temperature laboratory of the U.S. Geological Survey in 1900 and a permanent position in 1901. Arthur Day married Helene Kohlrausch in 1900; they would have four children. His work on the high temperature relations of plagioclase, and the extension of the gas thermometer scale to high temperatures was at the U.S. Geological Survey. This research caught the interest of the newly established Carnegie Institution of Washington, D.C., and it funded his research for several years. In 1906, the institution hired him as the director of the newly created Geophysical Laboratory. He remained in the position of director until his retirement in 1936. This streak was interrupted only with a two-year leave of absence from 1918–1920 to become vice president in charge of manufacturing at the Corning Glass Works in New York. His retirement did not curtail his research activities until 1946, when he had a physical breakdown. Arthur L. Day died suddenly of a coronary thrombosis on March 2, 1960.

Arthur L. Day led a very productive career with authorship on numerous scientific articles in international journals, professional volumes, and governmental and industrial reports. Most of his geological papers were on high temperature processes and especially on volcanoes. His papers on volatile components in igneous processes and seismology are benchmark studies. In recognition of these many contributions, he received numerous prestigious honors and awards. He was not just a member of the National Academy of Sciences but also home secretary and vice president. He was also a member of the American Academy of Arts and Sciences, as well as a member of the scientific academies in Sweden, Norway, and the USSR. He received honorary degrees from Columbia University, Princeton University, the University of Pennsylvania, and the University of Groningen. He also received the Penrose Medal from the Geological Society of America, the Wollaston Medal from the Geological Society of London, the William Bowie Medal from the American Geophysical Union, the John Scott Award from the City of Philadelphia, and the Bakhius Roozeboom Medal from the Royal Academy of Amsterdam, among others. Arthur Day also has awards in his honor from both the National Academy of Sciences and the Geological Society of America.

Day served the profession as well. He was president (1938) and vice president (1934) of the Geological Society of America. He also served as president of the Philosophical Society of Washington and the Washington Academy of Sciences.

⊠ DePaolo, Donald J.
(1951–)
American
Isotope Geochemist

The use of isotopes in geology began in earnest in the 1950s. The earliest evaluated systems were uranium-lead followed by potassium-argon, and rubidium-strontium. In the mid-1970s the new system of neodymium (Nd) to samarium (Sm) was investigated and Donald DePaolo was one of the true pioneers. He was one of the first to measure Nd isotopic compositions of terrestrial samples. He used these Nd compositions to convincingly constrain a fundamental process of Earth circulation. He compared Nd isotopic compositions of ocean island basalts with those from the mid-ocean ridges to show that the source of the mid-ocean ridge basalts cannot be from the entire mantle. It has to be purely shallow. As a result, DePaolo proposed a two-layer geochemical model for the Earth's mantle in which only the top layer participates in the melting and subduction process. This new model now appears in all introductory textbooks and has more recently been supported by seismic tomography studies by geophysicists. He was further able to define the Nd isotopic evolution for this shallow layer in the mantle. This work in turn allowed him to model the Nd ages for mantle separation from various segments of continental crust. These separation ages yield a more accurate age of the development of continents than the traditional radiometric methods. This work is described in the paper "Geochemical Evolution of the Crust and Mantle." DePaolo is the author of the definitive book on Nd isotopic systems entitled *Neodymium Isotope Geochemistry: An Introduction.*

But Nd is only one aspect of DePaolo's research. He has also investigated the fine scale strontium (Sr) isotopic evolution of seawater by using high-resolution stratigraphy of marine sediments. A careful study relating the composition of deep-sea carbonates with ocean water allowed him to develop a correlation scheme in order to trace this history. By careful fine scale investigations of sediments in San Francisco Bay and other areas, not only could he trace the isotopic evolution but he could also identify climate changes. As if all of these contributions are not enough, DePaolo also dabbled in other topics like groundwater-bedrock interactions using Sr isotopes as tracers, Sr isotopic zoning in garnets as a measure of metamor-

phic evolution, and Sr isotopes to document timing of large granitic magma systems. He was even involved in a project to drill through the crust in the Hawaiian Islands to conduct detailed geochemical and isotopic studies. The list of research topics continues to grow every year.

Donald DePaolo was born on April 12, 1951, in Buffalo, New York. He grew up in North Tonawanda, New York, on the Niagara Frontier. He entered college at Cornell University, New York, in 1969, intent on engineering, but quickly changed his mind to go into geology and transferred to the State University of New York at Binghamton. He graduated in 1973 with a bachelor of science with honors in geology and did his graduate studies at the California Institute of Technology in Pasadena. He graduated with a Ph.D. in geology with a minor in chemistry in 1978 as an advisee of Gerald Wasserburg. Upon graduation, he joined the faculty at the University of California at Los Angeles where he rose through the ranks to professor. In 1988, he accepted a faculty position at the University of California at Berkeley, where he remains today. DePaolo is the director of the Center for Isotope Geochemistry and a senior faculty scientist at the Lawrence Berkeley National Laboratory. He served as chair of the Geology and Geophysics Department from 1990 to 1993 and he is currently the head of the Geochemistry Department at Lawrence Berkeley Lab. He was also named Miller Research Professor in 1997 to 1998. He has been a visiting scientist several times during his career including as a Fulbright Senior Scholar at the Australian National University. Donald DePaolo has been married to Bonney L. Ingram since 1985 and he is the father of two children.

Donald DePaolo is amid a very productive career. He is an author of more than 130 scientific articles in international journals, professional volumes, and governmental reports. Many of these are benchmark studies of the isotopic evolution of seawater, isotopic evolution of the mantle, and the use of Nd/Sm systems. In recognition of his contributions to geology, DePaolo has received numerous honors and awards. He is a member of the National Academy of Sciences and a Fellow of both the American Academy of Arts and Sciences and the California Academy of Sciences. He received the F. W. Clarke Medal from the Geochemical Society, the J. B. MacElwane Award from the American Geophysical Union, the Mineralogical Society of America Award, and the Arthur L. Day Medal from the Geological Society of America.

DePaolo also performed significant service to the profession. He served on numerous committees in various capacities for the Mineralogical Society of America, the Geochemical Society, American Geophysical Union, and the Geological Society of America. He also served on panels and working groups for the U.S. Nuclear Regulatory Commission, the National Research Council, and the National Science Foundation. DePaolo served in numerous editorial roles including associate editor for *Isotope Geochemistry* and for *Journal of Geophysical Research*.

⊠ Dewey, John F.
(1937–)
British
Tectonic Geologist

After the pioneers of plate tectonics proved that the concept actually existed in the 1950s and 1960s, it was time to show how the Earth, both current and ancient, fit into this revolutionary paradigm. John Dewey went from a respected geologist to a household name among geologists worldwide with a single scientific article: "Mountain Belts and the New Global Tectonics." Written with J. M. Bird in 1970, it was a benchmark in this study that is still cited in scientific literature today. This initial article led to a flood of studies to place many of the mountain belts of the Earth into the plate tectonic context and fill in the various parts of the model. This research has

been soundly based on field observations coupled with any supporting evidence, be it geochemical, geophysical, or paleontological. This pioneering spirit and willingness to boldly address any problem within the field has most certainly earned John Dewey the respect of the profession. He can undoubtedly be considered the "father of modern plate tectonics."

The list of topics that John Dewey has addressed in his research reads like the chapters in a textbook on plate tectonics. He investigated continental breakup and dispersion, including triple junctions and hot spots. He investigated the obduction of ophiolites (pieces of ocean floor on land) during continental collisions and the complexities of the suture zones (where the old continents are stuck together) of those collisions. He investigated fracture zones (transform boundaries) on the ocean floor. He studied the distribution of relative strength profiles within the crust and upper mantle as the control on plate process and the collapse of orogens (mountain systems) as a result. He studied transpression (mixed compression) and transtension (mixed extension) in strike-slip fault zones among many other topics. Between all of these studies that defined a plate tectonic process, Dewey was constantly investigating specific areas worldwide and writing the seminal papers on the tectonics of them as well. His work has focused on the British Isles (Caledonides), but he has done detailed studies on the Alps, the Himalayas, the Appalachians (especially Newfoundland), Turkey, the Andes, and the Caribbean. His research always seems to guide the major direction of interest in the field of tectonics and is the topic of conversation around many universities worldwide.

John Dewey was born on May 22, 1937, in London, England. He attended Bancrofts School in Woodford Green, Essex, from 1948 to 1955 before entering Queen Mary College at the University of London, England, where he earned a bachelor of science degree in geology with first class honors in 1958. He continued his graduate studies at Imperial College at the University of London where he earned his Ph.D. in geology in 1960. His first academic position was at University of Manchester, England, where he was a lecturer from 1960 to 1964. He then joined the faculty at Cambridge University, England, in 1964 where he was a Fellow at Trinity College and a Fellow and associate dean of Darwin College. In 1970, Dewey accepted a position at the State University of New York (SUNY) at Albany. From 1980 to 1982, he was a distinguished professor at SUNY and a research professor after 1982. He became a professor and head of the department at the University of Oxford, England, in 1986. He joined the faculty at the University of California at Davis in 2000, where he remains today. During his career, he was a visiting scholar at Lamont-Doherty Geological Observatory of Columbia University, New York (1967), and at University of Calgary, Canada (1979). John Dewey was married on July 4, 1961, and he has two children. He is a serious cricket player and gymnast and enjoys skiing and model railroads.

John Dewey has led a very productive career. He is an author of some 134 articles in international journals and professional volumes. Several of these papers appear in high-profile journals such as *Nature* and many establish new benchmarks for the state of tectonics. Dewey has received many honors and awards for his contributions to tectonics from the geological community. He is a Fellow of the Royal Society of London and a member of the National Academy of Sciences. He was awarded two honorary doctorates from Memorial University of Newfoundland (1995) and the National University of Ireland (1998). He received the A. Cressy Morrison Medal from the New York Academy of Sciences in 1976. In 1983, he was awarded the T. N. George Medal from the Geological Society of Glasgow, Scotland, the Lyell Medal from the Geological Society of London, and the Award for Excellence in Journal Design for *Tectonics* from the Association of American Publishers. Additionally,

Portrait of John Dewey *(Courtesy of John Dewey)*

he received the Arthur Holmes Medal from the European Union of Geosciences (1993), the Wollaston Medal from the Geological Society of London (1999), the Penrose Medal from the Geological Society of America, and the Paul Fourmarier Prize and Medal from the Académie Royale de Belgique (1999), among numerous other honors.

The service John Dewey has contributed to the profession is unparalleled. He is a member or fellow of 12 geological societies from all over the world. He served on numerous advisory committees, including International Geodynamics Commission (secretary, 1972–1980), International Geological Correlation Programme, numerous committees for the Natural Environment Research Council of Great Britain, and the International Lithosphere Commission, among others. He served numerous committees for the Royal

Society and the Geological Society of London. All else pales, however, in comparison to his editorial work. He was founding editor and editor in chief for both *Tectonics* (1981–1984) and *Basin Research* (1989–present) and associate editor for *Geology, Geological Society of America Bulletin,* and *Journal of Geology.* He has been on the editorial board for nine journals and books. He has been an external evaluator for nine universities including Cambridge University, Oxford University, and University of Leeds.

⊠   Dickinson, William R.
(1931–)
American
Sedimentologist, Tectonics

When it comes to showing how plate tectonics controls sedimentation, there is no greater authority than William Dickinson. He devised a system based upon numerous observations to determine the plate tectonic setting of ancient sandstones. This system is presented in a paper entitled "Plate Tectonics and Sandstone Compositions," and includes a series of ternary discrimination diagrams based upon certain mineral and rock fragment proportions. By the relative percentage of these components, sandstones can be classified as to whether they originated in an island arc, stable interior, or in a rift setting with uplifted basement blocks, among others, even if the rocks had no other evidence of their settings. The term for the origin of these sediments is the provenance. The diagrams are now used regularly on a worldwide basis in the analysis of ancient sandstones.

Although far and away his most famous work, sandstone petrology is just the beginning of the contributions that Dickinson has made to geology. He has done extensive research on the evolution of sedimentary basins in all senses as well as that of island arcs, suture zones, transform plate boundaries, and foreland regions, as well as the accumulation of petroleum reserves in all these

settings. The studies built his concept of petrofacies for lateral relations of sedimentary rocks during a given time interval. Several details on these areas of research were to define the methods to evaluate volcaniclastic sedimentation related to magmatic arcs by carefully studying the field relations and sandstone petrology in California, Oregon, and Fiji. It was in this work that he defined his petrotectonic assemblages, which are characteristic groups of rocks and their structures to define plate tectonic setting. He studied lateral changes in sedimentation along the San Andreas, California, fault, in response to continuing movement along this transform margin. He defined the stratigraphy and structure of modern forearc and arc-trench systems in the western United States, as well as in Japan and New Zealand. Another area of research is the syntectonic sedimentation that accompanies severe Cenozoic crustal extension in the southwest United States. This research involves the development of sedimentary basins along active normal faults in Arizona. He also studied the tectonic and sedimentary development of Phanerozoic basins in the Cordillera and interior United States including Laramide basins, the Great Valley sequence, and the Ouachitas in Oklahoma.

Dickinson even studied the provenance of sand tempers in Melanesian and Polynesian potsherds. The results of these studies critically constrained the timing and directions of human migration among the south Pacific Islands. This work is deemed a breakthrough by archaeologists and anthropologists who study this area. Dickinson is by all means versatile in addition to his effectiveness.

William Dickinson was born on October 26, 1931, near Nashville, Tennessee. He grew up on a horse farm there. He attended Stanford University, California, and earned a bachelor of science degree in petroleum engineering in 1952. He then enlisted in the U.S. Air Force for a two-year hitch. He returned to Stanford University to earn master of science and Ph.D. degrees in geology in 1956

William Dickinson on a field trip in California in 1989 *(Courtesy of Arthur Sylvester)*

and 1958, respectively. He joined the faculty of Stanford University upon graduation where his students referred to him fondly as "Cowboy Bill." He was also dubbed the "perfect Prof" in a 1975 school newspaper for his inspired lectures. Although Dickinson was first married in 1953, he married his lifelong companion, Jacqueline Klein in 1970. In all, he has four children. In 1979, Dickinson moved to the University of Arizona in Tucson where he remained for the rest of his career. He served as department chair from 1986 to 1991 whereupon he retired to professor emeritus.

William Dickinson has led an extremely productive career. He is an author of more than 300 articles in international journals and professional volumes. Many of the definitive studies on the relations of sedimentation and petrology to plate tectonics (both descriptive and methodologies) are included in this group. Dickinson's contributions to geology have been well recognized by the profession in terms of honors and awards. He is a member of the National Academy of Sciences. He received both the Penrose Medal and the L. L. Sloss Award from the Geological Society of Amer-

ica and the Twenhofel Medal from the Society of Economic Paleontologists and Mineralogists. He was also a Guggenheim Fellow as well as having held numerous named lectureships.

Dickinson has performed significant service to the profession. He has held numerous positions of administrative responsibility for the Geological Society of America (including president and chair of the 1987 annual meeting), the National Research Council, and the Society of Economic Paleontologists and Mineralogists (including vice president), among others. He has also served in editorial positions for the *Geological Society of America Bulletin* and the *American Journal of Science,* among several others.

⊠ Dietz, Robert S.
(1914–1995)
American
Marine Geologist (Oceanographer)

Robert Dietz was one of the small group of revolutionaries who helped to turn the theory of plate tectonics into reality. His expertise was ocean floor mapping aboard literally dozens of research expeditions using the latest of technology. With the likes of H. WILLIAM MENARD, he studied mid-ocean ridges in the Pacific, submarine scarps (later called fracture zones and transform faults), and made the first map of the deep-sea fan at the mouth of the Monterey Submarine Canyon. This work showed that large amounts of sediment could be channeled into the deep sea from the continent. He contributed some of the seminal original work on the development of the continental shelves as well as the slopes. This geomorphic research added an important component to the plate tectonic paradigm.

Dietz was also interested in meteorite impact structures. In fact, he coined the now well-accepted phrase "astrobleme" to describe them. He studied craters both on Earth and on the Moon. He argued that the nickel-iron rich deposit of the

Sudbury Basin in Ontario, Canada, resulted from an extraterrestrial impact. He used some of his newly proposed features to further identify impact sites in the Ries and Steinheim basins in Germany and the Vredefort Ring in South Africa.

Dietz had a gift for finding adventure. He teamed up with Jacques Piccard to write a book about the deepest dive ever to the Challenger Deep at 35,800 feet. He visited and photographed Soviet oceanographic laboratories during the height of the cold war. He was present during the Russian invasion of Czechoslovakia in 1968 and even wrote slogans in chalk on Russian tanks. He sneaked a camera out onto the streets and photographed slaughtered Czechs and rebels fighting back. Some of those photos appeared in *Life* magazine.

Robert Dietz was born in Westfield, New Jersey, on September 14, 1914. He attended the University of Illinois at Urbana-Champaign from 1933 to 1941 during which time he earned bachelor of science, master of science, and Ph.D. degrees in geology with a minor in chemistry. Most of his research for his doctorate, however, was done at the Scripps Institution of Oceanography at the University of California at San Diego. He was in ROTC during his junior year and was called to active duty as a ground officer in the U.S. Army Air Corps with the 91st Observation Squadron in Fort Lewis, Washington, during World War II. He served as a pilot with many missions in South America. After active duty, he remained in the reserves for 15 years and retired a lieutenant colonel. After World War II, Dietz accepted a position at the U.S. Navy Electronics Laboratory in San Diego where he became the founder and director of the Sea Floor Studies Section. Through this position he participated in many marine expeditions, including Admiral Richard E. Byrd's last visit to Antarctica. His laboratory purchased the first Canadian Aqua-Lungs invented by Emile Gagnan and Jacques Cousteau. In 1953, the group who became expert with this equipment formed a private consulting company (Geological Diving Consul-

tants) to service the petroleum industry. Dietz was a Fulbright scholar at the University of Tokyo, Japan. He served with the Office of Naval Research in London, England, from 1954 to 1958. In 1963, Dietz accepted a position with the U.S. Coast and Geodetic Survey in Washington, D.C., which moved to Miami, Florida, and eventually became part of the U.S. National Oceanic and Atmospheric Administration (NOAA). He retired from NOAA to a series of visiting professor positions at University of Illinois, Urbana-Champaign (1974–1975), at Washington State University in Pullman (1975–1976), and Washington University in Saint Louis, Missouri (1976–1977), before accepting a permanent faculty position at Arizona State University. He retired to an emeritus professor position in 1985. Robert Dietz died of a heart attack on May 19, 1995, at his home in Tempe, Arizona.

Robert Dietz had a very productive career, producing numerous articles in international journals, professional volumes, and government reports. He was also an author of several books. Many of his works are seminal reading for plate tectonics and ocean bathymetry. He received numerous honors and awards for his contributions to geology. Among these awards are the Walter H. Bucher Medal from the American Geophysical Union, the Gold Medal of the U.S. Department of Commerce, the Alexander von Humboldt Prize from West Germany, and the Penrose Medal from the Geological Society of America. In addition, Dietz served numerous positions in professional societies as well as editorial positions for journals.

⊠ Dott, Robert H., Jr.
(1929–)
American
Sedimentologist

Robert Dott may be best known for making a strong stand against a growing creationist movement in the early 1980s based upon sound science. He wrote editorials and made numerous speeches attacking creationist claims for a young Earth and catastrophic change. He argued that episodic events like storms, volcanic eruptions, and earthquakes are the rule of the Earth rather than the exception. All could be explained by normal science. He also writes popular textbooks on Earth history that are widely adopted and read by geology majors and non-majors alike on college campuses around the country.

However, Robert Dott is known in the field for being one of the true pioneers of modern sedimentology. He did extensive research on ancient depositional systems in the southern Andes to Antarctica. The Dott Ice Rise was named for work done in Antarctica around 1970. At that time, he also conducted research on ancient systems in the western United States. Dott's timing was perfect, as he became one of the leaders in fitting sedimentology of mountain belts into the plate tectonic paradigm which was culminating at the time. He also attempted to fashion ancient plate reconstructions based upon paleomagnetic, paleontologic, and paleogeographic evidence. These ideas were applied to his next area of interest, which was the ancient sedimentary processes and environments within the stable craton interior during Proterozoic and early Paleozoic times. The research techniques were not innovative but his new eye for fitting the results into plate tectonic scenarios was innovative and many new ideas on the depositional systems of continental interiors arose as a result of Dott's work.

The impact that Dott made on geology with his sedimentologic research and defense of the science was enough for any career, but he also became renowned for his interest in the history of geology. He investigated the careers of several geologists who made significant impacts on the field. He studied the evolution of geological concepts and how episodic advances shaped the profession, similar to how such events shape the stratigraphy of an area. He has written numerous articles and several books on the subject.

Robert Dott shows the location of the Cambrian-Ordovician boundary in strata in a quarry in Madison, Wisconsin *(Courtesy of Robert Dott Jr.)*

Robert Dott was born on May 2, 1929, in Tulsa, Oklahoma. He attended the University of Michigan in Ann Arbor where he earned a bachelor of science degree in geology in 1950 and a master of science degree in 1951. He earned a Ph.D. from Columbia University, New York, in 1956. From 1954 to 1958, Dott worked part-time and full-time as an exploration geologist for Humble Oil and Refining Company (now Exxon Inc. of Exxon-Mobil, Inc.). In 1957 and 1958, he served as a first lieutenant for the U.S. Air Force Geophysics Research Directorate. In 1958, Dott joined the faculty at the University of Wisconsin at Madison where he remained for the rest of his career. From 1974 to 1977, he served as chair of the department. In 1984, he was named the Stanley A. Tyler Distinguished Professor of Geology. He retired in 1994 as a professor emeritus. During his years at University of Wisconsin, Dott was twice a National Science Foundation Visiting Fellow at Stanford University, California (1978), and the University of Colorado, Boulder (1979). He was a Cabot Visiting Professor at University of Houston, Texas (1986–1987), and an Erskine Visiting Fellow at Canterbury University, New Zealand (1987). He also served as chief scientist aboard the research vessel *Hero* to Cape Horn and Tierra del Fuego. Dott is married with five children.

Not only has Dott been the author of numerous articles in international journals and professional volumes, he is also author of several books, including *Evolution of the Earth,* probably the premier textbook of historical geology. His papers are also often-cited benchmark studies in the field of sedimentology. Dott's research contributions to geology have been well recognized by the profession as shown by his numerous honors and awards. He is a member of the American Association for the Advancement of Science. From the American Association of Petroleum Geologists, he received the President's Award for a young author (1956) and the Distinguished Service Award (1984). He received the 1992 Ben H. Parker Medal from the American Institute of Professional Geologists, the 1993 W. H. Twenhofel Medal from the Society for Sedimentary Geology, and the 1995 History of Geology Division Award and the 2001 L. L. Sloss Award from the Geological Society of America.

Robert Dott has performed extensive service to the profession. He was president of the Society of Economic Paleontologists and Mineralogists (SEPM) in 1981–1982 and of the History of Earth Sciences Society in 1990. He served on the U.S. Committee on the History of Geology (National Research Council) from 1981 to 1983 and the U.S. National Committee on Geology (National Academy of Sciences) from 1982 to 1986. He served on numerous committees for the Geological Society of America and was associate editor for *Geology.* He was also a distinguished lecturer on numerous occasions.

Drake, Charles L.
(1924–1997)
American
Geophysicist

In the tradition of several other renowned geophysicists from Lamont-Doherty Geological Observatory, New York, one of the main reasons that Charles Drake is so well known is for his advisory positions in politics. Like former colleagues FRANK PRESS and LYNN SYKES, Drake was a member of the Council of Advisors on Science and Technology to a president of the United States. In his case, it was President George H. W. Bush, from 1990 to 1992. In this role, Drake was a strong proponent of maintaining an active program of pure science as the emphasis shifted to applied science with the end of the cold war. This role was not the first experience in national-international politics for Drake. In 1986, he was asked to meet with the minister of geology of the USSR to discuss research directives.

In terms of geological research, Charles Drake established himself as one of the leading experts on the geology of continental margins. The margin marks the transition from continental crust to oceanic crust and is thus complex in terms of basement structure and coupling. It is also an area of active and varied sedimentation and is thus complex in terms of its cover geology as well. He studied this transition using geophysical techniques in addition to the results of deep-ocean drilling programs. Although he began with gravity studies, his most notable research involved the use of seismic reflection studies (like a sonogram of the Earth) on ocean sediments. In 1962, with Jack Nafe, he established a relation between the density of ocean sediments and the speed at which seismic waves travel through them known as the Nafe-Drake curve. It is still in use today. Drake also studied the development of these margins from rifted continent to mid-ocean ridge. He conducted a detailed study of the Red Sea, in addition to several seminal studies on the development of the Atlantic Ocean. Considering the timing of this research, primarily in the 1950s and 1960s, Drake became one of the important contributors of the plate tectonic paradigm.

Later in his career, primarily as the result of his interest in the geology of the Colorado Plateau, Drake became embroiled in the dinosaur extinction controversy. He took a strong stand in contrast to the popular opinion that the dinosaurs went extinct as the result of a giant meteor impact. He felt that terrestrial causes might be just as plausible.

Charles Drake was born on July 13, 1924, in Ridgewood, New Jersey, where he grew up. He enlisted in the U.S. Army and served in the South Pacific during World War II. After being discharged, Drake attended Princeton University, New Jersey, where he earned a bachelor of science degree in engineering in 1948. Upon graduation, he obtained a position with the U.S. Navy performing gravity measurements in submarines. This work sparked his interest and eventually led him to the Lamont-Doherty Geological Observatory of Columbia University, New York, where he earned a Ph.D. in geophysics in 1958. He remained at Lamont-Doherty, first as a research associate and later as a member of the faculty. In 1969, he accepted a position at Dartmouth University in New Hampshire and remained there until his retirement in 1992. He served as chair of the department in 1978–1979 and dean of graduate studies and associate dean of the science division from 1979–1985. Charles Drake married Elizabeth Ann Churchill on June 24, 1950; they had three children. In his spare time, he enjoyed boating, playing music, and woodcarving. Charles Drake died on July 8, 1997, at his home in East Thetford, Vermont.

Drake was one of the premier examples of just how much service to the profession and public one person can perform. In addition to that already mentioned, Drake served as president of the American Geophysical Union (1982–1984), the Geological Society of America (1976–1977), the

International Geological Congress (1989), and the International Council of Science Unions, Geodynamics Committee (1970–1975) in addition to serving on numerous committees and panels for each. He served on some 25 panels and committees for the National Academy of the Sciences-National Research Council, on many of which he was chair. He was a member of the National Advisory Committee on Oceans and Atmospheres and the governing board for the International Geological Correlation Program. He also served on many committees for the National Science Foundation, NASA, and the National Oceanic and Atmospheric Administration. In addition to all of this professional work, Drake somehow found time to serve as a trustee for the Village of South Nyack, New York, for eight years and even as deputy mayor from 1968–1969. Charles Drake received the G. P. Woolard Award from the Geological Society of America for his contributions to the profession.

⊠   **Dunbar, Carl O.**
(1891–1979)
American
Paleontologist

Carl Dunbar was an expert on fusulinids, football-shaped foraminifera that were common in the Paleozoic. He was especially expert in North American fusulinids from the late Devonian through Permian. Much of his effort was in the mid-continent in Illinois and Nebraska southward through Oklahoma and Texas, but also in British Columbia and Newfoundland. Because these fusulinids were so widespread and responsive to environmental changes, they make excellent fossil markers. Dunbar used them to study biostratigraphy and regional correlations of units as an extension. On this basis, he prepared several papers on major North American correlations of late Paleozoic biostratigraphic units. He was especially interested in major changes that took place across the profound Carboniferous to Permian boundary. To pursue this boundary required Dunbar to extend his studies outside of the United States, which he did with relish. He studied late Paleozoic fusilinids and related invertebrates worldwide in places like India, Central America, Mexico, South America (Peru and Argentina), the southern Urals in Russia, northwestern Yunnan, China, and east Greenland, among others. His correlations became worldwide and he identified subtle variations related to the shifting climates. Dunbar clearly established himself as the world expert on fusulinids and identified many new species in the process. His regional correlations, interpretations of paleoclimatic history, and evolutionary responses of fusulinids to these changes would later be utilized as key evidence in plate tectonic reconstructions.

Carl Dunbar was born on January 1, 1891, near Hallowell, in Cherokee County, Kansas. He worked as a wheat farmer in his youth and graduated from Cherokee County High School in 1909. He attended the University of Kansas in Lawrence where he earned a bachelor of science degree in geology in 1913 under the guidance of WILLIAM H. TWENHOFEL. He remained at the University of Kansas for one year of graduate studies before enrolling at Yale University, Connecticut, where he earned a Ph.D. in geology in 1917. He studied under Charles Schuchert. Carl Dunbar married Lora Beamer in September 1914; they had two children. Dunbar was an instructor at the University of Minnesota, Twin Cities, from 1918 to 1920. He returned to Yale University as a junior faculty member as well as the assistant curator of invertebrate paleontology at the Peabody Museum. He became curator in 1925 when Schuchert retired. Dunbar remained at Yale University through his entire career, retiring to professor emeritus in 1959. He was a visiting professor at the University of Kansas in 1962. Carl Dunbar

died suddenly on April 7, 1979, in Dunedin, Florida. His wife had recently predeceased him in December 1978.

Carl Dunbar was a productive geologist authoring more than 70 scientific publications including articles in international journals and professional volumes, chapters in books, and governmental reports. He was also an author of several popular textbooks, including *Textbook of Geology: Part II, Historical Geology* with Charles Schuchert and *Principles of Stratigraphy* with JOHN RODGERS. In recognition of his contributions to geology, the profession bestowed numerous honors and awards upon him. He was a member of both the National Academy of Sciences and the American Academy of Arts and Sciences. He received the Paleontological Society (United States) Medal, the Twenhofel Medal from the Society of Economic Paleontologists and Mineralogists, the Hayden Memorial Geological Medal from the Academy of Natural Sciences of Philadelphia, and several awards from his alma mater at the University of Kansas, including the Erasmus Haworth Distinguished Alumni Award in geology and the Alumni Distinguished Service Citation.

Dunbar's service to the profession was equally impressive. He was vice president (1952) and councilor (1940–1942) for the Geological Society of America. He was president (1952) and treasurer several times for the Paleontological Society. He was probably best known for his role as chairman for the Committee on Stratigraphy for the National Research Council (1934–1953), among many other panels and committees. He was also one of 27 scientists chosen to observe Operation Cross-Roads, the atomic bomb tests at Bikini Atoll. His editorial roles included associate editor of the *Geological Society of America Bulletin,* and the *Journal of Paleontology.*

E

Ernst, W. Gary
(1931–)
American
Geochemist (Plate Tectonics)

The pioneers of the plate tectonic paradigm were primarily concerned with large-scale physical forms on the planet. Geophysicists and theoreticians dominated this group. However, plate tectonics affects all aspects of geology. Gary Ernst is a true pioneer in the geochemistry of plate tectonics. He attacks these geochemical–plate tectonic problems from all angles, be it experimental, analytical, isotopic, or field-based and he travels to the ends of the Earth to find the best location. He became the world expert on high-pressure, subduction zone–related metamorphism and plate subduction processes gleaned from this research. His maverick approach, outstanding productivity in terms of both research and publication, and willingness to take on leadership roles in professional societies and on federal panels made him likely one of the top influential geologists in the world of the late 1970s to mid-1980s. In discussions among graduate students in departments nationwide, Ernst was regarded with awe as one of the heroes of the field. He still remains influential and respected.

Ernst's specific locations of research on subduction zone metamorphism include a complex of rocks of the Dabie-Sulu Belt of eastern China with Chinese, Japanese, and Russian colleagues. He has also begun work on the North Qaidam and North Qilian belts of northwest China. Previously he worked on the Maksyutov Complex in the southern Urals, the Kokchetav Massif of northern Kazakhstan and the Franciscan Complex of the California Coast Ranges, among others. Besides the high-pressure studies, Ernst has conducted field research in the White-Inyo Range of California for many years. This work involves petrogenesis of the Barcroft Granodiorite and contact metamorphism in the country rock. He even conducts environmental research in this area. He also continues his experimental studies on the synthesis of hydrothermal minerals, which he has done for many years.

W. Gary Ernst was born on December 14, 1931, in St. Paul, Minnesota, where he spent his childhood. He attended Carleton College, Minnesota, where he earned a bachelor of arts degree in geology in 1953. He attended the University of Minnesota, Twin Cities, for graduate studies and earned a master of science degree in geology in 1955. He then moved to the Johns Hopkins University, where he completed his graduate education by earning a Ph.D. in geochemistry. He won a postdoctoral fellowship from the National Science Foundation to the Geophysical Laboratory at the Carnegie Institution of Washington, D.C., in

Gary Ernst discussing Earth science aboard a research ship in California *(Courtesy of Arthur Sylvester)*

1959 and 1960. He joined the faculty at the University of California at Los Angeles in 1960, and remained until 1989. During that time he served as chair of the department from 1970 to 1974 and 1978 to 1982. From 1987 to 1989, he was the director for the Institute of Geophysics and Planetary Physics at UCLA. In 1989, he moved to Stanford University as dean of the School of Earth Sciences. Since 1999, Ernst has held the Benjamin M. Page Endowed Chair at Stanford University. During these years, he was a Crosby Visiting Professor at Massachusetts Institute of Technology in 1968, a National Science Foundation Senior Postdoctoral Fellow at University of Basel, Switzerland, from 1970–71, visiting professor at the Swiss Federal Institute, Zurich, from 1975–76, the William Evans Visiting Professor at Otago University, New Zealand, from 1982–83, the Universidade Federale de Pernambuco, Brazil, in 1988, and visiting professor at Kyoto University, Japan, in 1988.

Gary Ernst has been phenomenally productive, even if all of the administrative positions he has held are discounted. He has authored six books and research memoirs from popular science to textbooks to cutting-edge research works. He also served as editor of another 14 professional volumes. He was an author of more than 180 articles in international journals and professional volumes. Many of these papers are landmark studies that appear in some of the most prestigious journals and often-cited volumes in the field. His outstanding research has been recognized in the field in terms of honors and awards. He is a member of National Academy of Sciences and was chair of the geology section from 1979 to 1982 and 2000 to the present, as well as secretary from 1997 to 2000. He was a Fulbright Research Scholar at the University of Tokyo in 1963, a Guggenheim Memorial Fellow in 1975–1976 and a Japan Society for the Promotion of Science Fellow in 1995. He also won the Geological Society of Japan Medal in 1998.

Gary Ernst has performed extensive service to the profession. He was chairman of the board of Earth sciences for the National Research Council in 1984 to 1987 and a member from 1988 to 1993. He was president of the Geological Society of America in 1985 and 1986 and the president of the Mineralogical Society of America in 1980 and 1981. The rest of his service is in such abundance that it cannot all be listed here.

⊠ **Eugster, Hans P.**
(1925–1987)
Swiss
Geochemist

Hans Eugster was one of those rare people who can be given the title of "Renaissance Man" because he excelled at so many pursuits. He could have had a successful career as an artist or a musician or a chemist or a mathematician, among others. Fortunately for the Earth sciences, he chose to be a geochemist and an outstanding teacher. Unfortunately, he died far too young. His geochemical research was much like his life; he chose several directions and he excelled in all of them. Probably his greatest contribution to geochemistry was to demonstrate the control of gases, both in terms of presence and participation (called fugacity), on high-temperature chemical reactions among minerals in igneous and metamorphic rocks. These very minor components can actually control the minerals that will form with the major components. The first components considered were oxygen and hydrogen. With his colleague, DAVID R. WONES, Eugster investigated their participation in the formation of micas. Later, this work was extended to other minerals, as well as to other fluid components like carbon, fluorine, nitrogen, and sulfur species. He even looked at acids, bases, and metal chlorides in fluids, ultimately establishing a whole new field concerned with measuring the properties of fluids. It all led to a quantitative understanding of the role of fluids in the processes of mineral formation within Earth's crust and mantle.

Although this research may seem purely theoretical, it has practical applications as well. Eugster once began a project with a student through General Electric Corporation to devise a substance that was a perfect insulator in one direction and a perfect conductor in another. Eugster thought that synthesizing micas with gold lining the interstices would solve the problem. The student left and the project fell through. But if it had continued, they would have produced the first silicon chip well ahead of its time.

Eugster's second main research direction was in geochemistry as it applies to hydrogeology and sedimentology. Considering that his first interest was in high temperature applications, this second direction in surface reactions is surprising. In this research, he evaluated the hydrogeologic, chemical, and sedimentologic processes that lead to the formation of continental and marine evaporites. He discovered several new minerals and proposed a new origin for bedded chert including Precambrian banded iron formations. This research included experimental work, thermodynamic modeling, and geologically reasonable computer solutions to the evaporation of seawater, a feat that was attempted several times previously by other researchers without success. Two of the more important papers from this work include, "The Evolution of Closed Basin Brines" and "Minerals in Hot Water."

Finally, Eugster was also interested in the origin of ore deposits. He conducted experiments on the solubility of ore minerals to explain their deposition in hydrothermal systems. He explained several types of deposits with this work and he even investigated the source of these fluids in dewatering granites.

Hans Eugster was born in Landquart, Switzerland, on November 19, 1925, where he spent his youth. He gained an interest in geology climbing to the high Alps in the Grisons where he would lag behind the rest of his family because he was too busy collecting rocks. He attended the Swiss Federal Institute of Technology (ETH) in Zurich where he earned a diploma in engineering geology in 1948. He continued at ETH for his graduate studies and earned a Ph.D. in 1951 in geochemistry (his adviser was Paul Niggli). Eugster had accepted an eight-month postdoctoral research post at Massachusetts Institute of Technology, to return to ETH at its conclusion. However, the untimely death of Niggli led Eugster to accept a position at the Geophysical Laboratory of the Carnegie Institution of Washington, D.C., in 1952 with HATTEN S. YODER JR.. In 1957, he taught a course at the Johns Hopkins University as an adjunct and accepted a permanent position there the following year. He remained at Johns Hopkins for the rest of his life, serving as chair of the department from 1983 to 1987. He died suddenly of a ruptured aorta on December 17, 1987. His second wife, Elaine Koppelman, the James Beall Professor of Mathematics and Computer Science at Goucher College, and three daughters from his first marriage survived him. In addition to his talent as a geochemist, Eugster was also an accomplished violinist, painter, and potter.

Hans Eugster had a very productive career, publishing numerous articles in international journals and professional volumes. Many of them are considered benchmarks of geochemistry. The geologic profession has acknowledged his research contributions in terms of honors and awards. He was a member of both the National Academy of Sciences and the American Academy of Arts and Sciences. He received the Arthur Day Medal from the Geological Society of America in 1971, the Goldschmidt Medal from the Geochemical Society in 1976, and the Roebling Medal of the Mineralogical Society of America in 1983. He even had the new mineral eugsterite named after him in 1983. He served as the president of the Mineralogical Society of America in 1985.

⊠ Ewing, W. Maurice
(1906–1974)
American
Geophysicist

Maurice Ewing is one of the giants of Earth sciences and especially geophysics. He not only made significant discoveries about the ocean floor, he developed a good amount of the equipment to study it. In a field of some very impressive scientists, Maurice Ewing is still so prominent that he can be considered the "father of modern marine geophysics." He began this pioneering research in the 1930s, when he devised a method and developed the equipment to conduct the first seismic refraction profiles (like sonograms of the Earth) at sea. This work first established the shape of the continental shelf, slope, and rise, and showed how the sedimentary cover thickened oceanward over a highly faulted basement. That was just the beginning. He would later redesign the bathythermograph (temperature with depth) for use in moving ships, build equipment for continuous echo sounding (depth profiling) and precision depth recording, develop ocean bottom seismographs and advanced methods for marine seismic reflec-

tion and refraction surveying. He and his group at Lamont-Doherty Geological Observatory would develop the methods and protocol for piston core sampling of deep ocean sediments and Ewing would help found the Deep Sea Drilling Project, as well as serving as its first chief scientist in 1968. He would develop new methods for gravity and magnetic surveying at sea and with FRANK PRESS, he would establish the WorldWide Standardized Seismograph Network to monitor earthquakes as well as nuclear tests.

Naturally, all of these inventions and new methodologies were used to obtain the first data of their kind. Some of this research was physical-process based, such as how seismic waves travel through layered media or in ocean water, among many others; whereas other research was Earth-process based. He was the first to establish that there are fundamental differences between ocean crust and continental crust geophysically, geochemically, and petrologically. He studied the differences between them and formulated the basic features of all ocean basins. He determined that the mid-ocean ridges formed long chains of seismic activity and that they are unstable and ever-changing permanent features. In his sedimentary coring efforts, he identified a fundamental change in the sediment type and composition from ice ages versus those of interglacial periods and he modeled climatic oscillations on that basis. Even more fundamental was his publication of the first detailed maps of the seafloor in the North Atlantic, South Atlantic, Pacific and Indian Ocean basins. All of this classic fundamental research would later be expanded to spur discoveries in plate tectonics as well as climate change studies and it is still having major influence on many current research topics. It is difficult to overemphasize the impact that Maurice Ewing had on the Earth sciences.

William Maurice Ewing was born on May 12, 1906, in Lockney, Texas, near Amarillo, where he grew up. He was somewhat of a prodigy and graduated from Lockney High School at age 15.

Unfortunately, this high school was unaccredited and he had a difficult time getting into college. He was finally accepted at Rice Institute (now Rice University) in Houston, Texas, where he began as an electrical engineering major in 1922. He switched to mathematics and physics, and graduated in 1926 with a bachelor of science degree with honors and the title of Hohenthal Scholar. He remained at Rice Institute for graduate work and earned a master of arts in 1927 and a Ph.D. in 1931, both in physics. Maurice Ewing married in 1928 and was father to five children. Ewing was an instructor of physics at the University of Pittsburgh, Pennsylvania, from 1929–1930 before joining the faculty at Lehigh University, Pennsylvania, also in physics. However, he participated in marine geophysical surveying with geophysicists there and was moved to the geology department in 1940. That year, he also became a Research Associate at the Woods Hole Oceanographic Institution, Massachusetts. During World War II, Ewing worked with the U.S. Navy discovering and investigating the "Sofar" long-range submarine sound ranging and transmission system in which all underwater sound is naturally funneled in a 1,500 m depth horizon where it can be readily transmitted. He would receive the U.S. Navy Distinguished Service Award in 1955 for this research. In 1944, he moved to Columbia University, New York, where he would remain for the rest of his career. Ewing established and became director of the Lamont-Doherty Geological Observatory (now the Lamont-Doherty Earth Observatory) of Columbia University in 1949. It would become one of the world's premier geophysical laboratories, hosting such giants as Frank Press, JACK E. OLIVER, MANIK TALWANI, CHARLES L. DRAKE and LYNN R. SYKES, among others. He was named Higgins Professor of geology in 1959 and retired to professor emeritus in 1972. Maurice Ewing suffered a massive cerebral hemorrhage on April 28, 1974, in Galveston, Texas, and died in Palisades, New York, on May 4, 1974. His wife and five children survived him.

Maurice Ewing was an author of more than 340 scientific articles and reports in international journals, professional volumes, and government reports. Many of these publications are true classic works on all aspects of marine geophysics including numerous new and modified geophysical methods and processes as well as some of the basic work on the plate tectonics of ocean basins. His book, *Elastic Waves in Layered Media,* is still considered a seminal work. In recognition of his vast contributions to Earth sciences, Maurice Ewing was bestowed numerous honors and awards in addition to those already mentioned. He was a member of both the National Academy of Sciences and the American Academy of Arts and Sciences. He received the U.S. National Medal of Science in 1973 and numerous honorary degrees, including Washington and Lee University in Virginia. He received both the Agassiz Medal and the John J. Carty Medal from the National Academy of Sciences, both the Penrose Medal and the Arthur L. Day Medal from the Geological Society of America, both the William Bowie and the Walter H. Bucher Medal from the American Geophysical Union, the Gold Medal from the Royal Astronomical Society, the Sidney Powers Memorial Medal from the American Association of Petroleum Geologists, the Earl McConnell Award from the American Institute of Mining, Metallurgical and Petroleum Engineers, and the Vega Medal from the Swedish Society of Anthropology and Geography, among others. A medal from the American Geophysical Union is named for him.

Impossible as it seems, Ewing found time to perform significant service to the profession. He was president of the American Geophysical Union (1956–1959), president (1955–1957) and vice president (1952–1955) of the Seismological Society of America, and vice president (1953–1956) and councilor (1946–1948) of the Geological Society of America.

F

Fairbridge, Rhodes W.

(1914–   )
Australian
Geomorphologist (Climate Modeling)

Rhodes Fairbridge has several areas of expertise for which he is renowned. He studies coastal geomorphology, climate change and its control, and produces Earth science encyclopedias. His work in coastal geomorphology centers on eustatic sea level changes and their control, the role of gravitational processes in tectonic changes and resulting sedimentation patterns, and world geotectonics. His research concerns the emergence and submergence of coastlines, especially in response to ice ages and glacial loading of the crust. The crust in the far northern latitudes has undergone glacial rebound after the mile-thick continental ice sheet melted away after the last ice age. The plates were depressed into the mantle by the sheer weight of those continental glaciers by hundreds of feet in some cases. Fairbridge was the first to document and explain this depression and rebound and he showed the complex interaction of the rising sea level from the melting of the ice coupled with this emergence. The understanding of the shape and features of our coastlines is largely due to the work of Fairbridge. He also studied the effect of the sea level rise on coral reefs. He is truly the father of the modern science of coastal processes.

Fairbridge is even better known for his work on climate change. He uses his coastal geomorphology coupled with pollen analysis (palynology) and sedimentation patterns to chart climate changes. He correlates these climate changes with extraterrestrial influences. He proposed that planetary ephemeus, the alignment torque of the planets on the Sun, has an effect on solar particulate radiation mostly through sunspot activity. He showed that the carbon 14 flux rate has changed drastically and proposed that it controls the number and intensity of catastrophic droughts and floods. He looked at carbon dioxide abundances in the ocean through time with ROGER REVELLE by charting changes in the production of carbonates well before its current popularity. These studies led him to consider the interplay of sea level, greenhouse effect, and droughts on a worldwide scale through time.

Fairbridge is also the king of the Earth science encyclopedia. He has produced more than 24 high-quality encyclopedias on geomorphology, climatology, oceanography, environmental science, soil science, planetary science, geochemistry, sedimentology, hydrology and water resources, and world regional geology, among many others. He gathers the experts in the various fields to contribute to these works, but it is nonetheless phenomenal that he is so well versed in such a diverse variety of disciplines in Earth sciences to be able to attempt such endeavors.

Rhodes Fairbridge was born on May 21, 1914, in Pinjarra, Australia, son to the famous Kingsley Fairbridge. His family ran the Fairbridge Farm School, a world-renowned boarding school. He attended Queens University, Australia, where he earned a bachelor of arts degree in 1936. He then attended Oxford University, England, where he earned a bachelor of science degree in geology in 1940. During the period of 1938 to 1941 he was also a field geologist for the Iraq Petroleum Company. He returned to Australia to complete his graduate studies at the University of Western Australia, where he earned a doctor of science degree in 1944. He was a lecturer at University of Western Australia from 1946 to 1953 and a member of the faculty at the University of Illinois at Urbana-Champaign from 1953 to 1954. In 1955, he joined the faculty at Columbia University, New York, and remained there for the rest of his academic career, until his retirement in 1982, when he became professor emeritus. Throughout his academic career, Fairbridge served as a consultant to numerous companies and agencies, including the Hydroelectric Commission of Tasmania, Richfield Oil Company, Australian Bureau of Mineral Resources, Snowy Mountain Hydroelectric Authority, Pure Oil Company, and others. He also worked with several publishing companies, including *Life* magazine and Reader's Digest Books. After his retirement, Fairbridge became an associate of NASA-GISS, where he remains today. Rhodes Fairbridge married Dolores Carrington in 1943; they have one son.

Rhodes Fairbridge has led an extremely productive career. He is an author of more than 300 scientific publications. Many of these are seminal works on coastal geomorphology and climate change. As mentioned, he is an editor of 24 encyclopedias, in addition to being an author and editor of several professional volumes and books. He has received several honors and awards for his contributions to geology. He received an honorary doctorate from the University of Gothenburg,

Sweden. He was also awarded the Alexander von Humboldt Prize from the Humboldt Society, Germany, and the 1999 Mary B. Ansai Best Reference Work Award from the Geoscience Information Society.

Fairbridge has also performed significant service to the profession. He is a founder and the current vice president of the Coastal Education and Research Foundation. He was the president of the Shorelines Commission of the International Union for Quaternary Research. He also served on numerous committees and panels for the National Academy of Sciences, the National Research Council, Office of Naval Research, National Science Foundation, and others. He also served in an amazing number of editorial capacities including founder and editor of *Journal of Coastal Research,* series editor for 90 volumes of the *Geological Benchmark Collections* (Hutchinson-Ross Publishing Co.), and adviser for Random House, Fabbri Publishing Co., Milan, Italy, the *Van Nostrand Reinhold Encyclopedia of Earth Sciences,* and the *Chapman-Hall Encyclopedia of Earth Sciences.*

⊠ Folk, Robert L.
(1925–   )
American
Sedimentologist, Archaeological Geologist

Robert Folk became interested in classifying sedimentary rocks as a boy admiring his rock collection because the igneous rocks had such exotic names and the sedimentary rocks did not. Their names, sandstone, limestone, shale, were boring. He decided even then that he would remedy the situation and after his years at college, he did just that. If he had done nothing else with his career, Robert Folk would still be remembered for his classification system for carbonates, which still appears in many textbooks more than 40 years later, as well as a definitive textbook on sedimentology from

Robert Folk in a marble quarry in Lipari, Italy *(Courtesy of James K. Mather)*

about the same time. However, this work barely scratches the surface of a very successful career.

In addition to much more extensive work on carbonate deposition, Folk also became interested in Aeolian deposits. He researched grain roundness and coloring of sand in the Simpson Desert in Australia. He similarly investigated pebble shapes in rivers and on beaches in Tahiti. He then turned his attention to archaeological geology. He did research on environmental geology of classical Macedonia, limestone used in construction of the pyramids, and the archaeological geology of Israel, and especially Galilee. He located sources for building materials, isotopically determined the ages of mortar in structures, and determined how iron smelting was carried out in ancient Israel. He even began investigating how the ancient ruins were deteriorating.

An unquenchable love for Italy led to the discovery that bacteria mainly constructed Roman hot-water travertine (a kind of limestone in springs). Work on these rocks with the Scanning Electron Microscope (SEM) led to the discovery of dwarf bacteria (nannobacteria). About 10,000 of them will fit on a pinhead. He published on the topic with papers like "Nannobacteria in the Natural Environment and in Medicine" to little fanfare or even notice. Nonetheless, Folk began a full research program on the role of nannobacteria in both sedimentary and weathering processes. Then came the announcement by NASA that they may have found nannobacteria in Martian meteorites. The question as to whether there was life on Mars captivated the imagination of the public. Now Folk finds himself in the middle of a controversy and a frenzy of research and publication on terrestrial rocks, Martian meteorites, and human nannobacterial diseases.

Robert Folk was born on September 30, 1925, in Cleveland, Ohio. He graduated from Shaker Heights High School and enrolled in Pennsylvania State University in College Park in 1943. He earned all of his degrees at Penn State, including a bachelor of science degree in 1946, a master of science degree in 1950, and a Ph.D. in 1952. His adviser for all of his research was Paul Krynine, but he also spent a year at Columbia University, New York. He married Marjorie Thomas in 1947, and they had three children. He has enlisted them as field assistants on several projects. At the end of his graduate career he accepted a position as a geologist for Gulf Research and Development Co. in Houston, Texas, and Pascagoula, Mississippi. In 1952, Folk joined the faculty at the University of Texas at Austin and remained there for his entire academic career. He held several endowed chairs in the department including the Gregory Professorship in sedimentary geology (1977–1982) and the Carlton Professorship of geology (1982–1988). He retired as a professor emeritus in 1988 as well as accepting a position of senior research scientist at the Texas Bureau of Economic Geology, Austin. He was a visiting professor several times during his career at the Australian National University in Canberra (1965), at Università degli Studi in Milan, Italy (1973), and at Tongji University in Shanghai, China (1980).

Robert Folk has led a very productive career authoring more than 100 articles in international journals and professional volumes. Many of these articles are definitive works on carbonate petrology. He also wrote a successful textbook entitled, *Petrology of Sedimentary Rocks,* with six printings between 1957 and 1980. His research contributions and teaching ability have been well recognized by the profession as evidenced by his numerous honors and awards. For teaching, he received The Geology Foundation Outstanding Teacher Award and the Carolyn G. and G. Moses Knebel Distinguished Teaching Award, both from the University of Texas, the Neil Miner Award from the National Association of Geology Teachers (1989), and the Distinguished Educator Medal from the American Association of Petroleum Geologists (1997). For his research, Folk received three best paper awards from American Association of Petroleum Geologists, and Society of Economic Paleontologists and Mineralogists (SEPM). He was awarded the Twenhofel Medal from SEPM (1979), the Sorby Medal from the International Association of Sedimentologists (1990), and the Penrose Medal from Geological Society of America (2000).

⊠ Friedman, Gerald M.
(1921–　)
German
Sedimentologist

Two of the main benefits that geology provides to society are energy and environmental analysis. The principles behind these seemingly opposite fields overlap in the most important parts. The passage of oil and gas through sediments and sedimentary rock to a point of accumulation is analogous to the passage of groundwater and pollutants through the same materials. For this reason, there was a large migration of oil geologists to the environmental field during the oil bust of the 1980s. Unlike most academicians who simply refused to acknowledge the transition, Gerald Friedman moved from his position as one of the true leaders in petroleum geology to a position of prominence in environmental geology. Friedman's area of expertise is sedimentology of both clastic and carbonate rocks. He worked in the petroleum industry for some 10 years, discovering some of the major oil and gas fields mostly by using this expertise in sedimentology. When he moved on to academia in earnest, he brought his practical experience to research. Prior to his oil experience, he had done some of the groundbreaking research on carbonate diagenesis, which he continued upon his return. He performed primary research on the development of petroleum reservoirs in carbonate rocks and on sedimentology and depositional environments. He investigated both modern systems in the Gulf of Mexico, the Bahamas, Bermuda, Florida, the Red Sea, and the Dead Sea in Israel, and ancient systems in the Anadarko Basin in Oklahoma, the Permian Basin of west Texas, the Michigan Basin, the Appalachian Basin, and the Williston Basin.

Friedman truly came to the service of the world during the oil crises of the 1970s both with his research and by training large numbers of students to work in the petroleum industry or in academia related to petroleum exploration. He is also of service to the profession with his exceptional organizational ability.

Gerald Friedman was born on July 23, 1921, in Berlin, Germany. He attended the University of London, England, where he earned a bachelor of science degree in chemistry in 1945. He immigrated to the United States and accepted a position as an analytical chemist at E.R. Squibb and Sons, New Jersey. He married Sue Tyler Theilheimer in 1948 and entered the graduate program at Columbia University, New York. He earned a master of arts degree in 1950 and a Ph.D. in geol-

Gerald Friedman leads a field trip for the International Geological Correlation Project (IGCP) in the northern Appalachians in 1979 *(Courtesy of James Skehan, S.J.)*

ogy in 1952. In 1950, he joined the faculty at the University of Cincinnati, Ohio, but took a job as a consulting geologist in Sault Ste. Marie, Ontario, Canada, in 1954. From 1956 to 1964, Friedman was a research geologist and supervisor of research for Amoco Production Company in Tulsa, Oklahoma. In 1964, he joined the faculty at Rensselaer Polytechnic Institute, New York. He additionally became the president of the Northeastern Science Foundation, New York, in 1978, a position he holds today. He joined the faculty of Brooklyn College of the City University of New York (CUNY) and the CUNY Graduate School in 1985, where he continues today. In 1988, he was named Distinguished Professor of geology. He also served as deputy executive officer for the graduate program. Over the years, Friedman was a visiting professor at several schools, including He-

brew University, Israel; University of Heidelberg, Germany; Institute of Petroleum Research and Geophysics, Israel; the Geological Survey of Israel; and Martin Luther University in Halle-Wittenberg, Germany. Friedman and his wife have five children. In his leisure time, Friedman is a Judo master. He achieved third-degree black belt and was named a sensei.

Gerald Friedman is an author of more than 300 articles in international journals, professional volumes, and governmental reports. Several of the papers are seminal works on sedimentology and diagenesis. He has written or edited 16 books. One textbook, *Principles of Sedimentology,* sold more than 30,000 copies and won an award. Other highly regarded books by Friedman include *Depositional Environments in Carbonate Rocks* and *Exploration for Carbonate Petroleum Reservoirs,* among others. Friedman's achievements in research and professional leadership have been recognized with honors and awards too numerous to list fully here. Therefore, these are just the highlights. He has two honorary doctorates, one from the University of London, England, in which he was hooded by the late Queen Mother, and one from the University of Heidelberg, Germany, which issues them only once every 50 years. He received the Kapitsa Gold Medal of Honor from the Russian Academy of Natural Sciences. From the American Association of Petroleum Geologists (AAPG), he received the Distinguished Service Award in 1988, the Distinguished Educator Award in 1996, the Sidney Powers Medal in 2000, and the Environmental Teaching Award in 2001. From the eastern regional AAPG, he received the John T. Galey Memorial Award and a Certificate of Merit. He won the Twenhofel Medal from the Society of Economic Paleontologists and Mineralogists (SEPM), the Award for Outstanding Editing or Publishing Contributions from the Association of Earth Science Editors, and the James Hall Medal from the New York State Museum, among others. He also received a Best Paper Award and two

Honorable Mentions from *The Journal of Sedimentary Petrology,* as well as a Best Paper Award from an SEPM section.

Friedman's service to the profession rivals his honors and awards and as such can only be touched upon. He was president (1975–1978) and vice president (1971–1975) of the International Association of Sedimentologists. He was the president (1974–1975) and the vice president (1970–1971) of SEPM and vice president of AAPG in 1984–1985, among many other positions in both societies. He was president (1972–1973) and vice president (1971–1972) of the Association of Earth Science Editors and cofounder of the History of Earth Sciences Society. His editorial work is equally impressive. He served as editor for *Journal of Sedimentary Petrology, Northeastern Geology and Environmental Sciences, Earth Sciences History,* and *Carbonates and Evaporites,* and associate editor for *Sedimentary Geology, Journal of Geology, Geo Journal,* and *Journal of Geological Education* among others.

Fyfe, William S.
(1927–)
New Zealander
Geochemist

A summary of the accomplishments of William Fyfe in his geological career is very simple: he has done virtually everything and he has done it well. Originally, he was an experimental geochemist who was interested in problems of metamorphic petrology. Innovative techniques to conduct dehydration chemical reactions allowed him to define the formerly enigmatic appearance of the zeolite minerals. This research allowed him to define a new metamorphic facies (zeolite facies) that now appears in every petrology and introductory physical geology textbook. But he was not satisfied with considering only the low-temperature side of metamorphism, he also studied partial melting dynamics in the granulite facies within Archean

crust and high-pressure metamorphic rocks from the subduction zone complex in the Franciscan rocks of California. He was also interested in the formation of metal complexes and especially that of gold in hot fluid systems. This work truly revolutionized the field of hydrothermal systems and fluid flow, especially within fault zones, and also with regard to ore genesis. He pioneered the application of crystal field theory to the partitioning of trace elements among minerals. He also defined the application of stable isotope techniques to metamorphic problems among many other studies. These interests led him to write several classic memoirs and textbooks on the subjects including *Metamorphic Reactions and Metamorphic Facies, Geochemistry of Solids,* and *Fluids in the Earth's Crust.* He clearly established himself as one of the foremost authorities on metamorphism and perhaps the foremost authority on hydrothermal processes.

Later in his career, Fyfe became interested in environmental problems and he was no less productive in those studies. He pursued such topics as iron sulfide contents of coal from Ohio and its contribution to acidity in the environment, methods to determine soil erosion in the Arctic using smectite clay mineralogy, improvement of the crop potential of tropical laterite soils using geochemistry, manganese oxide precipitation on microbial mats within hot springs, the geochemistry of supratidal sediments in the Niger Delta, Africa, and many others in the general field of biogeochemistry. These studies are generally environmental and led Fyfe to become an outspoken advocate for science in the addressing of world problems. His research helped him to give sound advice on agricultural and environmental geochemistry, deep waste disposal, resource conservation, global climate changes, and assistance to Third World countries. In recognition of his untiring efforts in education and advocacy for a stronger role on the part of humankind in stewardship of the Earth, Fyfe was awarded the Companion of the Order of Canada, the nation's highest civilian award. Truly, the accomplishments of William Fyfe comprise several successful careers.

William Fyfe was born on June 4, 1927, in Ashburton, New Zealand. He attended Otago University, New Zealand, where he earned a bachelor of science degree in geology in 1948, a master of science degree in 1949, and a Ph.D. in 1952. From 1952 to 1954, he was a Fulbright Scholar in geology at the University of California at Berkeley. He had his first academic position at the University of California at Los Angeles for one year before moving back to Otago University as a reader in chemistry (1955–1958), and finally back to the University of California at Berkeley as a faculty member. In 1968, Fyfe was the recipient of a Royal Society Research Professorship in geochemistry at Manchester University in England. He was also a visiting professor at Imperial College in London, England. In 1972, he joined the faculty at the University of Western Ontario, Canada, in London as chair of the department, where he remained for the rest of his career. He was appointed dean of the faculty of science there from 1986 to 1990 and he is currently a professor emeritus. William Fyfe married Patricia Walker in 1981, and they have three children.

William Fyfe has led a phenomenally productive career. His tally of professional publications numbers in the vicinity of 800. He was typically an author of 30 or more books and articles per year. Many of these publications are benchmark studies on everything from hydrothermal metamorphism to the formation of soils. Fyfe received numerous honors and awards in recognition of his research contributions to the science. He was awarded six honorary doctoral degrees from colleges worldwide, including Memorial University in Newfoundland, Canada; University of Lisbon, Spain; Lakehead University, England; and Otago University, New Zealand, among others. He is a member of the Russian Academy of Sciences, the Indian Academy of Sciences, the Brazilian Academy of Sciences, the American Academy of Arts and Sciences, and a Fellow of the Royal Society of London and the Royal Society of New Zealand. He received the Logan Medal from the Geological Association of Canada, the Willet G. Miller Medal from the Royal Society of Canada, the Canadian Commemorative Medal, the Arthur Holmes Medal from the European Union of Geological Scientists, the Arthur L. Day Medal from the Geological Society of America, the Roebling Medal from the Mineralogical Society of America, the Canadian Gold Medal for Science and Engineering from the Natural Science and Engineering Research Council, the Queen's New Zealand Commemorative Medal, the Wollaston Medal from the Geological Society of London, the Sribrnou Medaili from the Czech Republic, and the Medal of the National Order of Scientific Merit from the country of Brazil, among many others.

Fyfe performed service to the profession too extensive to list here. His latest position was that of president of the International Union of Geological Scientists (IUGS) (1992 to 1996), where he has been especially active. He has served on dozens of important committees for the Natural Sciences and Engineering Research Council of Canada, commonly as chair. He also served in an editorial capacity for several international journals, including *Chemical Geology, Environmental Geophysics and Geochemistry, Geology,* and *Mineral Science and Engineering,* among others.

G

Garrels, Robert M.
(1916–1988)
American
Geochemist

Igneous and metamorphic rocks were treated as chemical systems early in the history of geology because there are diverse minerals involved and they are large enough to analyze. On the other hand, sedimentary rocks hold most of the economic reserves whether petroleum related or ore related. Robert Garrels did for the chemistry of sedimentary rocks what the likes of NORMAN L. BOWEN did for igneous rocks; he established the chemical systems. His 1952 paper, "Origin and classification of precipitants in terms of pH and oxidation-reduction," sums up much of his early research, which was specifically on rocks which formed as precipitants from water. One of his main areas of study at this time was the origin of iron deposits, which he would return to several times. However, he also worked on uranium and vanadium geochemistry. His later work looked at the rock-water interface geochemistry. This work involved both experimental research and advanced thermodynamics. The latter of these set him apart from many of the other researchers of the time and several of his students still maintain that position.

Garrels investigated the interaction of oceans and the sediments produced in them in chemical terms. He studied chemical mass balances between rivers, which carry on the chemical species, and oceans, which receive them. He set the standard for research on geochemical cycles with research on carbon, sulfur, and phosphorus. He modeled the interaction between oceans and the atmosphere with ROBERT BERNER to explain carbon dioxide abundances in the atmosphere in his famous "BLAG" model.

Much of this work was translated into books that became the handbooks for all geologists who ventured into this field. His book, *Mineral Equilibria at Low Temperatures and Pressures,* in 1960 showed how minerals form at surface and near surface conditions. His famous textbook, *Evolution of the Sedimentary Rocks,* published in 1971, set the standard for understanding the sedimentary cycle. It uniquely emphasized his research on the ocean-sediment interactions and clearly advanced the level at which students were introduced to the chemistry of sedimentary rocks.

Robert Garrels was born in Detroit, Michigan, on August 24, 1916, the second of three children. He spent some of his early years in Saltville, Virginia, before moving to Grosse Ile, Michigan, in 1928, where he attended high school. Garrels was a true athlete as well as a scholar, specializing in track and field. In fact, later in life he would hold the world high jump record for men over 57 years of age. Garrels entered the University of Michigan, Ann Arbor, at 17 years of age, vacillating between chemistry and literature. Instead, he turned to geology and graduated with a bachelor of science degree with honors in 1937. He entered graduate school at Northwestern University in Illinois the same year. He earned a master of science degree in 1939 and a Ph.D. in 1941. He then joined the faculty at Northwestern University, but quickly joined the Military Geology Unit of the U.S. Geological Survey for the duration of World War II. He returned to Northwestern University in 1945, but then returned to the U.S. Geological Survey in 1952. In 1955, he accepted a position at Harvard University, Massachusetts, where he remained for 10 years, including serving as chair. He moved back to Northwestern University in 1965, but only remained until 1969, when he accepted a position at Scripps Institution of Oceanography at the University of California at San Diego. There he was married to Cynthia Hunt in 1970. However, Garrels moved to the University of Hawaii in 1971, where he was named the James Cook Professor of oceanography. In 1974, he returned to Northwestern University, only to leave once again in 1980. He accepted the St. Petersburg Progress Chair in marine science at the University of South Florida, where he remained until his death. He contracted cancer of the spine in 1987 and succumbed to it on March 8, 1988. His wife, Cynthia, two daughters and a son by a previous marriage, and 13 grandchildren survived him.

In his very productive career, Robert Garrels produced numerous articles in international journals and volumes as well as several books. Many of these books and papers are the classical defining works for the field of sedimentary geochemistry. His work was well recognized and rewarded with honors and awards. He was a member of the National Academy of Sciences. He received honorary doctorates from the Free University of Brussels, Belgium, in 1969; the Louis Pasteur University of Strasbourg, Austria, in 1976; and the University of Michigan, Ann Arbor, in 1980.

He received both the Arthur Day Medal (1966) and the Penrose Medal (1978) from the Geological Society of America. He received the Goldschmidt Medal from the Geochemical Society (1973), the Roebling Medal from the Mineralogical Society of America (1981), and Wollaston Medal from the Geological Society of London, England. He served as the president of the Geochemical Society in 1962.

Gilbert, G. Karl
(1843–1918)
American
Geomorphologist

G. Karl Gilbert is one of the most famous American geologists of the 19th century and one of the founders of the U.S. Geological Survey. His geologic research of the American West during the late 19th century and early 20th century is historic. During his expeditions, Gilbert crossed Death Valley on foot and by mule, traveled along the steep, upstream terrain through the Grand Canyon, documented the Basin and Range Province of New Mexico, Arizona, and Utah, and explored the deserts of Nevada. He was even in the San Francisco earthquake of 1906 and fully documented the event. He is famous for the meticulous and detailed drawings and field notes that he kept on almost every geological, and on a lesser scale, biological feature he observed during his travels.

Karl Gilbert was a geomorphologist who developed many of the fundamental concepts that would define the discipline for many years. In his famous work, "Report on the Geology of the Henry Mountains," he determined that an intrusive body (laccolith) may deform its host rock. However, the real contribution was the expansion of John Wesley Powell's concept of subaerial erosion and base level into a fundamental theory. He emphasized lateral planation in this expansion. This idea would later be expanded into the theory of geographic cycles. Gilbert also studied graded streams. He showed that either by cutting down their beds or building them up with sediment, streams would always make room for themselves. On a long-term basis, they will transport exactly the load of sediment that is delivered to them from above. To quantify his observations, Gilbert conducted a series of flume experiments at the University of California at Berkeley from 1907–1909. This work was done largely to explain the sedimentary effects of hydraulic mining on the Sacramento River and San Francisco Bay and the great power of humans as geologic agents.

Gilbert was the first to describe the Basin and Range Province in terms of block faulting through extension. His real interest was the huge glacial Lake Bonneville, the ancestor of the Great Salt Lake, and the displaced shorelines as related to isostatic rebound. This interest in glaciation led him to participate in the Harriman Expedition to Alaska in 1899. He accompanied John Muir in studying the features of alpine glaciers. He described the effects of climate and topography on the motion of glaciers in his book, *Glaciers and Glaciation*. Gilbert even proposed an impact origin for the craters on the Moon in yet another famous study.

Grove Karl Gilbert was born on May 6, 1843, in Rochester, New York. He excelled in school in both academic achievements and social graces. Even with his family's limited resources, Gilbert graduated from high school at age 15, and went on to attend the University of Rochester, New York. While in college, his curriculum consisted of mathematics, Greek, Latin, logic, and one geology class. That one class was all that was needed to pique Gilbert's interest. In 1862, Gilbert graduated from college while the Civil War was beginning to tear the country apart. Either due to poor health or his dislike for violence, Gilbert did not enlist in the army. In 1863, with mounting student loans and no way to repay them, Gilbert accepted a position as a schoolteacher with the public school system in Jackson,

G. K. Gilbert studies a rock exposure at Monterey Formation in California in 1906 *(Courtesy of the U.S. Geological Survey)*

Michigan. He lived with his sister on the outskirts of Jackson. Gilbert did not adapt well to teaching unruly teenage schoolboys and returned to his hometown of Rochester before the school year ended. Gilbert found a position as a clerk with Ward's Cosmos Hall, a natural-science center, where he worked for the next five years (1863–1868). Even though Gilbert was inexperienced, he spent hours studying and documenting fossil samples.

The famous New York State geologist James Hall was leading an expedition to excavate a mastodon along the Mohawk River for Ward's. In late 1863, Hall injured his hip and Gilbert was given the opportunity to lead the expedition. As the expedition continued, it was evident that the skeleton was incomplete. Even though most of Gilbert's early work concentrated on paleontology, his real interest was in surficial geology. During the excavation process, he discovered potholes in the riverbed and began investigating their formation, as well as their association with a nearby retreating waterfall.

In 1869, Gilbert was hired by the Geological Survey of Ohio to conduct fieldwork. In 1871, Lieutenant G. M. Wheeler offered him a position as geologist with the newly formed Wheeler Geological Survey. The Wheeler Survey was one of four geological surveys (Hayden, King, Powell, Wheeler) that each had jurisdiction over a geographic area of the United States. The Wheeler Survey specialized in military and engineering goals. He met John Wesley Powell while completing the Wheeler Reports in Washington, D.C. In 1874, Gilbert moved to the Powell Survey. The four geological surveys were combined into the U.S. Geological Survey in 1878 and Gilbert and Powell were two of the original six geologists in charge. G. K. Gilbert spent the next three decades with the U.S. Geological Survey, including the position of the second director (1881–1892). Gilbert was married to Fanny Porter in 1874. They had three children, but his daughter died in 1883 and soon after, his wife became an invalid. She died in 1899. G. Karl Gilbert died on May 1, 1918, in Jackson, Michigan.

G. Karl Gilbert has some 400 scholarly publications of all varieties to his credit. Many of these are benchmarks of geomorphology, among other areas. He received numerous honors and awards in recognition of these contributions. He was a member of the National Academy of Sciences. He was awarded honorary doctoral degrees from the University of Rochester, the University of Wisconsin at Madison, and the University of Pennsylvania, among others. He received the Wollaston Medal from the Geological Society of London,

the Walker Grand Prize from the Boston Society of Natural History, and the Hubbard Medal from the National Geographic Society, among others. Gilbert is the only person ever to have been elected president of the Geological Society of America twice (1892 and 1909). He was also president of the Society of American Naturalists, the American Geographic Society, and the Philosophical Society and Geological Society of Washington, D.C., among others. He has awards named in his honor from the Geological Society of America, the U.S. Geological Survey, and the Association of American Geographers.

⊠ **Gilbert, M. Charles**
(1936–)
American
Petrologist (Geochemistry)

Polymorphs are minerals with the same chemical composition but different atomic bonding configurations. The best-known example of a polymorph is the transition of graphite to diamond, very different minerals but with the same chemical formula. In metamorphic rocks, the most important polymorphic transition involves an aluminum silicate mineral that changes based upon physical conditions. At high pressure it is kyanite; at high temperature it is sillimanite; and at low pressure it is andalusite. Charles Gilbert was involved in establishing the first accurate calibration of that transition published in the paper, "Experimental Determination of Kyanite-Andalusite and Andalusite-Sillimanite Equilibria; The Aluminum Silicate Triple Point." Until reliable geothermometers and geobarometers of common metamorphic minerals were established, this study provided the only real constraints on the physical conditions of metamorphism. Even today, before any analytical work is attempted, Gilbert's results are used as a first approximation based upon which polymorph is present.

The research career of Charles Gilbert can be divided into two parts, experimental geochemistry of minerals and assemblages of minerals as described above, and regional geology and petrology, mainly in Oklahoma. His experimental work was mostly performed on amphiboles and pyroxenes. He determined their stability under a variety of conditions both physical and chemical. These experimental studies provide the basic geochemical properties of the minerals as a basis for comparison with naturally occurring minerals and mineral assemblages. In terms of field-regional geology and petrology, Gilbert is mainly interested in the generation of magma during regional extension and especially in the North American mid-continent. There was a major extensional event in North America in the latest Precambrian and early Cambrian both along the East Coast and in the mid-continent. The Iapetus Ocean was created along the East Coast but the rifting in the mid-continent failed, creating a rift valley like that in East Africa. Extensive granite plutons intruded the rocks of southern Oklahoma. These granites are true classics for those formed during extension of the crust. Gilbert also studied some of the mafic plutons. With all of this petrologic work, he proposed models for the tectonic development of the southern mid-continent.

Charles Gilbert was born in Lawton, Oklahoma, on January 21, 1936. He graduated from Lawton High School in 1954. He enrolled in Cameron State Agricultural College, Oklahoma, before transferring to the University of Oklahoma at Norman in 1955. He earned a bachelor of science degree in geology with honors, Phi Beta Kappa, in 1958. Charles Gilbert married Mary Carol Leonard in 1958; they would have three children. Gilbert continued at the University of Oklahoma and earned a master of science degree in geology in 1961. He then attended the University of California at Los Angeles where he was the first graduate student of W. GARY ERNST. He earned a Ph.D. in 1965 and accepted a postdoc-

Charles Gilbert lectures to students in the Hunton quarry in the Arbuckle Mountains of Oklahoma *(Courtesy of M. C. Gilbert)*

toral fellowship at the Geophysical Laboratory at the Carnegie Institution of Washington, D.C. His first academic position was as a faculty member at Virginia Polytechnic Institute and State University in 1968. He served as department chairman from 1975 to 1980. During this time he was a visiting scientist at the Oklahoma Geological Survey on several occasions. Gilbert moved to Texas A & M University in 1983, where he served as department head from 1983 to 1985. During his time at Texas A & M University, he took a three-year leave to serve as a director in the Office of Basic Energy Sciences at the U.S. Department of Energy in Germantown, Maryland (1986–1989). Gilbert moved to his alma mater at the University of Oklahoma at Norman in 1990, where he remains today. He served as director of the School of Geology and Geophysics until 1998 and was named Eberly Family Professor from 1992 to 1998.

Charles Gilbert is an author of more than 60 articles in international journals, professional volumes, governmental reports, and field guides. Many of these are seminal papers on the geochemistry of minerals and rocks and the regional geology of Oklahoma. He is also an author or editor of three books and volumes. Gilbert has received extensive research funding from the National Science Foundation, NATO, NASA, U.S. Geological Survey, U.S. Department of Energy, and the Oklahoma Geological Survey.

Gilbert has performed extensive service to the profession. He served on numerous committees and panels for the Mineralogical Society of America and was elected secretary from 1979 to 1983. He also served as chair or member of numerous committees and panels for the American Geophysical Union, the Geological Society of America, the American Geological Institute, the National Research Council, the ILP Global Geoscience Transect Project, and the Basement Tectonics Association. He also served in numerous editorial positions including associate editor for the *Journal of Geophysical Research* and the *Geological Society of America Bulletin*.

⊠   **Glover, Lynn, III**
(1928–   )
American
Stratigrapher (Tectonics)

There is a division between pure geologic research and applied geologic research in that one is purely for the sake of knowledge and the other for a practical application. However, at certain times, applied research can drive the pure research. During the oil and gas crises of the 1970s, there was a tremendous effort to find alternative energy sources to petroleum. Although trained in classical geology, Lynn Glover III became an expert in balancing pure and applied geology to help with this search for alternative energy. This expertise was shown in his ability to obtain grants and contracts for applied energy research while still performing significant pure research within those

guidelines. Between 1974 and 1992, he organized several collaborators and obtained nearly $12 million in funding from the U.S. Department of Energy, the U.S. Nuclear Regulatory Commission, U.S. Geological Survey, and the National Science Foundation.

One project was to explore the geothermal potential of the granite plutons of the southeastern United States. Because granite is enriched in radioactive elements, it has an elevated heat flow. It is considered a "hot dry rock" in geothermal energy terms. By circulating water through deep wells, it can be heated enough to heat homes in the winter. The project involved drilling and coring numerous granite plutons, heat flow measurements, and analysis of radioactivity. There was more pure science done on the huge number of granites in the southeast in a short period of time than in all of the rest of the years of research combined. A successful well was drilled in the Atlantic Coastal Plain at Chrisfield, Maryland. This well would save much energy required by heat-pump technology to serve hospitals, apartment complexes, and commercial buildings.

Another project was to evaluate current earthquake activity in several areas where nuclear power plants are located or may be located in the future. The project involved detailed field mapping of these areas to identify the faults and possible surface expressions of them. Coupled with the fieldwork was seismic reflection profiling, similar to that which is used for petroleum exploration. A 20-ton truck is lifted on a pad, which then shakes, sending vibrations into the ground. Seismographs receive these vibrations after they have bounced off underground rock layers. The process is like a sonogram (ultrasound) of the subsurface geology. Such a seismic reflection profile was made all the way across the Appalachians in central Virginia similar to the COCORP profile farther south.

Although he participated in applied research, Lynn Glover's real passion is in the tectonics of the central and southern Appalachians.

However, unlike many researchers who jump on bandwagons of current popular ideas, Lynn Glover deals only with the geology rather than the models of others. His retirement symposium, sponsored by his former students, was entitled, "Renegade Tectonic Models and other Geologic Heresies . . ." because he was well known for strongly standing by his views. He proposed relatively straightforward plate reconstructions based on available data and steadfastly avoided the constant bombardment from people trying to apply the newest tectonic ideas to the southern Appalachians. He is well known for his willingness to battle out ideas with fellow researchers. Such an exchange of ideas forms a series of checks and balances for the science. A volume to summarize his contributions to Appalachian geology is entitled, *Central and Southern Appalachian Sutures: Results of the EDGE Project and Related Studies.*

Lynn Glover was born on November 29, 1928, in Washington, D.C. He spent most of his youth in Occoquan, Virginia, but moved around during World War II because his father was in the U.S. Navy. He earned his bachelor of science and master of science from Virginia Polytechnic Institute in 1952 and 1953, respectively. He was employed as a geologist for the U.S. Geological Survey from 1952 to 1967. He worked as a uranium exploration geologist in the southern Appalachians as well as a field geologist in Puerto Rico and the eastern Greater Antilles. He earned his Ph.D. from Princeton University, New Jersey, in 1967 as an advisee of HARRY H. HESS. He joined the faculty at Virginia Polytechnic Institute and State University in 1967, and remained there throughout the rest of his career. During that time he served as director of the Orogenic Studies Laboratory. He retired in 1998, and became an emeritus professor. Since his retirement he has developed and chaired an alumni relations committee for the department. Lynn Glover is married to Ellen Glover, to whom he credits most of his accomplishments.

Lynn Glover III in Spain with his wife, Ellen Glover *(Courtesy of Lynn Glover III)*

Lynn Glover has had a very productive career. In addition to his success with grants, he is an author of 43 articles in international journals and professional volumes. He edited seven volumes and guidebooks and wrote one monograph. Because of the unconventional nature of his work, he also produced 16 reports and maps. Glover has also performed service to the profession and especially to the Geological Society of America for which he was associate editor of the *Geological Society of America Bulletin* from 1988 to 1996, vice president of the Southeast Section (1994), and chair of the Southeast Section meeting in 1994, among others.

⊠ Goldsmith, Julian R.
(1918–1999)
American
Mineralogist, Geochemist

Many experimental mineralogists-geochemists specialize in a single mineral, performing research solely on that mineral throughout their career. Julian Goldsmith chose the most abundant and most important rock-forming mineral group in the Earth's crust, the feldspars. Following in the footsteps of his mentor, NORMAN L. BOWEN, he performed several groundbreaking experiments on alkali feldspars that concentrated on the ordering of silicon and aluminum atoms. With colleagues, he developed a new X-ray diffraction technique to determine the ordering in his samples that is still in use today. He also made a groundbreaking discovery on the formation of sodium feldspar at low temperatures. He was the first to synthesize ordered samples in this range although they are known in nature. Goldsmith never really figured out how he managed this synthesis, but he postulated that there was an unknown flux that drove the process. The flux was surmised to be hydrogen that penetrated his research vessels from dissociated water. Such an explanation would be consistent with observations of high water pressure in natural situations.

Goldsmith was not satisfied with becoming the foremost expert on the feldspars—he also performed extensive research on another important and abundant mineral group, the carbonates. He was interested in the mechanics of the substitution relations of calcium and magnesium with each other in the minerals calcite and dolomite. These experiments have implications for how certain marine animals make their shells. He also investigated the high temperature relations of these minerals and elements and in so doing, he established a method for determining the temperature of formation for metamorphic carbonate minerals. This "geothermometer" is still used today. Yet even conquering carbonates (as much as they can be conquered) was not enough for Goldsmith. He also investigated scapolite, admittedly a less important mineral to rock-forming processes. Nonetheless, this further branching into yet another mineral system attests to his versatility.

Even though Julian Goldsmith was certainly one of the pioneers in experimental mineralogy and geochemistry, the trait that he is most remembered for is his humor and friendliness. He literally rebuilt the department (and building) at

the University of Chicago and is likely the greatest influence on this great program. The easygoing personality of this great researcher is a rare combination that added greatly to all of his successes.

Julian Goldsmith was born on February 26, 1918, in Chicago, Illinois. He grew up in Chicago during the Great Depression but he was well provided for. He attended the University of Chicago, Illinois, and earned all of his college degrees there, including a Ph.D., which he earned in geochemistry in 1947. His dissertation adviser was N. L. Bowen. World War II interrupted his graduate career when he left the University of Chicago from 1942 to 1946 to do defense research at the Corning Glass Works in Corning, New York. He accepted a position of research associate at the University of Chicago upon graduation, and remained there for his entire career. Goldsmith inherited N. L. Bowen's lab when Bowen departed for the Carnegie Institution in Washington, D.C.; soon after, he joined the faculty. He was named Distinguished Service Professor in 1969, and he retired to professor emeritus in 1990. Goldsmith served as chair of the department from 1963 to 1971 and even served as associate dean for a short time. He married Ethel Frank and together they had three children. Julian Goldsmith died of leukemia in 1999.

Julian Goldsmith produced papers on the thermodynamics of feldspars in the late 1940s and 1950s that he was still being asked for copies of in the 1980s. Most articles are considered ancient and of marginal use if they are more than 10 years old, much less 40 years old. His contributions to the science were widely recognized and he received numerous awards and honors. He received the Mineral Society Award and the Roebling Medal from the Mineralogical Society of America in 1955 and 1988, respectively. He also received the Harry H. Hess Medal from the American Geophysical Union in 1987.

Goldsmith performed outstanding service to the profession. He served on the board of the National Science Foundation from 1964 to 1970.

He served as both vice president (1973) and president (1974) of the Geological Society of America. He served as both vice president (1968) and president (1970) of the Mineralogical Society of America, and both vice president (1955) and president (1965) of the Geochemical Society. Needless to say, he also served on and chaired numerous committees for all three of these organizations. He was also editor of the *Journal of Geology*, as well as several other editorial positions.

⊠   **Gould, Stephen Jay**
(1941–2002)
American
Paleontologist

Although Stephen Jay Gould was an outstanding invertebrate paleontologist and evolutionary theorist, he is best known for his popular books and media productions. In fact, these apparently separate careers are not so unrelated. Gould's scientific research on fossils led him to propose a punctuated equilibrium theory for evolution, probably the single most important modification to Darwin's theory of evolution. Animals tend to remain relatively unchanged for long periods of time (many generations) and then periodically undergo very rapid evolutionary changes. These changes are in response to some environmental stimulus and continue until the animal either adapts to that stimulus or becomes extinct. The speed of these adaptations between the otherwise slow changes accounts for the apparent abundance of "missing links" in the record of life. There would be many times more examples of the stable periods than the quickly changing ones by this theory. Human missing links are perfect examples. This idea was the impetus for his book, *The Panda's Thumb*. This quick turnover of science into ideas for public consumption exemplifies his apparent dual career. In fact, his fluency in paleontology, geology, and zoology fueled his media popularity. Other books that address this

topic of evolutionary adaptations, as well as extinctions and their causes, include *Hen's Teeth and Horse's Toes, The Flamingo's Smile, Wonderful Life: The Burgess Shale and the Nature of History, Ever Since Darwin* and *Eight Little Piggies*. These books are therefore further outgrowths of his scientific research.

Several of his other books more readily address the human condition and largely reflect his own experiences, but with a scientist's eye. The most famous of these, *The Mismeasure of Man*, argues convincingly that measures of human intelligence are inappropriate because they are not objective. The impetus for this book was Gould's eldest son Jesse's autism. His essay *Five Weeks* and his book *Questioning the Millennium* (1998) also deal with this subject. His book *Death and Horses: Two Cases for the Primacy of Variation, Case One 1996* documents his battle with the "invariably fatal" abdominal mesothelioma with which he was diagnosed in 1982. Later, Gould championed the battle against creationism as a replacement for evolution in public schools that has cropped up in various parts of the country.

Stephen Jay Gould was born on September 10, 1941, in New York City. He attended Antioch College in Yellow Springs, Ohio, where he earned a bachelor of arts degree in geology in 1963. He attended graduate school at Columbia University, New York, in evolutionary biology and paleontology. In 1965, he married Deborah Lee, an artist, and then accepted an instructorship at Antioch College for the summer of 1966. In 1967, he completed his Ph.D. and accepted a position at Harvard University as assistant professor and assistant curator of invertebrate paleontology. He remained at Harvard, where in 1982 he became the Alexander Agassiz Professor of zoology. In 1996, he became the Vincent Astor Visiting Research Professor of biology at New York University.

The productive Gould authored several hundred articles in international journals and professional volumes. In recognition of his research and popular books, Gould received an astounding number of awards and honors too numerous to list individually in this biography. He was awarded 44 honorary doctoral degrees from the United States, Canada, England, and Scotland. Included in this group are Duke University, Leeds University, England, McGill University, Canada, Rutgers University, New Jersey, and the University of Pennsylvania, among many others. He had 15 literary awards bestowed upon him, including the National Book Award in science (for *The Panda's Thumb*) in 1981, and the Rhone-Poulenc Prize (Britain's leading award for science books) for *Wonderful Life* in 1991. In 1992, he received the Golden Trilobite award for paleontological writing. In 1991, he was nominated and became a finalist for a Pulitzer Prize for *Wonderful Life* (nonfiction). He received 47 academic medals and awards. In 1975, he received the Schuchert Award for excellence in paleontological research (under age 40). He was named Scientist of the Year by *Discover* magazine in 1981. In 1983, he received the Neil Miner Award for excellence in teaching from the National Association of Geology Teachers and the Humanist Laureate from the Academy of Humanism. In 1985, the NOVA profile "S. J. Gould, This View of Life," won the Westinghouse Science Film Award. Stephen Gould received Distinguished Service Awards from American Geological Institute in 1986, American Institute of Professional Geologists in 1989, and the National Association of Biology Teachers in 1991, and Public Service Awards from the National Science Board of the National Science Foundation, as well as the Geological Society of America in 1999. He received two History of Geology Awards, one from the Geological Society of America in 1988 and the Geological Society of London (Sue T. Friedman Medal) in 1989. He received the James T. Shea Award for excellence in geological writing from the National Association of Geology Teachers in 1992.

Stephen Gould was a member of the board or council for numerous groups, institutions, and museums, including NASA, National Science Foundation, NOVA, Smithsonian, and the British Museum. He served in various editorial roles for several top journals including, *Science, Evolution, Systematic Zoology, Paleobiology* (including founder), and *American Naturalist*. He was a Fellow at most of the societies with which he was associated, including American Academy for the Advancement of Science, European Union of Geosciences, Geological Society of London, Linnaean Society of London, National Academy of Sciences, and the Royal Society of Edinburgh.

This short biography does not begin to do justice to the monumental accomplishments of Stephen Jay Gould as a top evolutionary theorist and paleontologist, a renowned writer of science, an inspired teacher, a contributor to the production of television science specials, and an adviser contributing his wisdom to countless organizations. He was also a historian of science.

Stephen Jay Gould died on May 20, 2002, of cancer. He is survived by his second wife, Rhonda Roland Shearer, with whom he had no children. He had two sons with his previous wife.

⊠ Grew, Priscilla C.
(1940–   )
American
Earth Science Advocate

Priscilla Grew may have started out in a very traditional geological career, but she changed course and quickly established the epitome of an astonishingly successful nontraditional career in the geosciences. Early in her career, she was a metamorphic petrologist studying metamorphism in subduction zones. However, she soon became interested in broader geoscience issues that are of importance to the public. This change occurred as she moved to California, one of the leading states where geoscience meets public policy. With some practical experience in applied geoscience issues like nuclear waste disposal, geothermal energy, earthquakes, and groundwater resources, Priscilla Grew moved into the public sector. This biography could have easily been recounting her career as an elected official, but instead she remained behind the scenes as a successful director and adviser. She has been one of the foremost public advocates in the United States for applying geology to public policy. Her influence has been not only at the state level in California, Minnesota, and Nebraska, but also has extended to the national arenas of energy, the environment, and education, where she has been a diligent adviser for the past three decades.

Priscilla Grew was born Priscilla Croswell Perkins on October 26, 1940, in Glens Falls, New York. She attended Bryn Mawr College, Pennsylvania, and earned a bachelor of arts degree in geology in 1962, graduating magna cum laude. That year she married Richard Dudley, currently a professor of mathematics at Massachusetts Institute of Technology. She earned her doctoral degree in geology from the University of California at Berkeley in 1967, and joined the faculty at Boston College, Massachusetts. She was divorced in 1972, and moved to California, where she became an assistant research geologist at the University of California at Los Angeles and executive secretary for the Lake Powell Research Project. In 1973, she had a one-year visiting assistant professorship at the University of California at Davis. In 1975, she married Edward Grew, currently a research professor of geology at the University of Maine at Orono. In 1977, then governor Jerry Brown appointed her the director of the Department of Conservation for the State of California, which includes the Division of Mines and Geology (State Geological Survey) and the Division of Oil and Gas. In 1981, the governor appointed her as one of the five commissioners of the California Public Utilities Commission. She then became the director of the Minnesota Geological Survey in 1986, and with it, a concurrent professorship at

Portrait of Priscilla Grew *(Courtesy of Peter Hasselbalch)*

the University of Minnesota, Twin Cities. In 1993, Priscilla Grew became the vice chancellor for research at the University of Nebraska at Lincoln, a position she held until 1999. She held a concurrent appointment of professor in the departments of geosciences and the Conservation and Survey Division of the Institute of Agriculture and Natural Resources. She has also been the Native American Graves Protection and Repatriation Act compliance coordinator since 1998.

Priscilla Grew has had a very productive career, however nontraditional. She is an author of some 57 publications but only about 15 are in international journals. Many are technical reports, treatises on geoscience and public policy, and policy evaluations. The profession has recognized her work in terms of honors and awards. She received the Distinguished Service Award from the Soil Conservation Society of America (1980), the Outstanding Service Award from the National Community Action Agency (1984) and a Certificate of Achievement from the Association of Environmental Professionals (1983). She also received the Tribute to Women Award from the Young Women's Christian Association (1994), two "People Who Inspire" Awards from Mortarboard (1997 and 1998), and the Ian Campbell Medal from the American Geological Institute (1999).

The amount of professional service in which Priscilla Grew has been involved covers six pages in her curriculum vitae and is far too extensive even to begin to summarize here. She was a member of the advisory board for the secretary of energy (1995–1997). She was usually chair on likely every board or task force relating to geology in California at one time or another. She was on several committees for the National Academy of Sciences and the National Research Council. She was on numerous important committees and panels for the National Science Foundation as well as chair of the national Geology and Geography Section of the American Association for the Advancement of Science. She served as chair of the Research Coordination Council for the Gas Research Institute. Yet she still served on numerous important committees and took several leadership roles in the Geological Society of America, American Geophysical Union, International Geological Correlation Project, and the Mineralogical Society of America. She has been on the evaluating committee and the advisory board for several universities including Stanford University, Chiba University, Japan, and the University of Colorado at Boulder. Priscilla Grew has been involved in advisory committees for many of the key geoscience initiatives at the national level, including continental drilling, global change research, and issues associated with energy and mineral resources.

⊠ Griggs, David T.
(1911–1974)
American
Geophysicist, Rock Mechanics

David Griggs had a true dual career, one in geophysics and one in national defense. He is truly the "father of modern rock mechanics." He greatly modified existing high-pressure experimental equipment that had been neglected for many years, or invented new equipment. One such apparatus, the "simple squeezer," achieved pressures up to 50 kilobars and temperatures up to the melting point. This equipment was unique in the world at the time. He then conducted exhaustive experiments on natural rock samples to determine their mechanical properties. He performed "creep" experiments in which he placed materials under high pressure and temperature conditions for periods up to nine months, allowing them to deform slowly. These conditions simulated the deep crust and mantle. The results and the processes that he defined are still the state of the art for the science. This research placed the first real constraints on deformational microtextures observed in naturally deformed rocks and the processes of developing alignment of minerals in metamorphic rocks. He defined the material science of many rocks and minerals in terms of strength and response, whether fracturing in a brittle response or stretching in a plastic response. In 1965, Griggs discovered a phenomenon he called "hydrolytic weakening" in minerals. The addition of very small amounts of water actually bonded into the quartz atomic structure caused significant weakening. The mechanism is not fully understood but it may be one of the most significant controls on deformation in the shallow to mid-crustal range.

This interest in rock deformation stemmed from Griggs's interest in mountain building processes. He wanted to better understand the deformational processes responsible. Before he started his groundbreaking experimental studies, he wrote a paper entitled "A Theory of Mountain Building" in which he proposed that thermal convection currents in the mantle were largely responsible for the distribution, structure, and periodicity of mountain building events. Even though the ideas were later proven sound during the plate tectonic revolution of the 1960s, at the time of publication they were roundly denounced and dismissed. It was Griggs's drive to vindicate his ideas that led him to undertake his later advanced experimental work.

David Griggs was born on October 6, 1911, in Columbus, Ohio. He spent his early childhood in Ohio, but his high school days in Washington, D.C. His father was a professor of botany at George Washington University and took expeditions to Puerto Rico, Guatemala, Texas, and Alaska where he discovered the famous Valley of Ten Thousand Smokes, near Mount Katmai. The highest mountain in the area was named Mount Griggs in his honor. David Griggs began his college career at George Washington University in 1928 but quickly transferred to Ohio State University and graduated with a bachelor of science degree in geology, Phi Beta Kappa, in 1932. He spent one year as a graduate student at Ohio State University before transferring to Harvard University, Massachusetts, and became a junior fellow the following year. During World War II, he served at the Massachusetts Institute of Technology Radiation Laboratory to develop microwave radar, and as an expert consultant to Secretary of War Henry Stimson. However, to introduce radar-guided bombing, he flew both training and combat missions in Europe, but after almost falling out the bomb bay doors on one mission and getting wounded on another (he received the Purple Heart), he was grounded. He then served with the Tactical Air Command. After VE day, he moved to the Pacific theater, where he was the liaison between General Douglas MacArthur and his staff, as well as prepared for the atomic bombing of Hiroshima. He received the Medal for Merit, the highest civilian honor, from President Truman for his efforts. Griggs married Helen Avery on May 4, 1946. They would have two children.

Griggs continued his activity in national defense even after the war. He worked with the Atomic Energy Commission and helped set up the RAND Corporation, serving as its first head of physics in 1947. He accepted a faculty position at the University of California at Los Angeles in 1948, but also served as chief scientist for the U.S. Air Force from 1951–1952. He became infamous in this role for supporting the government's desire to build a hydrogen bomb, in direct opposition to the position of J. Robert Oppenheimer, the head of the Manhattan Project, who was opposed to it. Many scientists regarded him as a Judas for this stand. Griggs got involved in an active combat situation when in 1967 he went to Vietnam to help General William Westmoreland design a scientific support structure. He improved the performance of new sensor technology to cut off enemy supply lines during the Tet Offensive.

David Griggs died of a heart attack during vigorous skiing in Colorado with former Secretary of Defense Robert McNamara. He had had a precedent heart attack but in typical Griggs style, he ignored it.

⊠ Gutenberg, Beno
(1889–1960)
German
Geophysicist

Beno Gutenberg gained fame for his work in the development of the Richter scale. In reality, this work was only a small part of his outstanding contributions to the science of seismology, and the interpretation of the deep structure of the Earth, both by natural and artificially generated seismic waves. He began this work while still in Germany and made many discoveries on the nature of the propagation of seismic waves there. The best-known volumes to which he contributed are entitled, *Handbuch der Geophysik* (Handbook of Geophysics).

Gutenberg's research during the second half of his career in the United States has led to some of the world's leading-edge scientific papers. He published a series of papers, entitled "On seismic waves," with CHARLES F. RICHTER between 1931 and 1939. The information provided by these exceptional papers included travel times of several seismic phases, information that researchers used to create models of the Earth's mantle and core. Also at this time, Gutenberg's keen observational talent led him to believe that a low-velocity zone existed in the upper mantle, now recognized as the asthenosphere.

In 1941, Gutenberg and Richter began working together again and published a book entitled *Seismicity of the Earth.* The theory of plate tectonics was developed from information obtained from illustrations of geographical patterns published in this book. Gutenberg and Richter also worked on establishing the famous "magnitude scales" using different types of seismic waves, so that magnitudes of earthquakes that had both shallow and deep foci and take place at different epicentral distances could be assigned by people observing their effects.

Beno Gutenberg was born in Darmstadt, Germany, on June 4, 1889. He attended the University of Göttingen, Germany, where he received a bachelor of science degree and later a Ph.D. in 1911, both in geophysics. His dissertation topic was on microseisms. He later used this topic during World War II in an attempt to track hurricanes and typhoons in the western Pacific. After receiving his Ph.D., Gutenberg joined the University of Strasbourg, Germany, in 1913, which at that time was the headquarters of the International Seismological Association. After a brief time, he left to serve with the Meteorological Service of the German army during World War I. When the war ended in 1918, Gutenberg accepted a professorship at the University of Frankfurt-am-Main, Germany. He also took a job as a business executive due to financial difficulties. In 1929, the Carnegie Institution of Washington, D.C., invited Gutenberg to participate in a meeting to discuss the future plans for the Seismological Laboratory in Pasadena, California. As a result, he was offered a

position at the laboratory in 1930. That same year, he joined the faculty at the California Institute of Technology as a professor of geophysics. In 1936, the Seismological Laboratory became integrated with Caltech and in 1947, Gutenberg became the director. Through Gutenberg's hard work and leadership ability, the laboratory became the leading center for the study of earthquakes and the deep Earth. Gutenberg retired to professor emeritus in 1958, but continued to conduct research. He contracted a virulent form of influenza that developed into fatal pneumonia. Beno Gutenberg died on January 25, 1960.

Beno Gutenberg led an extremely productive career. He was an author of close to 300 research articles in both German and English in international journals and professional volumes throughout his career. Many of these papers are true classics of earthquake seismology. In 1959, Gutenberg published his final book entitled *Physics of the Earth's Interior,* which summarized a lot of his views on the earthquakes and the physics of the Earth's internal structure. In recognition of his contributions to geophysics, Beno Gutenberg was awarded many scientific honors and awards during his career. He was a member of the National Academy of Sciences. He was awarded an honorary doctoral degree from the University of Uppsala, Sweden, in 1955. He was also awarded the Bowie Medal from the American Geophysical Union in 1933, the Lagrange Prize from the Royal Belgian Academy in 1950, and the Wiechert Medal of the Deutsche Geophysikalische Gesellschaft, among others. The American Geophysical Union named a medal in his honor.

Gutenberg also performed service to the profession. He served many committees and sections in the International Union for Geodesy and Geophysics, served on the board of directors and as president of the Seismological Society of America, and as a member of the Academia dei Lance and the Royal Society of New Zealand.

H

⊠ Handin, John W.
(1919–1991)
American
Structural Geology

John W. Handin was one of the first geologists to specialize in and dramatically improve many aspects of engineering geology including theory, application, and research. He was also instrumental in creating the world's leading laboratory in experimental rock deformation first at Shell Oil Company and later at Texas A & M University. Handin took an investigative approach to structural geology of the late 1950s and perfected many aspects, including theoretical, experimental, and observational methods. These improvements were readily adopted by his peers and are to this day considered the modern or new aspect of structural geology. A lot of the understanding and interest in the mechanical properties of rocks and how they apply to geological, geophysical, geoscience engineering, and general engineering problems has been attributed to Handin's research in the mid-20th century.

The problems that Handin addressed in his research include the effects of confining pressure, pore fluid pressure, and the effects of temperature on the strength and mechanical response of various types of rocks. Several of his special studies include the rate of strain, water phase changes, the surface energy of fractures, cracking on the surface and in the internal area of rocks due to temperature, analyses of rock fabric, and the flow and fracturing of folded rocks, among others. Handin developed many of the high-pressure and high-temperature techniques that made his laboratory work possible. Other studies that Handin began were the frictional sliding of rocks, and the study of remnant stresses in individual rocks as well as rock masses. He used the information that he gained during his research on the mechanical properties of rocks to model solutions for earthquake control.

Because this work is so applicable to engineering and earthquake studies, Handin was invited to serve on numerous committees and organizational boards. Several of these positions include the U.S. Geological Survey's advisory panel on earthquake studies, the policy board of the National Geotechnical Centrifuge Facility, and the JOIDES panel on sedimentary petrology and physical properties. He has also consulted for the U.S. Air Force, the U.S. Army Corp of Engineers, and the U.S. Environmental Protection Agency, in addition to the National Academy of Sciences and the National Research Council.

John Handin was born in Salt Lake City, Utah, on June 27, 1919. He grew up in West Los Angeles, where on he had the Santa Monica Mountains a short distance away and the Pacific

Ocean just a few miles in the other direction. During the 1930s, the country was in the middle of the Great Depression and college was the alternative to the sparse job market. In order to stay close to home, Handin enrolled in the University of California at Los Angeles (UCLA) in civil engineering. After taking only one course in geology, Handin decided to change his major from engineering to geology. In 1941, Handin graduated with a bachelor of science degree in geology, but World War II had just begun and the U.S. Coast Artillery Corps drafted him as a second lieutenant. His battalion wound up fighting in the liberation of the Philippine Islands, as well as in Okinawa, Japan. Handin remained in the U.S. Army Reserve after the war and he later retired a lieutenant colonel.

Upon discharge, Handin decided to return to UCLA to earn a master of science degree in geology and physics in 1947. That year, he married Frances Robertson; they would have two children. He finished his doctoral work on the source, transportation, and deposition of beach sands in 1949. Upon graduation, Handin accepted a postdoctoral fellowship with DAVID T. GRIGGS, at the Institute of Geophysics at UCLA. They worked at the high-pressure laboratory with Frank Turner of the University of California at Berkeley on dynamic petrofabric analysis of experimentally deformed marble.

M. KING HUBBERT developed a research program in structural geology at the Shell Exploration and Production Research Laboratory in Houston and hired Handin in 1950 to establish a high-pressure laboratory. Handin joined the faculty of Texas A & M University in College Station in 1967 as a distinguished professor of geology and geophysics. He established the Center of Technophysics and became its first director. Handin was also director of the Earth Resources Institute and also associate dean for the College of Geophysics at Texas A & M University all at the same time. He was appointed the director of the Earth Resources Institute from 1978 to 1982. He

retired to professor emeritus in 1984 and simultaneously established John Handin, Inc., geological consulting company. John Handin died in 1991.

During his career, John Handin was an author of more than 75 scientific articles in international journals, professional volumes, and governmental reports, in addition to numerous industry reports. Many of these papers are seminal studies on rock mechanics. In recognition of these contributions to geology, John Handin received several honors and awards. He received the Distinguished Achievement in Rock Mechanics award from the American Institute of Mining, Metallurgical, and Petroleum Engineers and the Walter Bucher Award from the American Geophysical Union. The John Walter Handin Laboratory for Experimental Rock Deformation at Texas A & M University was named in his honor.

⊠ **Harrison, T. Mark**
(1952–)
Canadian
Isotope Geochemistry

A relatively newer system of isotope geochemistry to determine the age of rocks and minerals involves the decay of radioactive parent potassium of atomic weight 40 to daughter argon 39. The problem is that potassium is a solid, whereas argon is a gas. The solution is that the samples are irradiated in a nuclear reactor to convert the potassium 40 to argon 40 so that both parent and daughter can be analyzed in a gas mass spectrometer. Because the daughter is a gas, at high temperatures it can escape from the mineral structure. At a certain temperature, the mineral will lock up the gas and prevent it from escaping, thus setting the clock going to record ages using this isotopic system. Different minerals have different lock-in or "closure" temperatures. Therefore, determining the age of a mineral using the Ar/Ar system in many cases yields the time when the mineral cooled through a certain temperature rather than

Mark Harrison (right) with University of California at Los Angeles colleague An Yin in the Himalayas with Mount Bauda behind them *(Courtesy of T. M. Harrison)*

the age of the rock. Therefore, it is termed thermochronology (age of a temperature) rather than simply geochronology. Mark Harrison has done much of the background research on the methods of this complex system as well as a good deal of the application. He has become one of the foremost experts on Ar/Ar geochronology and he did it in an astonishingly short time.

Determining the closure temperatures for each of the minerals that can be dated using Ar/Ar, namely biotite, muscovite, K-feldspar, and amphibole, and the factors which may affect that closure temperature, constitute the laboratory part of the research in which Harrison has participated. The methods for analysis and data reduction are also another analytical part of the research that he has reported on. However, he also applied

the Ar/Ar system to specific areas, including the Himalayas, in papers like "Raising Tibet," the New England metamorphic belt in "Pressure, Temperature, and Structural Evolution of West-Central New Hampshire: Hot Thrusts over Cold Basement," locations in California, southeast Asia (e.g., "Tectonic Evolution of Asia"), and other environments. With these regional studies, he addressed numerous topical problems that have also set the standard for Ar/Ar analysis. Some of these topics include dating of extraterrestrial impacts, dating detrital minerals in sedimentary basins, dating uplift and fault movements using thermal decay curves of metamorphic rocks, and the dating of plutons.

The research described above would constitute a successful career for anyone. However, for

Mark Harrison, it is not enough. He also participated with BRUCE E. WATSON of Rensselaer Polytechnic Institute on the role of accessory minerals in magma. He conducted studies on the granite generation, ascent, and emplacement. He participated in research on fission track dating of minerals and other isotope systems used in the dating of minerals. He collaborated in research on tectonic processes, especially in the Himalayas and New England. The amount and quality of this research out of the Ar/Ar field would constitute another successful career for most people but for Mark Harrison it is simply "additional interests."

Mark Harrison was born on November 2, 1952, in Vancouver, Canada. He attended the University of British Columbia, Canada, where he earned a bachelor of science degree in geology with honors in 1977. He earned a Ph.D. in geology from the Australian National University in 1981. Harrison was a postdoctoral research associate at the Carnegie Institution of Washington, D.C., in 1981 before joining the faculty at the State University of New York at Albany in 1982. In 1989, he accepted a faculty position at the University of California at Los Angeles (UCLA), where he remains today. He served as the chair of the department at UCLA from 1997 to 2000. He also serves a concurrent position of director of the Research School of Earth Sciences at the Australian National University from 2001 to the present, where he also was a visiting fellow from 1984–1985. Mark Harrison is married to Susan Annette Harrison; they have two children.

To date, Mark Harrison is the author of some 131 articles in international journals and professional volumes. Many of these studies are truly groundbreaking and appear in prestigious journals like *Nature, Science,* and *Geology.* He is the editor of one volume and the coauthor of the definitive textbook on argon geochronology/thermochronology entitled *Geochronology and Thermochronology by the Ar40/Ar39 Method.* Astoundingly, his books and articles are cited some 400 to 500 times per year in other scientific articles and books. He has received some $5.6 million in grant funding. This productivity and the wealth of his contributions to the profession have been well recognized in terms of honors and awards. He received the Presidential Young Investigator Award from the National Science Foundation in 1989. He also received the Outstanding Young Alumnus from the University of British Columbia (1989), the N. L. Bowen Award from the American Geophysical Union (1995), and an Outstanding Contribution in Geoscience Research Award from the U.S. Department of Energy (1996).

Harrison has performed significant service to the profession. He served on several committees and panels for the National Research Council, National Science Foundation, National Academy of Sciences, American Geophysical Union, Lawrence Livermore National Laboratory, and NASA. He was the associate editor for *Geochimica et Cosmochimica Acta* and on the editorial board for *Geology* and *Earth and Planetary Science Letters.*

⊠ Hatcher, Robert D., Jr.
(1940–)
American
Regional Tectonics

After the plate tectonic paradigm was established during the late 1950s and 1960s, the second order of questions about how ancient continental areas fit into this paradigm became the next frontier in the 1970s and early 1980s. This research involved reevaluating previously mapped areas within this new context. The field can be classified as regional tectonics and many new methods for unraveling complex relationships developed. Certain areas became hot spots for regional tectonic study. One major area was the central and southern Appalachians primarily because of the oil and gas potential of the Valley and Ridge province along the western side. This period coincided with the oil crisis of the 1970s. Many notable geologists and geophysicists participated in this evaluation, but without a doubt, Bob Hatcher was the leader. He is mainly interested in overthrust terranes, which are important to petroleum exploration. He established his leadership position by being the first to stick his neck out and propose an all-encompassing modern plate tectonic model for the entire southern Appalachian orogen. Every researcher thereafter had to address Hatcher's model, whether they supported it or refuted it. He has modified and revised this model several times and published additional models.

Based on this willingness to take such well-calculated risks in a high-profile venue, Bob Hatcher became the person with whom to collaborate in the southern Appalachians. Although his research is strongly field based, Bob is well versed in many supporting techniques in geophysics and geochemistry so he could participate in many projects. When JACK E. OLIVER's group at Cornell University decided to produce their world-famous COCORP seismic reflection line (like a sonogram of the Earth) across the southern Appalachians, it was Bob Hatcher they contacted. That research proved that much of the southern Appalachians had been thrust faulted westward some 200 kilometers and there were possible hidden oil and gas reserves under the crystalline rocks as reported in the paper, "Thin Skinned Tectonics in the Crystalline Southern Appalachians; COCORP Reflection Profiling of the Blue Ridge and Piedmont." When the concept of building orogens (and continents) with exotic pieces of crust from other parts of the Earth was born, the concept of suspect terranes, Bob Hatcher was coauthor of the definitive work with HAROLD WILLIAMS from Memorial University (e.g., *Appalachian Suspect Terranes*). When the significant strike-slip faulting was discovered, Hatcher was quick to bring it into context. He compiled a detailed geologic map in 1990 that succeeded the famous map of Harold Williams as the standard for the Appalachians (e.g., *Tectonic*

Portrait of Robert Hatcher *(Courtesy of R. D. Hatcher Jr.)*

Map of the United States Appalachians). Another high-profile project that Hatcher spearheaded was the National Science Foundation Appalachian deep hole project (ADCOH) which was to be drilled through the Blue Ridge overthrust and into the Valley and Ridge sedimentary rocks that are proposed to be hidden beneath. The project fell through in the end but the organization and work was outstanding and reported in the paper, "Appalachian Ultradeep Core Hole (ADCOH) Project Site Investigation Regional Seismic Lines and Geological Interpretation."

Robert Dean Hatcher Jr. was born on October 22, 1940, in Madison, Tennessee. He attended Northwestern High School in Springfield, Ohio, and he graduated in 1957. He earned bachelor's and master's degrees from Vanderbilt University, Tennessee, in 1961 and 1962, respectively, with majors in geology and chemistry and a minor in mathematics. His Ph.D. was in structural geology from the University of Tennessee at Knoxville in 1965. He then worked as an exploration geologist for Exxon USA for one year before accepting a faculty position at Clemson University in 1966. He achieved the rank of professor and remained there until 1978 when he accepted a position at Florida State University. He moved again in 1980 to the University of South Carolina at Columbia. In 1986, he returned to his alma mater at the University of Tennessee at Knoxville, where he occupied an endowed chair at the University of Tennessee/Oak Ridge National Laboratories.

Bob Hatcher has authored or coauthored some 140 journal articles, five books and monographs, and numerous field guides. One of these books is a popular textbook entitled, *Structural Geology, Principles, Concepts and Problems,* and another is a companion lab manual. His service to the profession is unparalleled. He served as editor for the *Geological Society of America Bulletin* from 1984 to 1988. For this work, he was awarded the first-ever Geological Society of America Distinguished Service Award in 1988. He served as president of the Geological Society of America in 1993 and the American Geological Institute in 1996. He served as science adviser to South Carolina governor Richard Riley for disposal of radioactive waste from 1984 to 1986. His interest in nuclear waste disposal resulted in Hatcher's serving a six-year term on the National Academy of the Sciences/National Research Council Board on radioactive waste management and a three-year term on the U.S. Nuclear Regulatory Commission nuclear reactor safety research review committee (1993–1996). Other awards include the 1997 I. C. White Award for his contributions to Appalachian geology and being made an honorary citizen of West Virginia in 1998 for the same reasons.

⊠ **Hayes, John M.**
(1940–)
American
Biogeochemist

Science has gone through something of a cycle over the years. At the dawn of modern science more than 150 years ago, researchers were simply scientists with no real affiliation. As science grew, it partitioned off into a whole series of highly specialized unrelated fields. However, this overspecialization was not conducive to solving the complex environmental problems that we face today. Now the new trend to study biodiversity, for example, has evolved into biocomplexity. John Hayes realized that a multidisciplinary approach was required to fully understand these complex interactions long before it was popular. John Hayes can be considered the "father of biogeochemistry," a new field of great interest and great opportunities for the future.

Hayes studies the Earth's "carbon cycle," the global network of processes in which plants and algae produce organic matter and animals and bacteria degrade that material to produce mobile, reactive substances like carbon dioxide and methane. Over time, these processes have built the atmospheric inventory of oxygen and controlled the abundance of atmospheric greenhouse gases, thus profoundly shaping conditions at Earth's surface. Hayes's particular specialty has been the measurement and interpretation of variations in the abundance of the isotopes of carbon, hydrogen, nitrogen, and oxygen. He has developed techniques that allow measurement of these variations using samples as small as one billionth of a gram. Such miniaturization has proven important because it allows isotopic analyses of individual organic compounds. As a result, intricate details of Earth's environmental machinery can be observed and mechanisms of control understood. For example, Hayes and his colleagues provided the first evidence for the origin of oxygen-producing photosynthesis at least 2.8 billion years ago.

They have also reconstructed pathways of carbon flow in ancient lakes, oceans, and sediments and shown that the concentration of CO_2 in oceanic surface water is a key factor controlling the abundance of the carbon 13 isotope in the organic matter produced by marine algae. Accordingly, isotopic analysis of ancient algal debris can aid estimation of former concentrations of CO_2.

John M. Hayes was born in Seattle, Washington, on September 6, 1940, and spent his childhood mainly in the American Northwest. He attended Iowa State University in Ames where he earned a bachelor of science degree in geology in 1962. He obtained his Ph.D. in chemistry at the Massachusetts Institute of Technology, Cambridge, in 1966. He was briefly a postdoctoral scholar at the University of Chicago, Illinois, at the Enrico Fermi Institute in 1966. He served in the U.S. Army from 1967–1968 in the Chemical Evolution Branch at the NASA Ames Research Center in California. After that, he held a one-year NATO–National Science Foundation postdoctoral fellowship in organic geochemistry at the University of Bristol, England, in 1969. He joined the faculty at Indiana University, Bloomington, in 1970, where he was named Distinguished Professor of biogeochemistry in 1990 and served as chairperson of the department from 1994 to 1996. In 1996, Hayes accepted the position of senior scientist and director of the National Ocean Sciences Accelerator Mass Spectrometry Facility at the Woods Hole Oceanographic Institution of the Massachusetts Institute of Technology, where he remains today. Hayes has also been professor of practice at Harvard University, Massachusetts, since 1997. He was a visiting scientist at the University of California at Los Angeles (1979–1980), and Australia Bureau of Mineral Resources, Geology and Geophysics (1988). Since 1962, John Hayes has been married to Janice Maria Boeke. They have three children.

John Hayes has been an author of 170 articles in international journals, four chapters in professional volumes, and two textbooks in the

John Hayes and technician attaching a sampling bottle to the hydro wire for a research project aboard the research vessel *Knorr* in October, 1977 *(Courtesy of John Hayes)*

fields of mass spectrometry, organic cosmochemistry, microbial biochemistry, isotopic and organic geochemistry, and chemical oceanography. He has received numerous honors and awards for his research. Hayes won an Eastman Prize as a graduate student at the Massachusetts Institute of Technology in 1962–1964. In 1987–88 he was a Fellow of the John Simon Guggenheim Memorial Foundation. He was a Bennett lecturer at the

University of Leicester in 1990, an Ingersoll lecturer with the Geochemical Society in 1994, and a Krumbein lecturer at the University of Chicago in 1994. In 1997, he was awarded the Harold C. Urey Medal of the European Association for Geochemistry. In 1998, he was awarded the Treibs Medal of the Geochemical Society and elected to membership in the National Academy of Sciences and the American Academy of Arts and Sciences.

Hayes has also performed significant service to the profession. He has been on editorial boards for *Precambrian Research* (1977–1998), *Organic Geochemistry* (1987–present), and *Biomedical Mass Spectrometry* (1975–1983). He was associate editor for *Geochimica et Cosmochimica Acta* from 1971 to 1975. He was the chair of two Gordon Conferences for the Geochemical Society from 1981–1983 and 1986–1988. He served on several committees for the Geochemical Society, as well as NASA.

⊠   **Head, James W., III**
(1941–   )
American
Planetary Geologist

How do we know anything about the earliest history of the Earth when tectonic and weathering processes have destroyed all of the early features? The answer is to study the development of the other planets and moons as analogs. None of them is exactly the same as the Earth so different planetary processes can be observed, but there are some general similarities. This is the type of research done by planetary geologists, and James Head III is one of the premier planetary geologists in the field. He has performed research on the Moon, Mars, Jupiter, moons of other planets, and several asteroids. He investigated processes of differentiation of the planets into layers and the formation of crust, volcanic processes, surface deformation (faulting), and impact structures.

Head's greatest achievements, however, have been through his research on Venus. He was one of the main participants on the Magellan mission, which mapped the surface of Venus using Synthetic Aperture Radar (SAR), as well as collecting gravity data on the planet. The images of the surface of Venus are superb with pixels at the 20–25 meter resolution for the whole planet. Therefore, very detailed and delicate features can be seen (imaged). Most of the discoveries made about Venus involved the participation of James Head. He identified mountain ranges formed by fold belts as reported in the paper, "Processes of Formation and Evolution of Mountain Belts on Venus," large normal faults in rift zones, large strike-slip faults, and strange deformation features called wrinkle ridges that we do not have on Earth. He did research on the volcanism on Venus, most of which is formed through hot spots. He mapped flows and their chronology by overlapping relations and radiating dike swarms on the planet. He also interpreted the large flat pancakelike volcanoes, which have no counterpart on Earth.

By looking at the density of impact structures on the planet surface relative to other planets and moons, Head realized that the surface of Venus must be relatively young. There was an apparent catastrophic resurfacing of the planet about 300 to 500 million years ago. There are highland areas that contain fragments of crust that are older but the majority of the surface is relatively young. This represents a planetary process that is not observed on Earth. Head theorizes that an abrupt change in the structure and chemistry of the mantle may have initiated this event. It likely involved massive deformation, which can still be observed in the highland structure, and large-scale volcanic flooding of the surface. Some of the crust may have been recycled in the mantle or else it was covered over by the massive eruptions. He has written several planetary interpretations of Venus including the paper, "The Geologic History of Venus: A Stratigraphic View."

James Head performed research on other planets and moons which produced such articles as *Oceans of the Past History of Mars: Tests for their Presence Using Mars Orbiter Laser Altimetry*, among others.

James Head III was born on August 4, 1941, in Richmond, Virginia. He attended Washington and Lee University, Virginia, where he earned a bachelor of science degree in geology in 1964. He attended graduate school at Brown University, Rhode Island, and earned a Ph.D. in geology in 1969. He accepted a position as a geologist with Bellcomm, Inc., in Washington, D.C., at NASA headquarters, in 1968. In 1973, he was the interim director at the Lunar Science Institute in Houston, Texas, as well as a faculty member at Brown University, Rhode Island, where he remains today as the James Manning Professor of geological sciences. James Head has two children and outside interests in music and literature.

James Head III has had an extremely productive career. He is the author of some 225 publications in international journals, governmental reports, and professional volumes. Many of these papers establish new benchmarks in lunar and planetary processes and evolution, especially that of Venus. His research contributions have been well recognized by the profession in terms of honors and awards. Head is a Fellow of the American Association for the Advancement of Science. From NASA, he received the Medal for Exceptional Scientific Achievement in 1971 and the Public Service Medal for his work on the Magellan mission in 1992. He received a Special Commendation from the Geological Society of America in 1973 and was named the CASE Professor of the Year for Rhode Island in 1990.

Head has performed an exceptional amount of service to the profession. He has served on numerous committees and panels for the National Academy of Sciences, Geological Society of America, Universities Space Research Association, and International Union of Geological Sciences. He has been an associate editor for *The Earth, Moon*

and Planets since 1974 and serves on the editorial board for *Planetary and Space Science* for Pergamon Press, Ltd. However, his real service has been to NASA. In addition to serving on most of the major committees and panels, he was a member of the Viking mission to Mars, Galileo mission to Jupiter, Magellan mission to Venus, the Lunar Scout II mission, and the USSR Venera 15/16 and Phobos missions.

Helgeson, Harold C.
(1931–   )
American
Geochemist

What could geology have to do with the origin of life? The obvious answer is "nothing," until we consider the conditions of the Earth at the time life originated. There was a lot of rock in various states of decomposition and fluids and gas carrying by-products of that decomposition. Not only were minerals (like clays) involved in this transformation but the thermodynamics of the biochemical molecules in this geochemical system must also have been a controlling factor in the first life. These truly revolutionary ideas are those of Harold Helgeson, who has established himself as the leader in this new field of biomolecular geochemistry that he originated. The existence of living microbes in deep oil field brines or around the vents deep under the ocean at the mid-ocean ridge, among others, tell us that the experiments done by microbiologists at conditions of standard temperature and pressure may not be completely representative. After all, the surface conditions of the Earth some 3.5 to 4 billion years ago or more were far from the standard conditions of today. As a result, Helgeson conducted thermodynamic studies of biomolecules at elevated temperatures. He also collected samples and observed the environments of these hyperthermophilic (high temperature) microbes on the island of Vulcano in Sicily to characterize their natural habitats. The steps in this research process are to determine what stabilizes the biomolecules that stabilize cells, as well as the conditions and role of enzymes. To accomplish this, Helgeson became interested in protein chemistry and even began working with the thermodynamics of DNA, and ultimately RNA. This new approach is basically to look at the physical chemistry of these organic biomolecules. This combining of microbiology with physical chemistry and encapsulating it in geology to provide natural constraints almost defines a whole new science rather than simply an arm of existing sciences. It is truly a field of the future that may have profound implications for life.

This hybrid biogeochemistry is not the only area of research of Harold Helgeson. He first established his expertise by solving geochemical problems using elegant advanced mathematical solutions. By using advanced thermodynamics, Helgeson was able to explain old and puzzling problems for which there had been solutions that explained only part of the observations. Much of this work was on aqueous geochemistry and water-rock interactions including hydrothermal deposits, but mineral thermodynamic solutions were also addressed. This mathematical approach, using any and all variables taken together, really took the profession by storm. As a result, many geologists have come to view Helgeson's work with a combination of awe and trepidation.

Harold (Hal) Helgeson was born on November 13, 1931, in Minneapolis, Minnesota. He spent his youth in the Saint Anthony Park suburb of Saint Paul, Minnesota. He attended Michigan State University in chemical engineering through ROTC, but switched to geology and graduated with a bachelor of science degree in 1953. He accepted a job with the Technical Mine Consultants, Ltd., Toronto, Canada, as a uranium geologist that year. He was called to active duty in 1954 and became a second lieutenant in the U.S. Air Force, first for training in Denver, Colorado, and later with the 497th Recon Tech-

nical Squadron in Schierstein am Rhein, Germany. He met his first wife Velda there; they were married in 1956. The same year, Helgeson took a job as a mining exploration geologist searching for diamonds with the Anglo American Corporation in South Africa (owner of De Beers). He enrolled in Stanford University, California, for graduate studies in the spring of 1959, but changed his mind and switched to Harvard University, Massachusetts, instead. He graduated with his Ph.D. in 1962 as an advisee of ROBERT M. GARRELS and accepted a position as a research chemist with Shell Oil Company in Houston, Texas, that year. In 1965, Helgeson accepted his first academic position at Northwestern University, Illinois, at the request of Robert Garrels. He joined the faculty at the University of California at Berkeley in 1970, and has remained there ever since. He was a Miller Research Professor in 1974–1975. Hal Helgeson was married twice more, currently to France Damon, and he has three children.

Hal Helgeson has contributed to numerous articles in international journals and professional volumes. Many of these papers establish new benchmarks in applying mathematical solutions (mostly using physical chemistry) to geochemical problems. He received the Goldschmidt Medal from the Geochemical Society in 1988 in recognition of his research contributions to the profession. He was also a Guggenheim Fellow in 1988.

⊠ **Herz, Norman**
(1923–   )
American
Archaeological Geologist

Although Norman Herz began his career as an economic geologist who also addressed problems of regional geology, he made a decision midway through his career to apply his experience to archaeological problems and has firmly established himself as one of the leading archaeological geologists in the world. Because archaeological geology involves the use of geological materials for economic purposes, the transition from economic geology to archaeological geology was a smooth one. Herz's most important contribution to the science is to establish methods to determine the sources of marble used in ancient building, statuary, and other applications. He uses stable isotope geochemistry in combination with petrology and trace element geochemistry to identify the rock units and perhaps even the quarries where the marble in these pieces was obtained. Such work can help identify the trade routes and trading partners in the ancient world. Herz is an author of the definitive book on this topic, entitled *Classical Marble: Geochemistry, Technology and Trade.* Herz has worked on problems from Neolithic–Early Bronze Age, classical Greece and Rome, and even through Renaissance and modern times. This work has resolved several puzzling issues regarding ancient trade patterns, especially in Greece. It has also resolved questions regarding association of broken fragments for reconstruction and the authenticity of artifacts. Herz has been consulted by numerous museums, including the Louvre, the British Museum, the J. Paul Getty Museum, the Ny Carlsberg Glyptotek (Copenhagen), the Metropolitan Museum of Art, the Art Institute of Chicago, the Walters Art Gallery, and the National Gallery of Art to help with such problems. He is especially well known for his work on the Getty Museum kouros, an article about which appeared in the *New York Times.*

Norman Herz's other major contribution to geological archaeology has been organizational. In addition to establishing his own reputable center, he has consulted worldwide, helping to establish state-sponsored surveys and professional societies.

Norman Herz was born on April 12, 1923, in New York, New York, where he spent his youth. He attended the City College of New York and earned a bachelor of science degree cum laude in geology in 1943. Upon graduation he enlisted in

the U.S. Air Force, where he advanced to second lieutenant between 1943 and 1946. After he was discharged, Herz attended the Johns Hopkins University and earned a Ph.D. in geology in 1950. Between 1950 and 1951, he was employed as an instructor at Wesleyan University, Connecticut, and a part-time geologist for the Connecticut Geological Survey. From 1951–1952, Herz was a Fulbright Senior Research Scholar in Greece before accepting a position as a research geologist with the U.S. Geological Survey. From 1956 to 1962, he worked in the Brazil office. Herz left the U.S. Geological Survey in 1970 to become a professor and head of the department at the University of Georgia in Athens. He stepped down as department head in 1977, but would serve again from 1991 to 1994. He also helped set up the Center for Archaeological Sciences at the University of Georgia, and served as its director from 1984 to 1994. Norman Herz retired to professor emeritus in 1994. He was a visiting professor several times during his career to the University of São Paulo, Brazil, George Washington University, Université d'Orléans, France, American School of Classical Studies, Greece, as well as the Romanian and Bulgarian Academy of Sciences. Norman Herz married his current wife, Christine M. Suite, in 1993. He has three children by previous marriages.

Norman Herz has been very productive throughout his career, serving as an author of some 200 scientific articles in international journals, professional volumes, and governmental reports. These papers run the gamut from regional geology to economic geology to archaeological geology, where he has produced several benchmark studies. Since 1988, he has been an author or editor of five books, mainly on archaeological geology, including the definitive textbook, *Geological Methods for Archaeology.* In recognition of his contributions to the field, Norman Herz has received several honors and awards. He is a foreign member of the Brazilian Academy of Sciences. He received the Pomerance Award from the Archaeological Institute of America and the Creative Research Medal from the University of Georgia, among others.

Herz has also performed service to the profession and the public. He has served as adviser to the U.S. National Park Service, the Archaeological Survey of India, and the National Research Council, mainly for preservation and dealing with acid-rain problems. Herz is also the U.S. member for the International Council on Monuments and Sites. He was the founding member and president of the Association for the Study of Marble and Other Stones used in Antiquity (ASMOSIA), originally a NATO-sponsored program, from 1988 to 1998.

⊠ **Hess, Harry H.**
(1906–1969)
American
Plate Tectonics

Harry Hess was one of the mavericks who pioneered the plate tectonic paradigm. His name stands prominently among the giants of the Earth sciences. Using the bathymetric data he collected while a commander of a naval vessel during World War II and gravity and magnetic data from other work, he identified mid-ocean ridges and their importance. He hypothesized that they were the location where new ocean crust was being formed and proposed the mechanism of large thermally driven circulation (convection) cells in the mantle to drive the plates. He proposed that the ocean crust was therefore youngest at the mid-ocean ridge and was progressively older away in both directions. He even identified oceanic trenches as the location where old ocean crust was returned to the mantle in a conveyor-beltlike system. This 1960 theory that he called an "essay in geopoetry" sparked the research efforts of such later giants as ALLAN V. COX, Fred Vine, and DRUMMOND H. MATTHEWS, among many others who would confirm his hypotheses. This work forms the fundamental basis for our current understanding of plate tectonics and places Hess on par with AL-FRED WEGENER as one of the two fathers of the theory.

Plate tectonics, however, came late in Harry Hess's career. He first achieved prominence for his work on peridotites, serpentinites, and pyroxenes. His papers, "Pyroxene of Common Mafic Magmas" and "A Primary Peridotite Magma," first established him in the field of geology. This interest would take him to study alpine peridotites but also to the oceans. He became interested in geophysical surveys over ocean crust while still a graduate student but later he combined the two apparently disparate areas to consider the generation of his beloved rocks. This attempt to combine ideas that had not been previously related led to his breakthroughs on oceanic processes. He modeled the serpentinization of mafic and ultramafic rocks at the mid-ocean ridges and formulated hypotheses for the origin of the Hawaiian Islands, as well as oceanic guyots, among others. He devoted much effort to explaining the circulation of seawater into the newly formed, dual-layered hot rocks at the mid-ocean ridge and the chemical reactions that took place as a result. These studies were unique in their addressing features at so many scales from so many directions and would all eventually contribute to the plate tectonic work. This expertise in mafic and ultramafic rocks would later lead NASA to invite Hess into a prominent position in the Apollo program to explore the lunar surface. Unfortunately, he did not live long enough to see this project to completion.

Harry H. Hess was born on May 24, 1906, in New York, New York. He attended Asbury Park High School, New Jersey, before entering Yale University, Connecticut, in 1923 to become an electrical engineer. Along the way, he switched to geology and graduated with a bachelor of science degree in 1927. Upon graduation, he accepted a position with the Loangwa Concessions, Ltd., as an exploration geologist in Rhodesia. He returned to the United States in 1929 to attend graduate school at Princeton University, New Jersey, as an advisee of ARTHUR F. BUDDINGTON. He earned a Ph.D. in 1932 and accepted a teaching position at nearby Rutgers University. In 1934, he returned to Princeton University to join the faculty. Harry Hess married Annette Burns on August 15, 1934; they would have two sons. Hess was a U.S. Navy reserve officer and was called to active duty after the attack on Pearl Harbor in 1941. He was first stationed in New York City, where he oversaw the detection of enemy submarine operation patterns in the North Atlantic. His work resulted in the virtual elimination of the threat within two years. He even tested the effectiveness of the program by serving on the decoy vessel USS *Big Horn*. He then took command of the attack transport USS *Cape Johnson* and took part in four major combat landings, including Iwo Jima. Hess carefully chose his travel routes around the Pacific Ocean and used his echo sounder continuously to map the bathymetry of the ocean floor. Through this surveying, he discovered flat-topped submarine volcanoes which he named guyots after the geology building at Princeton University. Hess remained in the active reserves after the war and was called up during the Cuban Missile Crisis, the loss of the submarine *Thresher,* and the *Pueblo* affair. He held the rank of rear admiral at the time of his death. Hess returned to Princeton University after the war and was named the Blair Professor of geology in 1964. He also served as department chair from 1950 to 1966, when he retired to professor emeritus. He was a visiting professor at the University of Capetown, South Africa (1949–1950), and Cambridge University, England (1965). Harry H. Hess suffered a fatal heart attack on August 25, 1969, in Woods Hole, Massachusetts, while serving as chair of a meeting of the Space Science Board of the National Academy of Sciences.

Harry Hess was an author of more than 110 monographs, articles, and discussions in international journals and professional volumes; his papers on "Serpentinites, Orogeny and Epeirogeny," "The Ocean Crust," and "Sea Floor Spreading" are true classics. In recognition of his vast contributions to geology, Hess received numerous honors and awards. He was a member of the National Academy of Sciences and the American Academy of Arts and Sciences. He was awarded an honorary doctorate from Yale University. Hess also received the Penrose Medal from the Geological Society of America, the Distinguished Service Award from NASA, and the Feltrinelli Prize from the Academia Nazionale dei Lincei, among others. He now has an award of the American Geophysical Union named in his honor.

Equally impressive was Harry Hess's service to the profession and the public. Amazingly, he served as president of the Mineralogical Society of America (1955), the Geological Society of America (1963) and two sections of the American Geophysical Union (geodesy, 1951–1953, and tectonophysics, 1956–1958).

⊠ Hochella, Michael F., Jr.
(1953–)
American
Geochemist, Mineralogist

It is commonly the case that the Earth sciences lag behind the other sciences in terms of perceived impact in the future. It is viewed more as a historical discipline because it deals mainly with events that occurred many years ago; many of which are unlikely to result in new applications. The glamour of genetic engineering and the development of new energy sources, medicines, and superconductors are not typically in the realm of the Earth scientist. Michael Hochella, however, is involved in unique research on that level. Working with microbiologists, he studies how microorganisms attach themselves to the surface of minerals and then use the chemicals in those minerals to live. For example, Hochella and colleagues have studied how the common microorganism Shewanella attaches itself to the widespread mineral in soils called goethite. Shewanella uses a weak attractive

Michael Hochella demonstrates some of the analytical equipment in his laboratory at Virginia Tech *(Courtesy of Michael Hochella Jr.)*

force to attach itself and then makes a special protein that allows it to use the iron in the goethite to respire when oxygen is not available. This research has profound implications for a variety of applied sciences. Not only will it demonstrate how microorganisms degrade rock and soil, but also it could let us take a critical step forward in environmental remediation. It even has implications for food production. How do the basic chemicals in rock and soil wind up within the food we eat? Combined with genetic engineering, the implications for this research are as important as any science today.

To perform such research, Michael Hochella has become one of the foremost experts on the surface chemistry of minerals. His papers, "Atomic Structure, Microtopography, Composition and Reactivity of Mineral Surfaces" and "Mineral Surfaces: Characterization Methods and their Chemical, Physical and Reactive Nature," are classics in the field. Chemical reactions occur on the surface of the minerals, so the understanding of the surface is critical to all geochemistry whether environmental at surface conditions or in igneous rocks at 1,000

degrees Centigrade. Surfaces can be analyzed as deep as many layers of atoms, or the analysis can be restricted to the top layer. All of these analyses require the most advanced of high-tech analytical techniques. The instruments he uses employ scanning tunneling and atomic force microscopies and spectroscopies, transmission electron microscopy (TEM), X-ray and ultraviolet photoelectron spectroscopies, scanning Auger microscopy and spectroscopy, and low energy electron diffraction. Hochella has written several papers describing the application of these new techniques including, "Auger Electron and X-ray Photoelectron Spectroscopies." These methods use electrons under various states and trajectories and electromagnetic radiation to image the actual individual atoms on the mineral surface. The applications of these techniques to minerals alone are a whole field of research, much of which Hochella defined. He also applied them to the surfaces of minerals, including plagioclase and degraded plagioclase, sulfides (pyrite and galena), calcite, hematite, barite, gypsum, goethite, asbestiform riebeckite, and even gold. Like the microbial research, this advanced mineral surface chemical analysis is one of, if not the most, significant and pioneering research that is being conducted today. The potential for important discoveries is immense.

Michael Hochella was born on September 29, 1953, in Yokohama, Japan. His father was a highly decorated B-25 pilot in World War II who was flying missions in the Korean War at the time. He remained in the U.S. Army after the war and the family moved to New Jersey, France, Germany, Arizona, and Bel Air, Maryland, where his father retired and Hochella was able to spend sixth through twelfth grades in the same town. He attended Virginia Polytechnic Institute and State University and earned a bachelor of science degree in geology in 1975. He completed his graduate studies at Stanford University in California, where he earned a Ph.D. in geochemistry in 1980 as an advisee of Gordon Brown. Hochella began his career in industrial science at Corning Glass, Inc., in

Corning, New York, as a research chemist in 1981. Realizing that academia was more to his liking, he returned to Stanford University in 1983 as a senior research associate and later as an associate research professor. He joined the faculty at his alma mater of Virginia Polytechnic Institute and State University in 1992, where he remains today. Michael Hochella is married to fellow geologist and faculty member at Virginia Tech, Barbara M. Bekken. They have two children. In his spare time, Hochella is an avid amateur aviator.

Michael Hochella is amid a strongly productive career having been an author of some 100 scientific articles in international journals and professional volumes. Several of these papers establish new benchmarks in mineral surface chemistry and the use of the latest in high-energy analytical techniques in their study. He is also an editor of an important volume, *Mineral-Water Interface Geochemistry.* He has been very successful in obtaining funding for his research, with more than $4.6 million to date. In recognition of his research contributions to geology, Hochella has received several honors and awards. He received the Dana Medal from the Mineralogical Society of America, the Alexander von Humboldt Award from the von Humboldt Society, Germany, and was named a Fulbright Scholar and a Mineralogical Society of America distinguished lecturer.

Hochella served as president of the Geochemical Society (2000–2001) as well as a member of several committees there and for the Mineralogical Society of America, and a member of the advisory committee for geosciences for the National Science Foundation.

⊠ Hoffman, Paul
(1941–)
Canadian
Stratigrapher (Tectonics)

A relatively recent but high-profile concept that Paul Hoffman has championed is that of the "Snowball Earth" hypothesis. This idea, first proposed by Joe Kirschvink in 1992 but dismissed as a "wild idea" by the geologic community, has been proven through the work of Paul Hoffman. There has always been a problem with Neoproterozoic stratigraphy (just before Cambrian, or about 700 to 550 million years ago). It appears that there was glaciation on a worldwide basis, even at low latitudes. Hoffman has found evidence of this odd situation in Namibia in the Damaran fold belt, in the Svalbard Archipelago in the Barents Sea and the Anti-Atlas Mountains of Morocco. It has been long known in many other areas (Scandinavia and Canada, for example) including the Appalachian Mountains of the United States. It seems that the whole Earth went into a deep-freeze condition, causing continental glaciation all over. This glaciation is not only evident in glacial deposits of this age but also in geochemical signatures of the sediments. The glaciation accompanied one of the most profound extinction events on record, the development of shells on animals right after the event, and the deposition of the largest terrestrial sand oceans ever in geologic history. It is clear that this was a very momentous time in Earth history; Paul Hoffman has provided a reason for it.

Although Paul Hoffman is now associated with the snowball earth idea, he did not even address it until age 57 and would have still qualified to be in this book based upon his accomplishments until that time. He is probably the foremost expert on Proterozoic plate tectonics of the Canadian shield. He took an extremely complex geology and put it in the context of modern plate interactions, particularly the Wopmay orogen. He recognized geometries and interactions in these ancient rocks that were barely understood in modern frameworks such as conjugate strike-slip systems through plate collisions. He figured out sediment transport directions on 2-billion-year-old rocks. He used modern geochemical techniques on carbonate rocks to predict the chemistry of ancient oceans. His recognized authority led him to be asked to construct a tectonic synthesis of the Precambrian evolution of Laurentia (North America and Greenland) as part of the Decade of North American Geology project for the Geological Society of America in 1984. He compiled a geologic map of the northern half of the Canadian shield from original sources.

Paul Hoffman was born on March 21, 1941, in Toronto, Canada. He was attracted to geology through classes and field trips at the Royal Ontario Museum. He attended McMaster University in Hamilton, Ontario, and earned a bachelor's degree in geology in 1964. Summer employment with the Ontario and Canadian Geological Surveys gave him 15 months of practical field experience before he started graduate school at the

Johns Hopkins University in Baltimore, Maryland. Paul's dissertation advisers were FRANCIS J. PETTIJOHN and Robert Ginsburg. He did a field research project on a Paleoproterozoic fold belt in Great Slave Lake of the Northwest Territories. He graduated in 1968 and taught for one year at Franklin and Marshall College in Lancaster, Pennsylvania, before accepting a permanent position at the Geological Survey of Canada (GSC) in Ottawa. Paul Hoffman married Erica Westbrook in 1976; they would have one child. Hoffmann left the GSC in 1992 to become a professor of geology at the new School of Earth and Ocean Sciences at the University of Victoria, British Columbia. In 1994, he joined the faculty at Harvard University where he is the current Sturgis

Paul Hoffman shows features of a research sample at Harvard University *(Courtesy of John Chase, Harvard News Office)*

Hooper Professor of geology and associated with NASA's Astrobiology Institute as well as the Canadian Institute for Advanced Research.

Paul Hoffman is a foreign associate of the National Academy of Sciences and the American Academy of Arts and Sciences. He is a Fellow of the Royal Society of Canada and the American Association for the Advancement of Science. He is an Alfred Wegener Medalist of the European Union of Geosciences, a William Logan Medalist of the Geological Association of Canada, and a Henno Martin Medalist of the Geological Society of Namibia.

⊠ Holland, Heinrich D.
(1927–   )
German
Geochemist

Dick Holland working in his office at Harvard University *(Courtesy of H. Holland)*

Every student in historical geology classes worldwide learns about the evolution of the atmosphere. They learn about the chemical changes that have taken place, the buildup of oxygen, for example, and their causes. These students and those in environmental geology classes learn about the close interaction and chemical exchanges between the atmosphere and hydrosphere. Indeed, the evolution of both is tightly associated. Heinrich "Dick" Holland is the main reason for this understanding. As far back as the mid-1960s, long before it was fashionable, Holland was investigating the exchange of gases and chemical interdependence of the two, including geochemical cycles. This important research culminated in his award-winning book, *The Chemical Evolution of the Atmosphere and Oceans* in 1984, although his research continues. Now all of the climate modelers seeking to determine our fate apply Holland's groundbreaking research on a daily basis. Holland can really be considered the "father of the climate modeling movement" which is now by far the most vigorous and well-funded field in Earth sciences.

Holland has also been called the "father of modern geochemistry of hydrothermal ore deposits" because that is his "other" research life. He was originally trained as an economic geologist and unlike many of his peers, he was well versed in all aspects of hydrothermal ore deposits including theoretical, experimental, and analytical aspects. Holland was the first to use thermodynamic theory and data in order to estimate the conditions of formation of the ore deposits. He was among the first to perform experimental studies to determine solubilities of sulfides and carbonates at elevated temperatures, as well as the partitioning of elements between fluids and minerals or melts. He also determined stability relations of minerals in hydrothermal fluids and water-rock interactions at elevated temperatures in his experiments. Holland was the first to use the analytical techniques of fluid inclusion analysis on ore deposits as well as stable isotope studies on ores. He was also the first geochemist on ALVIN dives to investigate hydrothermal mineralization at mid-ocean ridges. He was even involved in reporting the oldest traces of life on

Earth in the paper, "Evidence for Life on Earth more than 3,850 Million Years Ago." With all of his "firsts" it is no wonder that the profession holds him in such high esteem.

Heinrich Holland was born on May 27, 1927, in Mannheim, Germany. He received his primary education in Germany, but his secondary education was completed in England and the United States. He attended Princeton University, New Jersey, and graduated in 1946 with a bachelor of arts degree in chemistry. He graduated magna cum laude, Phi Beta Kappa, and with the Physical Chemistry Prize. He was in the armed forces in 1946 and 1947, where he held the rank of technical sergeant assigned to work with Wernher von Braun's group. He attended Columbia University, New York, for graduate study and earned master of science and doctoral degrees in 1948 and 1952, respectively. His first faculty position was at his alma mater, Princeton University, where he remained from 1950 to 1972. He then moved to Harvard University in 1972, where he remains today. He was named H.C. Dudley Professor of economic geology in 1996, and he still holds that endowed chair.

Heinrich Holland wrote or edited five books and professional volumes and published some 147 articles in international journals, chapters in professional volumes, and professional reports. An astonishing number of these papers are seminal works in the field of geochemistry, appearing in the most prestigious of journals and often-cited volumes. Holland has received numerous honors and awards in recognition of this work. He was awarded a National Science Foundation postdoctoral fellowship at Oxford University, England, in 1956–57. He was a Fulbright lecturer at Durham University and Imperial College, London, England, in 1963–1964. He was a Guggenheim Fellow in 1975–1976. He received the Alexander von Humboldt Senior Scientist Award in 1980–1981 and the 1984 Best Physical Science Book Award by the Association of American Publishers for *Chemical Evolution of the Atmosphere*

and Oceans. He received the 1994 V. M. Goldschmidt Award of the Geochemical Society and the 1995 Penrose Gold Medal from the Society of Economic Geologists.

Holland has also performed notable service to the profession. He is a member of the National Academy of Sciences and a Fellow of the American Academy of Arts and Sciences. He has held positions of member of council, vice president, president, and chair of several important committees for the Geochemical Society.

⊠   **Holmes, Arthur**
(1890–1965)
British
Isotope Geochemist, Geophysicist,
Geomorphologist, Petrologist

Arthur Holmes has been called the greatest geologist of the 20th century. He made some of the greatest contributions to geology on the whole, as well as to numerous individual disciplines. He even made the first steps toward the plate tectonic paradigm. His most famous contribution was his reevaluation of *The Age of the Earth* published in the book of the same name. During the 19th century, many scientists attempted to derive an absolute time scale for our planet. The age of Earth was a question that had been plaguing geologists for a long time. Lord Kelvin published a scientific paper that proposed geologic time to span 20 to 40 million years. His calculations were based on the assumption of a uniformly cooling Earth and gravitational and chemical sources for terrestrial and solar energies. Holmes decided to initiate his own research into the subject and after collecting enough data, he showed that Kelvin's conclusions were not valid by the availability of radioactive heat. He compared the amounts of uranium and thorium in rocks with their decaying products (daughter products) of lead and helium, respectively, and was able to make an assumption of a constant half-life for each element. Based on this information, Holmes felt the age of the Earth was at least 1.6 billion years old, which he later revised to 4 billion years. He also placed the beginning of the Cambrian at 600 million years. These ages were refined later but considering the primitive technology of the time, they are remarkably accurate.

The research Arthur Holmes conducted on the age of the Earth added insight into other questions that were plaguing geologists at the time. By recognizing that radioactive heat was available from the breakdown of uranium, thorium, and potassium, it could no longer be believed that the Earth was cooling and contracting. Holmes believed that the tectonic movements of the crust were caused by cyclical expansion alternating with contraction of the crust. He also agreed with ALFRED WEGENER that continents are drifting, in stark contrast to the popular opinion of the profession. Holmes was also the first geologist to deduce that convection currents were present in the mantle of the Earth and likely drove the continents. HARRY HESS would later expand on this idea to form the basis of the plate tectonic paradigm.

The work of Arthur Holmes in the field of petrology was also groundbreaking. He was a strong believer in the idea that both extrusive and intrusive igneous rocks originated from liquid magma. However, due to the limited amount of physiochemical and thermodynamic data at the time, he began to question the idea. Holmes began to work with the Geological Society of Uganda due to the excellent specimens of alkalic volcanic rocks that were found in the West African Rift Valley and began to think about solid state metasomatism and transformation of preexisting rocks by differential introduction of fluxes of hydrothermal fluid. He also made important links between geophysics and petrology, with regard to the origin of kimberlite rocks (diamond pipes) and his belief in eclogite as a high-pressure equivalent of basalt. Holmes's wide range of scientific research can be found in detail in his highly regarded book, *Principles of Physical Geology.*

Arthur Holmes was born on January 14, 1890, in Hebburn-on-Tyne, England. His first insight into geology was discovered while he attended Gateshead High School. After graduating from high school, Holmes enrolled at Imperial College in London, England, in 1907. He earned a bachelor of science degree in physics under R. J. Strutt (later Lord Rayleigh), but changed to geology as an advisee of W. W. Watts. Holmes graduated with a second degree in geology as an associate of the Royal College of Science in 1910. Holmes did his graduate studies with Strutt in investigating the area of radioactivity with geology. He also took a position as a prospector to Mozambique, Africa, to earn money, where he learned the art of fieldwork and then began studying Precambrian metamorphic rocks and Tertiary lavas. He contracted malaria there. After graduating with a Ph.D. in 1912, he accepted the position of demonstrator at Imperial College, where he taught petrology. In 1920, he decided to work in industry and accepted the position of chief geologist of an oil exploration company in Burma. He lost his young son to dysentery at that time. He returned to England in 1925 to become professor of geology and chair at the University of Durham. Holmes reorganized the entire department and conducted some of his most extensive research there. He transferred to the University of Edinburgh, Scotland, where he was appointed regius chair of geology in 1943. He remained until his retirement to professor emeritus in 1956. Arthur Holmes was married twice, first to Margaret Howe in 1914, and after her death to petrologist Doris Reynolds in 1939. Arthur Holmes died in London on September 20, 1965.

Arthur Holmes was extremely productive in terms of numbers of scientific articles in international journals and professional volumes. These papers include numerous benchmark studies in a variety of topics ranging from geochronology and the age of the Earth to geomorphology, petrology, and Earth history. In recognition of his outstanding contributions to the many fields of geology, Arthur

Holmes received numerous honors and awards. He was a Fellow of the Royal Society of London. Among many other additional awards, Holmes received the Penrose Medal from the Geological Society of America and the Vetlesen Prize from Columbia University. He also has a medal named in his honor at the European Union of Geosciences and a society in England named after him.

⊠ **Hsu, Kenneth J.**
(1929–)
Chinese
Tectonics, Sedimentologist, Structural Geologist

Surprisingly, Kenneth Hsu is probably best known by the general public for the popular and sometimes controversial books and papers that he has written. Most of these are the scientific research with which he was involved, translated into material that is understandable by the general public. Books like *The Great Dying: Cosmic Catastrophes, Dinosaurs, and Evolution; The Mediterranean was a Desert;* and *Challenger at Sea* are examples of geology for the public. On the other hand, he wrote books like *Applied Fourier Analysis* and *Circle Graphs in Polynomial Time* and papers like "Fractal Geometry and Music" and "Is Darwinism Science?" (some written with his son that are only marginally related to geology). Many of these wind up in the newspapers and on multiple websites for a variety of reasons.

This popular work is surprising because Kenneth Hsu is one of the giants of the profession who has led a full career in geology as well. This versatility in apparent multiple careers characterizes his geologic career as well, thus the difficulty in titling his specialty. To tectonics researchers, Hsu is the discoverer and definer of the melange, a chaotic mixed deposit that forms in a subduction zone. It contains sediment and volcanics scraped from the ocean floor mixed with metamorphic rock squirted back up the subduction zone and

into a mass. If found in the mountains, these rocks define the line marking the "suture zone" between two ancient plates. He is also the editor of the groundbreaking book, *Mountain Building Processes*. To structural geologists, Hsu is the inventor of a diagram bearing his name in which strain features in rocks may be plotted and compared. He is also a contributor to the understanding of thrust fault movement through his work in the Alps. To climatologists and oceanographers, he is the leader of an historic cruise of the *Glomar Challenger* through the Mediterranean Sea, which resulted in the Mediterranean salinity crisis idea. It showed that large sea-level changes could affect climate and biota and was the impetus for his later popular book. To sedimentologists, he is the definer of Alpine carbonate and flysch sedimentation by investigating modern analogs again aboard the *Glomar Challenger* from the stormy Atlantic to the sun-scorched Persian Gulf. He developed new ideas on how evaporation affects the deposition of carbonate rocks that appear in all textbooks today. He is a great contributor to the understanding of sedimentary facies (lateral changes in deposits). To the Chinese, he is a returning hero who mentored students and breathed new life into the geologic research there. He even wrote the book *Geologic Atlas of China*. Hsu's great success for each of these endeavors is his rare combination of an "eye for detail" but a "mind for the whole." He is also willing to travel to the ends of the Earth to track down a problem; he has visited every continent except Antarctica. Hsu is a true renaissance man in every sense of the term.

Kenneth Jinghwa Hsu was born on July 1, 1929, in Nanking, China. He attended the National Central University at Nanjing, China, and earned a bachelor of science degree in geology in 1948. He moved to the United States and attended Ohio State University, where he earned a master of arts degree in geology in 1950. He earned a Ph.D. in geology from the University of California at Los Angeles in 1954 as an advisee of DAVID T. GRIGGS. Upon graduation, he accepted the position of geologist and later as project chief and research associate at Shell Development Company in 1954, and he remained there until 1963. He joined the faculty at Harpur College in 1963 and moved to the University of California at Riverside in 1964. Ken Hsu was married to Ruth Hsu. They had three children but a tragic auto accident took her life in 1964. Hsu was remarried in 1966 to Christine Eugster, a native of Switzerland; they had one child. He accepted a faculty position at the Swiss Federal Institute of Technology (ETH) in Zurich in 1967. Hsu remained at ETH until 1994 and twice served as chair of the department. He was a visiting professor at Scripps Institution of Oceanography at the University of California at San Diego in 1972 and at California Institute of Technology in 1991. After he left ETH, he was a visiting professor at National University of Taiwan in 1994–1995, at Hebrew University in Jerusalem in 1995, at the Berlin Institute for Advanced Studies, Germany, in 1996, and at Colorado School of Mines in 1997.

Kenneth Hsu has had an amazingly productive career, producing more than 400 articles in international journals, professional volumes, and governmental reports, and 20 books and volumes of international importance. Many papers are published in high-profile journals and establish a new benchmark for the state of the science. His research contributions have been recognized by the geologic profession as shown by his numerous honors and awards. Hsu is a member of the National Academy of Sciences. He received an honorary doctoral degree from Nanjing University, China, in 1987. He was awarded the Wollaston Medal by the Geological Society of London (1984), the Twenhofel Medal from the Society of Economic Paleontologists and Mineralogists (SEPM)(1984), and the Penrose Medal from the Geological Society of America (2001). He was also a Guggenheim Fellow in 1972.

Hsu has also performed outstanding service to the profession. He was the co-chief scientist for the Deep Sea Drilling Project, Atlantic and

Kenneth Hsu (center) studying a deep sea drill core with colleagues aboard a research cruise *(Courtesy of Kenneth Hsu)*

Mediterranean, a panel chair for the Joint Oceanographic Institute for Deep Earth Sampling (JOIDES), and a section chair for the International Union for Geological Sciences (IUGS), as well as serving on numerous committees for each. He was president of the International Association for Sedimentology in 1978 to 1982. He served as editor in chief for the journal *Sedimentology* and the associate editor for *Journal of Sedimentary Petrology* and *Marine Geophysical Research*.

⊠ **Hubbert, M. King**
(1903–1989)
American
Geophysicist

In its classical form, Earth science is largely descriptive in nature. But it is also a composite science, drawing upon methods of other sciences to explain phenomena of the Earth. The integration of ideas from quantitative fields into geology caused a major revolution in each of the geologic disciplines as it was realized. M. King Hubbert was one of the true pioneers of this integration. He had training in physics and mathematics, but a strong interest in rock mechanics as well. FRANCIS J. PETTIJOHN called him "a student of nobody" even while he was a graduate student. Nobody had the knowledge that he was after. His first assault on the science was a 1937 paper entitled "Theory of Scale as Applied to the Study of Geologic Structures," in which he used dimensional analysis and continuum mechanics to scale-model geologic structures. He derived scaling laws to model familiar geologic systems based upon the length, mass, and time constants of the systems. The work was considered controversial and raised a stir in the profession. He later applied this work

to all of Earth in a paper entitled "The Strength of the Earth," which would later form the basis for deriving more quantitative plate tectonic models.

In his second major assault, M. King Hubbert addressed the process of fluid flow. He verified Darcy's law of flow through experimentation and then derived field equations for the movement of fluids through the permeable media of the Earth's crust in his paper, "Theory of Groundwater Motion." He introduced gravity as the major controlling factor, but showed that fluids did not necessarily flow from higher to lower pressure. This work caused the previously feuding hydrogeologists and petroleum geologists to join forces against him because it made all of their work obsolete. But Hubbert prevailed and later applied this work to the migration and subsequent entrapment of oil and gas. He modeled the interactions of fluids with unlike densities in a dynamic continuum which produced several counterintuitive outcomes,

King Hubbert (front left) with colleagues on a resistivity survey in Franklin County, Alabama *(Courtesy of the U.S. Geological Survey, E.F. Burchard Collection)*

at least with regard to accepted ideas. It altered the course of petroleum exploration.

Hubbert took this fluid research and applied it to rock mechanics. He showed that increased fluid pressure would decrease the strength of rock, ultimately causing fracturing in unexpected orientations. This work was directly applied to oil exploration by pumping fluids under high pressure into oil wells to cause the rock around the well to fracture, thus increasing permeability. But it was also applied to problems of overthrusting, which occurs at angles that were previously unexplainable. He was involved with the classic "beer can experiment" in which a warming empty beer can scoots along a virtually flat piece of glass on a cushion of air. Thrust sheets were shown to move in the same manner but on a cushion of fluid in a classic Hubbert paper.

M. King Hubbert was born on October 5, 1903, in San Saba County, Texas, where he grew up on a farm. He attended Weatherford College, a nearby two-year school from 1921 to 1923. He enrolled at University of Chicago, Illinois, but had to perform grueling work as a wheat harvester and to replace track for Union Pacific just to obtain travel money. He finally arrived at the University of Chicago in 1924 and earned a bachelor of science degree in geology and physics with a minor in mathematics in 1926. He remained at the University of Chicago for graduate studies and earned a master of science degree in 1928 and a Ph.D. in 1937 in geophysics. Hubbert worked over the summers from 1926 to 1928 as an exploration geologist for the Amarada Petroleum Company in Tulsa, Oklahoma. He became an instructor at Columbia University, New York, in 1931 while working summers for the Illinois Geological Survey. He met and married Miriam Graddy Berry in 1938. He left Columbia University in 1940 to write and conduct his own research. In 1942, Hubbert joined the World War II effort as a senior analyst for the Board of Economic Warfare in Washington, D.C. He joined Shell Oil Company in 1943 as a geophysicist and

held various positions. He retired from Shell Oil Co. in 1963 to assume concurrent positions as a geophysicist at the U.S. Geological Survey in Washington, D.C., as well as a member of the faculty at Stanford University, California. In 1968, Hubbert retired to professor emeritus from Stanford University. After his second retirement, he was a visiting professor at the Johns Hopkins University in 1968, and a regents professor at the University of California at Berkeley in 1973. In 1976, Hubbert retired for a third and final time from his position at the U.S. Geological Survey. M. King Hubbert died in his sleep of an embolism on October 11, 1989.

M. King Hubbert's busy career can be measured in many ways. His written contributions spanned governmental reports, industrial reports, nearly 100 articles in scientific journals and professional volumes and presentations. The subjects he addressed were just as varied, ranging from petroleum exploration to geophysical techniques to rock mechanics, among others. They were typically innovative and nontraditional, and therefore pioneering. He also wrote a popular textbook, *Structural Geology*. In recognition of his contributions to the science, numerous honors and awards were bestowed upon him. Hubbert was a member of the National Academy of Sciences and a Fellow of the American Academy of Arts and Sciences. He was awarded honorary doctoral degrees from Syracuse University, New York, and Indiana State University. He received both the Arthur L. Day Medal and the Penrose Medal from the Geological Society of America, the William Smith Medal from the Geological Society of London, the Elliott Cresson Medal from the Franklin Institute of Philadelphia, the Rockefeller Public Service Award from Princeton University, the Anthony F. Lucas Gold Medal from the American Institute of Mining, Metallurgical and Petroleum Engineers and the Vetlesen Prize from the Vetlesen Foundation at Columbia University.

In terms of professional and public service, Hubbert was equally notable. He served as president of the Geological Society of America in 1962, among many other committees and panels. He served on numerous committees and panels for the National Research Council, the National Academy of Sciences, the U.S. Office of Naval Research, U.S. delegations to the United Nations, and the U.S. Nuclear Regulatory Commission. In terms of editorial work, he was editor of *Geophysics* and associate editor of the *American Association of Petroleum Geologists Bulletin* and the *Journal of Geology.*

I

⊠ **Imbrie, John**
(1925–)
American
*Invertebrate Paleontologist,
Paleoclimatologist*

One of the "hottest" areas of the Earth sciences today is climate change, and one of the main spokespersons for its study is John Imbrie. His primary approach to research is to study the history of climate, carefully documenting its variability with time. Much of this work involves looking at the changes which have taken place in the marine environment using biological, sedimentological, and chemical markers mostly found in deep-sea cores. He then models these changes in an attempt to identify the main mechanisms of climatic change and control. These variations range in duration from as short as one month to as long as 100,000 years or more. The models identify the cyclicity in these variations and attempt to tie them to astronomical sources like distance of the Earth to the Sun (Milankovitch cycles) or terrestrial sources. He has even translated his research into a more popular forum by publishing an award-winning book entitled, *Ice Ages: Solving the Mystery.* This book provides a summary of his research findings on the controls and processes in moving into and out of ice ages that can be understood by laypersons.

Imbrie has truly had two careers in geology: in paleoclimatology as described, as well as an earlier career as a premier invertebrate paleontologist. His areas of expertise include paleoecology and biometrics. Imbrie developed methods of studying assemblages of fossils and their relative abundances to predict the paleoenvironment. Small changes in relative abundances can indicate significant ecological changes. Clearly, this work has been applied to the climate research. Biometrics is a highly quantitative treatment of paleontological changes using probability and statistics. These quantitative results can detect small changes in features of animal populations that might otherwise go unnoticed using standard observational techniques. It was these numerical methods that established John Imbrie as a leader in paleontology and established a whole new direction of research.

John Imbrie was born on July 4, 1925, in Penn Yan, New York. He enlisted in the U.S. Army in World War II and served in combat in Italy as an infantryman in the 10th Mountain Division. He was wounded in the Po Valley. He later wrote a book about his experiences there. Imbrie attended Princeton University, New Jersey, upon discharge where he earned a bachelor of arts degree in geology in 1948. He attended graduate school at Yale University, Connecticut, and earned a master of arts degree in geology in 1951 and a

Ph.D. in 1952. He accepted a position at the University of Kansas at Lawrence in 1951, but moved to Columbia University, New York, the following year. He served as department chair in 1966 and 1967. In 1967, he joined the faculty of Brown University, Rhode Island, where he spent the rest of his career. In 1976, he was named the Henry L. Doherty Professor of oceanography and in 1990 he retired to professor emeritus. He remains active, but more with regard to his work on the historical records of World War II. Imbrie was a visiting scientist at Lamont-Doherty Geological Observatory of Columbia University and at the University of Rhode Island. John Imbrie married Barbara Z. Imbrie in 1947, and they have two children.

John Imbrie has had an extremely productive career. He was an author of some 60 articles on climate change and numerous others on invertebrate paleontology in international journals and professional volumes. Both groups include many seminal works that are required reading in their respective fields. He has also written four books, two of which are popular rather than purely scientific. His contributions to science have been well recognized by the profession in terms of honors and awards. He is a member of the National Academy of Sciences and the American Academy of Arts and Sciences. He was awarded honorary doctoral degrees from the University of Edinburgh, Scotland, and from the Christian-Albrechts University of Kiel, Germany. He received the Vega Medal of the Swedish Society of Anthropology and Geography, the Vetlesen Prize from the Vetlesen Foundation, the Lyell Medal from the Geological Society of London, the Leopold von Buch Medal from the Deutsche Geologische Gesellschaft, the Maurice Ewing Medal from the American Geophysical Union and the U.S. Navy, the Award for the Advancement of Basic and Applied Science from the Yale Science and Engineering Association, and the MacArthur Prize Fellowship. He also won the Phi Beta Kappa Prize for his book, *Ice Ages: Solving the Mystery.*

Imbrie has also given professional service too extensive to list here. In short, he served on numerous committees and panels as both a member and chair for the National Science Foundation, the National Academy of Sciences, and the National Research Council, among many others.

J

⊠ **Jahns, Richard H.**
(1915–1983)
American
Mineralogist, Petrologist

Granite pegmatites are formed from the last liquids in a crystallizing magma or the first melts in
a metamorphic rock undergoing partial melting.
Because of this fringe position relative to normal
igneous processes, the liquid tends to be enriched
in anything that will not readily fit into the common rock-forming minerals. These components
include incompatible elements and water and result in many economic mineral deposits. It was
his World War II assignment with the U.S. Geological Survey to find sources of strategic (incompatible) elements and minerals that sparked
Richard Jahns' interest in pegmatite dikes. Pegmatites can contain economic sources of beryl-
lium, tantalum, and lithium, in addition to
minerals like mica, feldspar, and gemstones like
beryl, emerald, tourmaline, and others. Early in
his career he established himself as arguably the
foremost authority on pegmatites in addition to
training himself as a renowned field geologist. He
collaborated with some of the top mineralogists
and geochemists like Wayne Burnham and O.
FRANK TUTTLE to produce the classic studies on
pegmatites. These studies include both experimental work as well as observational and thermo-

dynamic. He explained an old and perplexing
phenomenon of the layering of pegmatites with
aplites with the cycling of fluid during crystallization. He explained the giant crystals up to 40 feet
long that he observed using a fluid model.

Even as Jahns completed these pure science
studies, he never drifted far from the applied aspect of the science. Later in his career, he took on
administrative roles and used this applied interest
to greatly energize the departments and schools in
which he served. Several departments owe their
current success to his effectiveness and foresight.
He also used this interest to form a very successful
consulting business to industry, largely for mineral
exploration and mining. In addition, he successfully applied his geologic expertise to public service. He was one of the outspoken leaders on land
use management as well as mining and earthquake preparedness. He served in leadership capacities for several state agencies.

Richard Jahns was born on March 10, 1915,
in Los Angeles, California, but he grew up in
Seattle, Washington. He graduated from Seattle
High School as class valedictorian. He entered
California Institute of Technology at age 16 intent
on chemistry but switched to geology after one
year and graduated with a bachelor of science degree in geology with honors in 1935. He played
varsity baseball as an undergraduate. Jahns attended graduate school at Northwestern Univer-

sity, Illinois, and earned a master of science degree in 1937 in petrology. He married Frances Hodapp while a graduate student. They would have a son and a daughter. He accepted a position with the U.S. Geological Survey upon graduation and had numerous assignments, including mapping granites in New England, mica deposits in the southeastern United States (Piedmont), and pegmatites in New Mexico. He completed his doctoral work through this period and was awarded a Ph.D. from the California Institute of Technology, Pasadena (Cal Tech), in 1943. He joined the faculty at Cal Tech in 1946. In 1960, Jahns moved to the Pennsylvania State University at State College where he served as chair of the Division of Earth Sciences and dean of the College of Mineral Industries. Jahns returned to California to become the dean of the School of Earth Sciences of Stanford University in 1965, where he spent the rest of his career. He was named the Welton J. and Maud L' Anphère Crook Professor of geology and applied Earth sciences at Stanford. He retired to professor and dean emeritus in 1980. He died of a heart attack on December 31, 1983.

Richard Jahns had a very productive career producing numerous articles in international journals, collected volumes, and governmental reports. These articles define the state of knowledge for pegmatites and their genesis. His publication, *The Study of Pegmatites,* remains the classic in the field. Jahns's research was well received by the geologic profession and as a result, he received numerous honors and awards. He received the Ian Campbell Award from the American Geological Institute (1981), the Public Service Award from the American Association of Petroleum Geologists (1982), the Distinguished Achievement Award from the American Federation of Mineralogical Societies (1972), and the Distinguished Alumnus Award from California Institute of Technology. He also received the Outstanding Teaching Award from the Stanford School of Earth Sciences.

Jahns performed outstanding service to the profession. He served as president of the Geological Society of America (1970–1971), among many committees and panels. He served on many committees and panels for the Mineralogical Society of America. For NASA, he was in the astronaut training program for Apollo 15 and 16, as well as part of the Lunar Exploration Planning Group. He was chairman of the Earth Sciences Advisory Panel for the National Science Foundation. He was president of the California Academy of Sciences and served as chair, president, and/or member of most of the California state boards that relate to geology (earthquakes, mining, licensing, etc.). Jahns also served editorial roles too numerous to list here.

⊠ **Jordan, Teresa E.**
(1953–)
American
Stratigrapher

The Earth Science department at Cornell University has undertaken a broad multi-investigator, multidisciplinary research objective called the Cornell Andes Project. This project focuses on modern mountain building in relation to plate tectonics and climate. Teresa Jordan contributes the expertise of physical stratigraphy and sedimentology to the project. She is a strongly contributing member of this team because she traditionally looks broadly into problems integrating other geologic research with her stratigraphic studies. She collaborates with other diverse Earth scientists like earthquake seismologists, gravity-magnetic geophysicists, groundwater hydrologists, stable isotope geochemists, structural geologists, igneous petrologists-geochemists, climatologists, and plate tectonic modelers. Jordan investigates stratigraphy and related structures in the field and combines them with satellite image analysis of the surface features, seismic stratigraphy which uses seismic reflection profiles (like sonograms of the Earth) to determine the geometry of the strata underground, and magnetic polarity stratigraphy

which yields ages of the rocks based upon the preserved orientation of the Earth's magnetic field at
the time of deposition compared to a calibrated
chart. She combines these data with the textural
and compositional distribution of the sediments
as studied microscopically to determine the history of the stratigraphic sequences, the geometry
and timing of subsidence of the basin, the source
of the sediments making up the rock and consequently the nearby uplift history of the surrounding rocks. She also determines the environmental
conditions and evaluates whether the observed
changes are episodic or simply continuous slow
change. With all of this information in hand, Jordan then models the history of the basin. This
process allows her to isolate the controlling factors
in the changes she observes. Some factors can include plate tectonics, sea level change, properties
of the source region for the sediments, and climate change. Such a multidisciplinary approach
to large-scale geologic problems sets a new standard for research and Teresa Jordan is one of the
leaders in this field.

This move to undertake research in the Argentine and Chilean Andes is a natural outgrowth of
Teresa Jordan's background. She began her research
on the development of sedimentary basins in
highly deforming mountainous areas by studying
the ancestral Rocky Mountains in western North
America (Idaho and Wyoming). She calculated the
flexing of the crust downward under the tremendous loads imposed during mountain building
which consequently forms basins in the foreland
(just ahead of the main mountains) that fill with
sediment. She took this experience to the Andes
Mountains where she investigated the Bermejo
foreland basin in western Argentina. She looked at
the fluvial drainage patterns, as well as the basinal
distribution of sediment grainsizes and the development of evaporite deposits. This climatic and
hydrologic information was then correlated with
concurrent structural studies to determine the sedimentary response to deformation versus climate
changes. Because the area is a desert, small changes

Teresa Jordan explains the geology of the rock exposure
in the background *(Courtesy of Carrie Allmendinger)*

in water supply have a great impact on sediment
supply and distribution as well as organic component resulting from changes in plant growth. Several important papers have resulted from this work
including, "Andean Tectonics Related to the Geometry of Subducted Nazca Plate" and "Retroarc
Forearc Basins," among others.

Teresa Jordan was born on April 14, 1953, in
rural Dunkirk, New York. She attended Rensselaer
Polytechnic Institute in Troy, New York, where she
earned a bachelor of science degree in geology in
1974 and received the Joseph L. Rosenholtz
Award. She completed her graduate studies at
Stanford University, California, on a National Science Foundation graduate fellowship, where she
completed a Ph.D. in geology in 1979. She also
received an American Association of University
Women doctoral fellowship and worked as a geologist for the U.S. Geological Survey upon graduation. That year Jordan accepted a position as a
postdoctoral research scientist at Cornell University in Ithaca, New York, but joined the faculty in
1984. She remains at Cornell University as a full
professor today. She was a visiting scientist at the
Colorado School of Mines in Golden. Teresa Jordan is married to structural geologist and fellow

Cornell University professor Richard Allmendinger. They have one daughter.

Teresa Jordan is amid a very productive career. She is an author of more than 90 scientific articles in international journals and professional volumes. Many of these are seminal papers on the tectonics and stratigraphy of the Andes Mountains. In recognition of her contributions to geology, Jordan received the National Science Foundation Award to Women Faculty in Science and Engineering. She performed several instances of service to the profession, including membership on several committees for the Geological Society of America, as well as on several panels for the National Research Council. She has also taught numerous short courses to professional societies and oil companies in Argentina and Chile.

Karig, Daniel E.
(1937–   )
American
Marine Geologist, Geophysicist

Island arcs are one of the major discoveries by the giants of plate tectonics. These convergent zones of ocean plate consumption, however, exhibit many complex relations and processes that are unique on Earth. If a student researched the literature to learn about these features, he or she would quickly learn that one of the foremost authorities on them is Daniel Karig. The interaction of rapid deformation complexly interacting with rapid sedimentation forms a radically diverse deposit sitting within a deep-sea trench. This pile of complex sediment plus metamorphic and igneous rock forms the forearc or accretionary prism or wedge and is analogous to the wedge-shaped pile of snow scraped up by a snowplow. Because it is such a dynamic system, erosion and sedimentation work in concert. In addition to normal sedimentary processes, there is tectonic erosion and addition of material and the regular earthquakes in these zones play a role in shaking sediments loose. Daniel Karig studied these sediments in several ways. He studied them on land where the forearc prism emerges above sea level; he studied them in deep-sea drill cores taken within these areas; and he studied them aboard submersible vehicles like the famous

ALVIN launched from the research vessel, *Glomar Challenger*. He studied exposed forearc prisms in Japan, Iran, the Philippines, and Indonesia. He studied these areas directly on Deep Sea Drilling Program (DSDP)/Ocean Drilling Program (ODP) projects in several areas in the western Pacific Ocean (Marianas, Philippines, Sumatra, etc). These studies resulted in such papers as "Ridges and Basins of the Tonga-Kermadec Island Arc System" and "Structural History of the Marianas Island Arc System," among others. He later modeled the features he observed by building an analog experimental subduction zone apparatus to study the paths of deforming objects in forearc prisms.

The articles that Karig produced from this research include some descriptions of specific areas but many define the processes that occur in subduction zones. They include definitive papers like "Remnant Arcs, Tectonic Erosion at Trenches" and "Initiation of Subduction Zones: Implications for Arc Evolution and Ophiolite Development," among many others. He has shown how deformational fabrics form in sediments from forearc prisms as well as how the fluids (mostly water) within the sediments are squeezed out under high pressure as a result of the extreme deformational pressures. Sedimentary structures are formed as a result of this process. In addition to the forearc work, Karig also researched (and defined) the back-arc-basin areas and their processes. He inves-

Dan Karig on a research cruise aboard the research vessel *Glomar Challenger* *(Courtesy of Daniel Karig)*

tigated island arcs where the two plates do not collide head-on but instead at an angle in a process called oblique convergence. He also showed how fragments of ocean crust (ophiolites) could be lifted (obducted) from their plate and emplaced onto the island arc. His explanation was a new and sensible approach to an old problem.

Daniel Karig was born on July 20, 1937, in Irvington, New Jersey. He attended the Colorado School of Mines in Golden, where he earned a bachelor of science degree in geological engineering in 1959 and a master of science degree in geology in 1964. From 1960 to 1961, Karig served as a second lieutenant in the U.S. Army Corps of Engineers with combat training and even attended ranger school. He completed one year of research at Victoria University in Wellington, New Zealand, before enrolling in a doctoral program at the Scripps Institution of Oceanography at the

University of California, San Diego, in 1965. He earned his doctoral degree in 1970 in Earth sciences as an advisee of H. WILLIAM MENARD. He remained at Scripps Institution for a year after graduation as a postdoctoral fellow before joining the faculty at the University of California at Santa Barbara in 1971. He moved to Cornell University, New York, in 1973, and remained there for the rest of his career. He served as department chair from 1991 to 1995 and retired to professor emeritus in 1998. Since his retirement, Karig pursued athletic events, including winning three national cross-country ski championships and a national championship in tandem marathon canoeing. He has also been involved in local environmental issues. Daniel Karig is married to Joanne Molenock; they have one son.

Daniel Karig has had a very successful career. He is an author of some 120 articles in international journals, professional volumes, and major professional reports. Many of these papers are often-cited seminal studies on the mechanics of subduction zones and development forearc prisms. Karig's research has been well received by the geologic community as evidenced by his honors and awards. He received the D.C. Van Dienst Medal from the Colorado School of Mines and was a Fulbright Fellow, among other honors.

Karig has also performed significant service to the profession. He served on numerous panels and committees for the Ocean Drilling Project (ODP). He was an associate editor for the *Bulletin of the Geological Society of America, Journal of Geophysical Research,* and *Neotectonics* as well as *Ophioliti.*

⊠ **Kay, Marshall**
(1904–1975)
Canadian
Stratigrapher

Although primarily a stratigrapher, Marshall Kay gained fame as the originator and champion of the geosynclinal theory. In this theory, deep subsiding sedimentary basins, called geosynclines, evolve into mountain ranges. The geosynclinal theory was widely accepted until the origination of the plate tectonic theory in the 1960s. Kay's 1951 publication, "North American Geosynclines" (and later "Geosynclines in Continental Development"), was the high point of the theory. He concluded that the nature of the sedimentation within geosynclines depends upon the complex interrelationships of uplift, contemporaneous subsidence, weathering, and the presence or absence of volcanism. These observations described deposition within active basins where sediment sources and rates of input are constantly shifting. These sources could be volcanic, recycled sedimentary rocks or uplifted basement blocks. The geosynclines were divided into three major groups based upon their association with tectonically active margins, passive margins, or on continental cratons. Each of these categories was subdivided into one to three geosynclinal terms with a prefix. For example, within tectonically active (convergent) zones there could be a Eugeosyncline if volcanic strata are interlayered or a Miogeosyncline if the sequence is volcanic-free, though these were not Kay's terms. Kay was not overbearing with his system and yet it became almost sneered at during the plate tectonic revolution of the 1960s, even though the distinctions and processes have validity. Unfortunately, Kay commonly took the brunt of the criticism.

Marshall Kay was considered one of the leading stratigraphers in the world. He gained recognition during the biostratigraphic revolution of the 1930s, as he was a recognized paleontologist as well. Out of this beginning, he became a pioneer in physical stratigraphy, especially with regard to basin reconstructions. Three of his papers, "Paleogeographic and Palinspastic Maps" (1945), "Analysis of Stratigraphy" (1947), and "Isolith, Isopach, and Palinspastic Maps" (1954), established the new methods by which stratigraphy would be analyzed in the future. These benchmark works are among the most important in sedimentary geology.

Kay was born in Paisley, Ontario, Canada, on November 10, 1904. He was raised in a scientific household with a father, George F. Kay, who was a distinguished geologist in his own right. Kay's father moved his family to the United States in 1904 to accept a position as a professor at the University of Kansas. He taught Pleistocene (2 million–8,000 years ago) geology. He became head of the department and state geologist in 1911, and then dean of liberal arts from 1917 to 1941.

Marshall Kay graduated with a bachelor of science degree cum laude and Phi Beta Kappa from the University of Iowa in 1924, and with a master of science degree in 1925, both in geology. He received the Lowden Prize for his thesis work. Kay decided to attend Columbia University, New York, where he received his Ph.D. in 1929 on a Roberts Fellowship and as assistant curator of paleontology. After receiving his Ph.D., he accepted a teaching position at Barnard College, New York, before joining the faculty at Columbia University in 1931, where he taught geology for 44 years. He served as chair of the department from 1953 to 1956 and again from 1971–1973, after which he retired to professor emeritus. He achieved his greatest appointment as Newberry Professor of geology in 1967. From 1944 to 1946, Kay was administrator of Columbia University's Division of War Research Program, which was part of the Manhattan Project. Marshall Kay married Inez Clark in 1935; they would have four children, three of whom would go into geology. Marshall Kay died on September 3, 1975.

Marshall Kay served as an author of some 110 scientific articles in international journals and professional volumes. He also produced three books. Several of these are classics on stratigraphic processes, as well as the geosynclinal theory. In recognition of his outstanding contributions to geology, Marshall Kay received several honors and awards. He was awarded an honorary doctoral degree from Middlebury College, Vermont. He received the Penrose Medal from the Geological Society of America, the Distinguished Service Award from the University of Iowa, and the Kunz Prize from the New York Academy of Sciences, among others.

Kay was very active in terms of service to the profession. He served as vice president of the Paleontological Society (United States), and of the New York Academy of Sciences. He also served in leadership roles on numerous committees for the International Geological Congress, the International Commission on Stratigraphy, the American Association of Petroleum Geologists, and the New York Botanical Gardens, among others.

Keller, Edward A.
(1942–)
American
Geomorphologist

When an earthquake occurs in the eastern or central part of the United States there are seismic waves that shake buildings and other structures, but rarely is there evidence on the surface as to where it occurred. The only way to locate the earthquake is with seismographs and patterns of seismic activity, which are as uncommon as the surface features. For that reason, geomorphology is regarded as a rather gentle branch of geology there. In the western United States, on the other hand, earthquakes and other tectonic movements leave scars, induce landslides, and generally wreak havoc on buildings and people. In stark contrast to the East, tectonic geomorphology is a dynamic and dangerous study in the West. Edward Keller is one of the foremost experts on tectonic geomorphology especially with regard to earthquake hazard reduction and prevention. By studying relative uplift and subsidence both in terms of rates and elevation changes, tectonic movements and their extent and intensity may be revealed. The beautiful wave-cut terraces of the California Pacific coast are excellent examples of the types of features that Keller studies. They reveal sequential tectonic uplift of the land surface with erosion during the quiet periods. Such studies can reveal information on recurrence intervals for earthquakes, potential for blind faults, as well as landslides and other hazards. They have great implications for building codes and disaster preparedness plans. Keller primarily studies the geomorphology and Quaternary deposits related to active faults and folds that result from faults.

Edward Keller's other main area of interest is fluvial geomorphology. He studies the development of channels in streams, as well as the controls on where pools and riffles develop and how they change with time. This research involves an attempt to explain and even quantify a process that is otherwise chaotic in appearance. In addition to determining location of the features of a stream, Keller studies the processes involved in the transport of material as well as the seasonal changes in streams. This research has profound implications for studies of drainage basins and planning especially with regard to flood control. Currently, as an offshoot of this research, he has been studying the hydrologic processes in the chaparral ecosystem of southern California and role of wildfire in the recurrence of high magnitude flood deposits and debris flow deposits.

Edward Keller was born on June 6, 1942, in Los Angeles, California. He attended California State University at Fresno where he earned a bachelor of science degree in mathematics in 1965. However, he decided that he was really better suited to geology and returned to California State University to earn a bachelor of arts degree in geology in 1968. Keller was married in 1966. He then earned a master of science degree in geology from the University of California at Davis in 1969. He earned a Ph.D. from Purdue University, Indiana, in geology in 1973. He joined the faculty at the University of North Carolina at Chapel Hill the same year. In 1976, he accepted a position at the University of California at Santa Bar-

Edward Keller aboard a research boat in California
(Courtesy of Edward Keller)

bara and has remained there ever since. He has served as chair of both the Environmental Studies and the Hydrologic Science programs several times. Keller and his wife have two children.

Edward Keller has had a very productive career. He is an author of some 90 articles in international journals, governmental reports, and professional volumes. Many of these are seminal works on fluvial processes and tectonic geomorphology. Even more impressive are the books he has written. He is the author of the most successful environmental geology textbook, *Environmental Geology,* in its eighth edition in 2000. He also wrote the definitive textbook on tectonic geomorphology, *Active Tectonics,* in its second printing. He is also an author of four other books including *Environmental Science.* Keller has received several honors and awards for his contributions to the profession. He received a Hartley Visiting Professor Award from the University of Southampton, England, in 1982–1983 and the Quatercentenary

Fellowship from Cambridge University, England, in 2000. He received two Outstanding Alumnus Awards from Purdue University, Indiana, one from the department (1994) and one from the School of Science (1996). He also received a Distinguished Alumnus Award from California State University at Fresno in 1998. He received the Outstanding Research Award from the Southern California Earthquake Center in 1999.

⊠ **Kent, Dennis V.**
(1946–)
American/Czechoslovakian
Geophysicist (Paleomagnetics)

When a volcanic rock that contains magnetite cools through the Curie temperature of 578°C, the magnetite crystals capture the direction of the prevailing magnetic field of the Earth as thermoremnant magnetism. In contrast, small magnetite grains that settle in water as sediment will spin and align with the prevailing geomagnetic field of the Earth like a compass needle as they settle aligned to the ocean floor, thus preserving depositional remnant magnetism. Dennis Kent helped to define this processes of paleomagnetism as described in his paper, "Post Depositional Remnant Magnetism in Deep Sea Sediments." He studied paleomagnetism on the layer-by-layer basis in marine sediments taken from deep-sea piston cores from locations throughout the world. Kent discerned the recurrent reversals of the Earth's magnetic poles in these sediments. The patterns of these observed reversals were correlated with known magnetic reversals from studies of paleomagnetism in the ocean crust recorded by the thermoremnance of oceanic basalts. Between the thermoremnant and detrital remnant magnetism, a detailed "magnetostratigraphy" for the past 180 million years was established. This research provided the framework for the integration of a fossil and isotopic dating system that virtually all modern geologic time scales for the late Mesozoic and Cenozoic now incorporate. Papers describing this work include "Cenozoic Geochronology" and "A Cretaceous and Jurassic Geochronology." Kent is one of the pioneers in establishing and applying the magnetic polarity time scales.

Kent has been extending the application of magnetostratigraphy for time correlation to more remote geologic periods. This fine resolution is especially useful in continental sediments that have few index fossils or appropriate rock types for dating. He collaborated with PAUL E. OLSEN of Lamont-Doherty Earth Observatory to study climate cycles during the late Triassic as a baseline for comparison with proposed climate change of today. They drilled a continuous core more than 5,000 meters long through the Mesozoic Newark Basin of the Middle Atlantic, which contains one of the most continuous non-marine sedimentary sections on Earth. The complete magnetostratigraphy recorded by these sediments and the fine age control to model provided by the climatic cycles allowed construction of a new geomagnetic

Dennis Kent drills a rock core from the Paleocene-Eocene Esna Formation in Egypt for paleomagnetic studies in 2001 *(Courtesy of Dennis Kent)*

polarity time scale for more than 30 million years of the late Triassic and early Jurassic. Kent and his colleagues have been using this template to link the history of the early Mesozoic rift basins throughout North America, Greenland, and North Africa.

Kent also did research on some of the late Paleozoic units mainly in the central, but also in the northern Appalachians. This work was dominated by studies on the Catskill and Helderberg sequences of New York and the relative positions of Laurentia and other continents during the late Paleozoic. These studies evolved into tracking Alleghenian deformation, large-scale rotations, and remagnitization during orogeny. Kent's land-based studies also included plutons from New England, as well as collaborations on more exotic areas like the Antarctic, the Yangtze Platform of China, West Africa, the southern Alps, the Greek Islands, and the Colorado Plateau, among others.

Dennis Kent was born in Prague, Czechoslovakia on November 4, 1946, and after residing in London, England, he came with his family to the United States in 1953. He attended the City College of New York and earned a bachelor of science degree in geology in 1968. He completed his graduate studies at the Lamont-Doherty Geological Observatory of Columbia University in marine geology and geophysics, earning a Ph.D. in 1974. He accepted a research position at Lamont-Doherty upon graduation and moved through the ranks of research scientists, ultimately becoming the director of research for the renamed Lamont-Doherty Earth Observatory in 1993. He joined the faculty at Rutgers University in New Brunswick, New Jersey, in 1998, and remains there today. During his tenure at Lamont-Doherty, he twice accepted a visiting professorship at the Institute for Geophysics at ETH (Swiss National Institute) in Zurich. Dennis Kent married Carolyn Ann Cook in 1971 and they have one daughter.

A prolific author, Dennis Kent is a contributor to nearly 200 articles in international jour-

nals, professional volumes, and professional reports. Many of these articles are in high-profile
journals like *Nature* and set new benchmarks for
paleomagnetism and especially magnetostratigraphy. He is also an editor of four professional volumes. He has performed significant service to the
profession. He served on numerous panels and
committees for the Joint Oceanographic Institutions for Deep Earth Sampling (JOIDES). He
served on the U.S. Continental Scientific
Drilling Program, the ICS/IUGS, and many
committees for the Lamont-Doherty Earth Observatory. Kent served on several committees for
the American Geophysical Union and was section
president from 1994 to 1996. He served as associate editor for several journals including *Journal
of Geophysical Research, Geophysical Research Letters, Paleoceanography, Terra Nova,* and the electronic journal *G3.* He also participated in five
scientific cruises including the *Glomar Challenger*
in 1979.

⊠ **Kerr, Paul F.**
(1897–1981)
American
*(Applied) Mineralogist, Economic
Geologist*

There is the study of mineralogy for theoretical
purposes to understand the chemistry, physics,
and processes of formation of minerals and there
is applied mineralogy to determine the processes
involved in economic and environmental applications. Paul "Pappy" Kerr is considered the "father
of applied mineralogy in the United States."
Among his research directions was an interest in
refining analytical techniques. He began with X-
ray diffraction techniques to identify minerals in
his graduate career and pioneered their use in
mineral identification as described in his paper,
"The Determination of Opaque Ore Minerals by
X-ray Diffraction Patterns." His work on opaque
minerals and clay minerals was unparalleled and is

still used today. He would later pioneer X-ray flu-
orescence and infrared and ultraviolet spectroscopy applications to mineralogy as well. He is
probably best known for his systematic organization and compilation of optical techniques and
properties for the study of minerals using an optical microscope. The technique of differential thermal analysis (DTA) was also taken from an
interesting observation to a cutting-edge analytical
method especially for clays as the result of Kerr's
innovations and adaptations of the instruments.

All of this analytical organization and adapta-
tion was applied toward economic minerals. The
systematic nomenclature and classification of clay
minerals that is used today is the result of efforts
by Paul Kerr. He was especially interested in
"quick clays" and their role in landslides and slope
stability. Expanding clays that take on large
amounts of water have generated numerous devastating landslides and other mass movements. He
was also interested in the clay mineralogy in alteration of rocks around ore deposits. The interest in
these "alteration haloes" around ore deposits
stemmed from his interest in ores themselves. He
did an enormous amount of research on tungsten
mineralization in the western United States and
published a comprehensive study on the nature of
tungsten mineralization in general, entitled *Tungsten Mineralization in the United States.*

Kerr was also interested in uranium mineralization which brought him a great amount of notoriety. As part of the Manhattan Project, Kerr
investigated the availability of raw materials for
atomic weapons. In addition to the western
United States he traveled to the Belgian Congo in
Africa and the Northwest Territories in Canada,
among others, which continued for many years as
an association with the U.S. Atomic Energy Commission. Because of this expertise, in 1945 he was
chosen by the Carnegie Endowment for International Peace to chair a commission to investigate
problems associated with inspection of atomic
materials. In 1955, on behalf of the United Nations, Kerr set up a program on raw materials for

the First International Conference on Peaceful Uses of Atomic Energy in Geneva, Switzerland.

Paul Kerr was born on January 12, 1897, in Hemet, California. He worked in the citrus orchards and bean fields in San Jacinto Valley in his youth to save enough money for college. He attended Occidental College, California, where he earned a bachelor of science degree in chemistry and mathematics in 1919. His undergraduate career was interrupted by a brief period of military service during World War I. He continued with his graduate studies at Stanford University, California, where he earned a Ph.D. in 1923. He was a visiting assistant professor at Stanford University for one semester as a sabbatical replacement before joining the faculty at Columbia University, New York, where he remained for the rest of his career. He became the Newberry Professor of mineralogy at Columbia University in 1959. Kerr served as department chair from 1942 to 1950. During that time he was largely responsible for the acquisition of the site for the future Lamont-Doherty Geological Observatory, as well as for naming it. He retired to professor emeritus in 1965, whereupon he moved back to California to become a consulting professor until 1977. Paul Kerr died of a heart attack on February 27, 1981, in Palo Alto, California. His wife of 54 years, Helen Squire Kerr, died several years earlier in September of 1978. They had three children.

Paul Kerr led an extremely productive career. He is an author of some 250 scholarly publications in international journals, professional volumes, and governmental reports, including several books and monographs. He is probably best known for his widely used textbook, *Optical Mineralogy,* originally entitled *Thin-Section Mineralogy* and published in 1933. However, he also published an astounding number of benchmark papers on X-ray techniques, clay mineralogy, uranium mineralogy, tungsten mineralogy, and applied mineralogy. In recognition of his many contributions to geology he received numerous honors and awards. He received an honorary doctorate from his alma mater, Occidental College. He also received the K.C. Li Medal, the Distinguished Member Award from the Clay Minerals Society, and he was made an honorary member of Great Britain's Mineralogical Society.

Kerr also performed extensive service to the profession. He was president (1946) and secretary (1934 to 1944) of the Mineralogical Society of America, as well as serving on numerous committees. He was vice president of the Geological Society of America in 1947, as well as serving on numerous committees. He was also vice president for the American Association for the Advancement of Science and held several positions for the New York Academy of Sciences, among others.

⊠   **Kerrich, Robert**
(1948–)
British
Geochemist

The formation of many of the deposits of economic minerals occurs through hydrothermal processes. Hot chemically reactive fluids dissolve certain mineral species and transport them to chemically favorable areas to precipitate them. These favorable areas can be in lithologic units of a certain chemistry, but commonly they are faults and fractures. By this process, minerals can be naturally concentrated to economic abundance. Robert Kerrich is one of the foremost experts on metamorphic hydrothermal processes. One of his main interests is gold deposits. He devised what is regarded as the standard model for the formation of the layering of gold within seams produced by metamorphic-hydrothermal processes. By studying isotopic systematics of zircons within gold veins produced under low-temperature hydrothermal conditions, Kerrich was able to prove a fractionation model for the segregation of gold. However, this is not his only contribution to gold exploration. By studying numerous other isotopic systems in lode gold deposits in addition to their relation to metamorphism, magmatism, deformation, and plate tectonics he has evaluated the processes of gold deposition from Archean to present. Through this exhaustive study he identified a timing paradox from the processes that derive the gold to those that deposit it. There can be lags of many millions of years between them. This work on the timing of deposits relative to tectonism in an area allowed Kerrich to place the process of gold emplacement to the supercontinental cycle. He also found odd lamprophyric magmatism that includes gold in the igneous rock. This all-encompassing research makes Kerrich one of the top few experts on gold in the world.

Robert Kerrich did not start out with gold as his primary interest. Rather, he was more interested in the geochemistry and role of fluids in the development of crustal features. He looks at the development and diffusion of fluid reservoirs within the crust. These reservoirs take on certain chemical characteristics based upon physical conditions as well as the composition of the rocks in which they are contained. Kerrich has become interested in the development of the Archean crust in his adopted country of Canada. He concentrates his efforts on the Superior Province of the Canadian Shield. This work relies on isotopic systems to track the role of fluids in the development of the lithosphere. He evaluated Archean mantle chemical reservoirs by studying 3.0-to 2.7-billion-year-old ocean plateau basalts and other basalts. By looking at trace elements in these rocks as well as plutons that intrude them in conjunction with unconventional ratios of isotopes like niobium/uranium and thorium/lanthanum he had shed light on how this ancient crust was formed. The continuing research is adding an important aspect to models of the early formation of continents.

Robert Kerrich was born on December 15, 1948, in England. He attended the University of Birmingham, England, where he earned a bachelor of science degree in geology in 1971. He completed his graduate studies at Imperial College in London, England, earning a master of science degree in 1972 and a Ph.D. in geology in 1975. Kerrich immigrated to Canada, where he was awarded a NATO postdoctoral Fellowship at the University of Western Ontario from 1975 to 1977. He remained at the University of Western Ontario as a member of the faculty until 1987. He then moved to the University of Saskatchewan, Canada, where he was named to a George L. McLeon Chair in Geology. In 1996, Kerrich was awarded an earned doctor of science degree from the University of Saskatchewan. He remains at the University of Saskatchewan today.

Robert Kerrich is amid a very successful scientific career. He has been an author of some 156 articles in major international geoscience journals and professional volumes. Several of these are seminal papers on hydrothermal geochemistry, gold mineralization, and global tectonics, and appear in top-quality journals. In recognition of his research contributions to geology, Kerrich has received several honors and awards in addition to those mentioned. He was the youngest person ever to be elected a Fellow of the Royal Society of Canada. He received the W.H. Gross Medal from the Geological Association of Canada, a Steacie Fellowship from the National Environmental Research Council of Canada, the Willett G. Miller Medal from the Royal Society of Canada, and the Distinguished Researcher Award from the University of Saskatchewan.

Klein, George D.
(1933–   )
Dutch
Sedimentologist, Petroleum Geologist

Geology plays a prominent role in everyday life. Major sources of energy are oil, gas, and coal, all of which originate in sedimentary rocks. George Klein is one of the foremost experts on the application of sedimentology to petroleum exploration.

Portrait of George Klein *(Courtesy of George Klein)*

Although he has actively participated in exploring for, finding, and developing oil and gas fields, his main contribution is the development of depositional models to help geologists predict where petroleum can be found. He taught oil geologists how to use sedimentology in their exploration by teaching numerous short courses. In his research, he developed the concept of "Tidalites", which are sediments deposited by tidal currents. He also developed an evolutionary model for the development of cratonic (on continental crust) basins. He applied the tidalite concept to predict tidal circulation on ancient craton platforms (shelves) based on observations of modern processes. Klein showed that increasing shelf width also increased tidal range and thus tidal circulation dominated cratonic seaways. Other work on tidal flats includes the documentation of vertical sequences of rocks and sedimentary structures developed in carbonate banks. He developed new criteria for recognizing features and sediment distribution within tidal flats and the tidal reach in coastal areas.

In addition to his tidal work, Klein documented the control of the bedrock source on the composition of sandstone in rift margins. Such deposits form in basins over granites that are formed during continental breakup and a whole model was proposed for this scenario. This model is especially applicable in the breakup of supercontinents. Klein also developed new field methods to identify ancient lake deposits.

George Devries Klein was born on January 21, 1933, in s'Gravenhage, Netherlands. He immigrated to the United States and attended Wesleyan University, Middletown, Connecticut, where he earned a bachelor of arts degree in geology in 1954. He attended the University of Kansas in Lawrence and earned a master of arts degree in geology in 1957. He worked for the Kansas State Geological Survey while completing his degree. He earned a Ph.D. from Yale University, Connecticut, in 1960. Sinclair Research Inc. (petroleum) employed him as a research sedimentologist in 1960 and 1961. He joined the faculty at University of Pittsburgh, Pennsylvania, in 1961, but moved to the University of Pennsylvania in Philadelphia in 1963. In 1970, Klein accepted a position at the University of Illinois at Urbana-Champaign, where he remained for the rest of his academic career. He retired to professor emeritus in 1993. From 1993 to present, Klein has been the president of the New Jersey Marine Sciences Consortium and director of the New Jersey Sea Grant College. He has also run a geologic consulting business (George D. Klein and Associates, and SED-STRAT Geoscience Consultants, Inc.) part-time from 1970 to 1996, and has been a full-time consultant in the petroleum field since 1996. He has been a visiting professor several times to Oxford University, England; University of Tokyo, Japan; University of Utrecht, Netherlands; Seoul National University, Korea; University of Chicago, Illinois; Scripps Institution of Oceanography, California; and several others.

George Klein has led an extremely productive career. He is an author of some 137 articles in international journals, governmental reports, field guidebooks, and professional volumes. Several of these establish new processes in sedimentology. He is the author or editor of eight books and volumes. His book, *Sandstone Depositional Models for Exploration for Fossil Fuels,* has been reprinted in three editions. Other books and volumes include *Tidal Sedimentation* and *Clastic Tidal Facies.* He has also written some 41 technical reports for companies, societies, and trade magazines. Klein received numerous awards for his contributions to the profession. He received the Outstanding Paper Award from the Society of Economic Paleontologists and Mineralogists (SEPM) in 1970 and honorable mention in 1971. He received the Laurence L. Sloss Award from the Geological Society of America in 2000. He was awarded a Citation of Recognition from the Illinois House of Representatives and the Erasmus Haworth Distinguished Alumnus Award from the University of Kansas, both in 1980. He was a Senior Fulbright Research Fellow in 1989 and Senior Research Fellow for the Japan Society for the Promotion of Science in 1983. At the University of Illinois, he was twice an associate at the Center for Advanced Study, and he received an Outstanding Faculty Award.

Klein has performed outstanding service to the profession. He has served as member and chair of numerous committees and panels for the American Association of Petroleum Geologists, Geological Society of America, Society of Sedimentary Geology of SEPM, International Association of Sedimentologists, Global Sedimentary Geology Program, JOIDES (Joint Oceanographic Program), DOSECC (Deep Drilling of Continents), and the Society for Exploration Geophysicists, among others. He served in numerous editorial roles including associate editor for the *Geological Society of America Bulletin* and on the editorial board for *Geology, Sedimentary Geology,* and *Journal of Geodynamics,* and numerous advisory boards for publishing companies.

Kuno, Hisashi
(1910–1969)
Japanese
Igneous Petrologist

Hisashi Kuno was one of the greatest volcanologists of the 20th century. He had to overcome a language barrier to publish his studies in international literature. Perhaps his most famous work was the study of Japanese volcanoes, in which he discovered an association between how the source of various basaltic rocks is distributed relative to the depths of earthquake foci. He found that tholeiite basaltic magma is produced at less than 200 kilometers and alkali olivine basaltic magma is produced at greater than 200 kilometers. With additional research, he also found that there is a high alumina basalt magma, which lies between the two magmas in composition and forms at an intermediate depth. This hypothesis sparked interest among fellow petrologists. Several petrologists decided to test this model. HATTEN S. YODER JR. and ALFRED E. RINGWOOD, among others, conducted several different experiments at high temperatures and pressures on laboratory constructed systems that contained olivine, pyroxene, and natural rocks. Many of these experiments concluded that Kuno's hypothesis was correct.

One of his earliest papers, entitled, "Petrological Notes on Some Pyroxene Andesites from Hakone Volcano," was well received. This earlier work and subsequent papers were evidence of his remarkable talents as a field petrologist. Kuno's field data and optical data were evidence of his hard work and natural ability to observe geological characteristics. He researched and discovered how the mineral pyroxene crystallized from magmas. The key to this important discovery is the groundmass minerals, which are the materials surrounding the phenocrysts (a relatively large, conspicuous crystal) of a volcanic rock.

Kuno also produced the world-famous and often-cited *Catalog of Active Volcanoes.* His 1954 book, *Volcanoes and Volcanic Rocks,* was used throughout Japan as a standard textbook. He traveled to Hawaii to conduct research on Hawaiian magmas. Kuno's research paper on his travels, "Differentiation of Hawaiian Magmas," illustrates his idea that a possibility exists that granitic magma can be generated from tholeiitic magma.

Hisashi Kuno also conducted research on many other aspects of volcanology, including such topics as the development of the craterlike calderas that sit atop volcanoes, volcanic eruptions based on pyroclastic materials, and the origins of andesite and petrographic provinces. During the last years of his career, Kuno became extremely interested in the petrology of the Moon. He worked with NASA as a principal investigator on the acquisition of lunar samples.

Hisashi Kuno (front left) on a visit to Boston College, Massachusetts *(Courtesy of James Skehan, S.J.)*

Hisashi Kuno was born on January 7, 1910, in Tokyo, Japan, where he grew up. His parents sent him to Sendai to complete his studies at the Second High School. Even though he was very interested in geology, he spent much of his time on extracurricular activities. In 1929, he enrolled in the Geological Institute of the University of Tokyo and studied petrology. He earned a bachelor of sciences degree in geology in 1933, and remained at the University of Tokyo for graduate studies. In 1939, he was appointed to the faculty at the University of Tokyo, but was drafted in the armed forces in 1941 to fight in World War II. He was stationed in northeastern China throughout the war. He returned to academia in 1946, but did not finish his Ph.D. until 1950. In 1951 and 1952, he was invited to collaborate with HARRY H. HESS at Princeton University, New Jersey, on the study of pyroxenes, and as a result became internationally known. Kuno was promoted to full professor of petrology in 1955, a position he would hold until his untimely death resulting from cancer on August 6, 1969, in Tokyo. His wife Kimiko and two children survived him.

Hisashi Kuno was the author of numerous scientific articles in both English and Japanese in international and national journals and professional volumes. Several of his papers on igneous processes in volcanoes and volcanic rocks are benchmarks in igneous petrology. In recognition of his contributions to the profession, Hisashi Kuno had several great honors and awards bestowed upon him. He was a member of the U.S. National Academy of Sciences and an honorary member of numerous professional societies. His most prestigious award was the Japan Academy Prize, which he received in 1954. He also served in a leadership role in many societies, including president of the Volcanological Society of Japan, the Geological Society of Japan, and the International Association of Volcanology and Chemistry of the Earth's Interior, as well as vice president of the International Union of Geodesy and Geophysics.

L

Landing, Ed
(1949–)
American
Paleontologist, Stratigrapher

Coming of age in the 1960s, Ed Landing maintained parallel interests and energy for social change and the study of paleontology and its use in biostratigraphy (relative time correlation of sedimentary rocks). He is one of the foremost experts of one of the most profound transitions in the geologic record, the Cambrian-Precambrian boundary. This boundary marks the demise of a rich diversity of invertebrate animals that lacked shells. These soft-bodied Ediacarian animals, named for a sequence of strata particularly rich in fossils in Australia, include soft corals, jellyfish and wormlike forms. They were replaced by a whole new group of hard-shelled invertebrate animals in the Cambrian. Many of these organisms were the ancestors of the marine invertebrates of today. The current global reference rock section for this boundary is located at Fortune Head, eastern Newfoundland, Canada, and Landing and associates established it. This area is part of the Avalon Terrane, which records part of the building of the supercontinent Rodinia, but the type section is in the overlying breakup sequence. Landing and his colleagues studied evolutionary changes in metazoans through this period and de-

termined the actual beginning of the Cambrian by a change in the behavior of the animals. The geologic community deemed this careful work on a very complete section in the Avalon terrane the best of its kind. Other top researchers, like SAM A. BOWRING collaborated with Landing on this work to tightly define the actual age of the strata by performing uranium-lead geochronology on interlayered volcanic units. Ed Landing has produced some of the seminal works on this important time interval and he is in great demand for talks and papers for volumes.

The Precambrian-Cambrian boundary in ancient Avalon is not the only area of expertise for Ed Landing. He has also done extensive work in the Cambrian to Ordovician-age rocks of the shallow shelf and in the outboard Taconic sequence of New York, Vermont, Quebec, and west Newfoundland. These rocks were formed offshore of the ancient Laurentian (North American) continent. They are mostly from deep water. He studied enigmatic animals like graptolites and conodonts within these rocks and shed some light on this mysterious group of rocks. He proposed that the black shale in this sequence reflects periods of global warming and stagnation of the world's oceans whereas green shale reflects falling sea levels and glaciation. Landing also documented a unique early Paleozoic reef made completely of snails.

New York State Paleontologist Ed Landing in his office *(Courtesy of Ed Landing)*

Ed Landing was born on August 10, 1949, in Milwaukee, Wisconsin, and he grew up in Port Washington, to the north of the city. He attended University of Wisconsin at Madison, where he graduated with a bachelor's of science degree with honors in geology in 1971. He earned a master's of science degree and a Ph.D. in geology from the University of Michigan, Ann Arbor, in 1975 and 1978, respectively. His theses were both in early Paleozoic paleontology and biostratigraphy. Ed then completed several postdoctoral research positions. His first position was in 1978–79, at University of Waterloo, Ontario, Canada, where he worked on early Ordovician conodonts from Devon Island, Canadian Arctic Archipelago. He next held a prestigious National Research Council postdoctoral assistantship in 1979–80, at the U.S. Geological Survey in Denver, Colorado, where he worked on late Cambrian-early Ordovician conodonts and stratigraphy of the Bear River Range, Utah–Idaho. Finally, he was a postdoctoral fellow

at the University of Toronto, Canada, in 1980–81 where he worked in Jasper National Park, southern Canadian Rocky Mountains, Alberta. He was hired as a senior scientist-paleontology by the New York State Geological Survey in 1981, and became the seventh New York State paleontologist, following such notables as Don Fisher, Rudolf Ruddeman, and James Hall. In 1996, he was promoted from senior scientist to the only principal scientist in the New York State Geological Survey. He lives in Albany, New York, with Jeanne Finley.

Ed Landing has edited 10 professional volumes and been an author of 99 peer-reviewed professional articles. He received the 1990 Best Paper Award from the Paleontological Society. He was also recognized in the State of New York Legislative Assembly (number 479) for establishing the Precambrian-Cambrian boundary global stratotype. Ed Landing has had continuous funding for his research from the National Science Foundation for some 20 years. He is a voting member and co-vice chair of the Cambrian Subcommission of the International Stratigraphic Commission and a corresponding member of the Ordovician Subcommission.

Lehmann, Inge
(1888–1993)
Danish
Geophysicist

Inge Lehmann was not only the first true woman geophysicist, she was also the first Danish geophysicist, among other firsts. She made fundamental contributions to geophysics and to the understanding of the Earth's structure. She ensured her place in the history of geophysics with her 1936 paper simply titled, "P'" (P prime). This study suggested a new discontinuity in the seismic structure of the Earth, now known as the Lehmann Discontinuity. This discontinuity was based on the diffraction of seismic waves and marks to boundary between the solid inner core and liquid outer core of the Earth. Before that the inner core was not known; now it is a basic piece in the architecture of the Earth. However, Lehmann did not stop at the Earth's core. She subsequently began studying body-wave amplitudes and travel times in the upper mantle and became an expert there as well. She proposed a 220-km discontinuity (also named after her) based upon seismic velocities and would figure prominently in plate tectonic models and more recent seismic tomography.

Because Inge Lehmann was a real organizer in establishing seismic networks, she was invited to become significantly involved in nuclear test monitoring. She helped design and implement the Worldwide Standardized Seismographic Network, which was extensively used during the cold war. She played a pivotal role in both the monitoring and interpretation of the seismic expression of these nuclear tests. Standards were established to estimate timing, location, and strength of the tests as well as test-ban treaties. Her contributions were therefore not only to the Earth sciences, but the public as well.

Inge Lehmann was born on May 13, 1888, at Osterboro by the Lakes in Copenhagen, Denmark, where she grew up. Her father, Alfred Lehmann, was a professor of psychology at the University of Copenhagen. Inge Lehman attended an enlightened coeducational school that was run by Hannah Adler, an aunt of Niels Bohr. She entered the University of Copenhagen in 1907 to study mathematics and passed the first part of the examination in 1910. She was admitted to Newham College in Cambridge University, England, where she spent one year before dropping out of school. Instead of college, she worked as an actuary for the next six years. She reentered the University of Copenhagen in 1918, and graduated with a master of science degree in mathematics in 1920. By 1923, Lehmann was an assistant to the professor of actuarial science at the University of Copenhagen, but in 1925 she switched to assisting in setting up the first seismic networks in Denmark and then Greenland. In 1928, she earned a second master of science degree in geodesy and was appointed as the chief of the seismological department in the newly established Royal Danish Geodetic Institute. Lehmann remained in this position until her retirement in 1953. She described herself as the "only Danish seismologist." In 1952, she was first invited to be a visiting scientist at the Lamont-Doherty Geological Observatory. She would visit several other times in 1957–1958, 1960, 1962–1964 and 1968. She was also a visiting scientist at the Dominion Observatory in Ottawa, Canada (1954, 1957, 1965, 1968), and the University of California at Berkeley (1952, 1954, 1965, 1968), as well as the California Institute of Technology. Inge Lehmann died in 1993 at 105 years old.

Inge Lehmann was an author of numerous scientific articles in both Danish and English. Several of these are benchmarks on the deep structure of the Earth as well as the travel of seismic waves. She published with some of the most notable geophysicists ever, including Sir Harold Jeffreys, BENO GUTENBERG, FRANK PRESS, and W. MAURICE EWING, among others. In recognition of her vast contributions to geophysics and the understanding of the internal structure of the Earth, Lehmann received numerous honors and awards. She was awarded honorary doctorates from Columbia University and the University of Copenhagen. She received the Gold Medal from the Royal Danish Academy of Sciences and Letters, the William Bowie Medal from the American Geophysical Union, the Emil Weichert Medal from the Deutsche Geophysikalische Gesellschaft, the Medal of the Seismological Society of America (first woman), two Tagea Brandt Awards from Denmark, the Harry Oscar Wood Award in seismology, and she was named an Honorary Fellow of the Royal Society of Edinburgh, Scotland. The American Geophysical Union named an award in her honor.

Lehmann was also very active in service to the profession. She was one of the founding members of the Danish Geophysical Society of which she was chair in 1941 and 1944. She was the first president of the European Seismological Federation (1950) and the vice president of the International Association of Seismology and Physics of the Earth's Interior (1963–1967), as well as a member of the executive committee.

⊠ Liebermann, Robert C.
(1942–)
American
Geophysicist, Mineral Physicist

Geophysicists who study the travel of seismic waves through the mantle of the Earth have found that the velocities of these waves vary with depth and location. Seismic imaging of the mantle indicates that processes appear to vary spatially as well. Considering that we cannot visit or sample these places, how can we tell what these variations represent? The answer is to study physics and chemistry of minerals under simulated high pressure and temperature conditions. Robert Liebermann is among the foremost experts in this field. He and two colleagues established a Center for High Pressure (CHiPR) beginning in 1985 with facilities that are the first of their kind in North America. There are two large volume high-pressure devices. One of these generates 2,000 tons of force and the other is used for synthesizing materials at pressures to 250 kilobars and temperatures to 2,500°C. The second is a cubic anvil that is installed at the National Synchotron Light Sources, a high-energy source of X rays at the Brookhaven National Laboratory. The research that Liebermann conducts includes the transformation of minerals in response to pressure. Just as graphite converts to diamond with pressure, so do other minerals change their form, becoming denser with each change. The new higher-pressure minerals and

their transformations are studied both with X rays and electron microscopy to determine the actual mechanisms by which the transformation takes place. The velocity of seismic waves of the new minerals and even the transformation states is then measured and compiled as a function of pressure and temperature even under these extreme conditions. Liebermann then compares the observed velocities of seismic waves in the deep Earth with this experimental information. By this method, he can interpret the mineralogy and composition of the Earth as a function of depth and discuss the implications for large-scale dynamic processes of the Earth's interior. The research involves the study of numerous different mineral species typically with colleagues who investigate the same systems under less extreme conditions. This research even has application to superconductive materials, as many are formed or exist under these extreme conditions. Although he has done more typical geophysical research as well, Liebermann's real contributions to the science are more uniquely to provide geophysicists with guidelines to more accurately interpret the results of their observations on mantle processes. In this important function, he is among the top few authorities in the world. Several of the papers by Liebermann on these topics include "Elastic Properties of Minerals, Mineral Physics and Geophysics" and "Material Sciences of the Earth's Deep Interior."

Robert Liebermann was born on February 6, 1942, in Ellwood City, California. He attended the California Institute of Technology where he was an Alfred P. Sloan Scholar. He earned a bachelor of science degree in geophysics in 1964. He attended graduate school at the Lamont-Doherty Geological Observatory of Columbia University, New York, where he earned a Ph.D. in geophysics in 1969. He was a research scientist at Lamont-Doherty Observatory in 1969–1970 and a research fellow at California Institute of Technology in 1970. He served as a research fellow and a senior research

Robert Liebermann (right) at the Center for High Pressure Research at State University of New York at Stony Brook with Dr. Gwanmesia (center) and a summer scholar *(Courtesy of Robert Liebermann)*

fellow at the Australian National University in Sydney from 1970 to 1976, where he worked with ALFRED E. RINGWOOD. He joined the faculty at the State University of New York at Stony Brook in 1976, and he remains there today. Liebermann was named a distinguished service professor in 1996, a title which he still holds. He has been chair of the department from 1997 to 2001 and associate director of the Mineral Physics Institute at Stony Brook since 1993. He now serves as interim dean of the College of Arts and Sciences at Stony Brook. He has been a visiting professor numerous times throughout his career at schools like University of Tokyo, Japan, University of Paris VII, France, Australian National University, and others. Liebermann is married with three children.

Robert Liebermann has led a very productive career. He is an author of some 134 articles in in-

ternational journals and professional volumes. Many of these papers are benchmarks in mineral physics. As a result of the respect he has earned in the profession through his research contributions, Liebermann has been invited into several of the most prestigious groups in the world. He has been a member of several committees for the National Academy of Sciences and the National Research Council. He represented the United States in several U.S.–Japan High Pressure Research Seminars. He has served on numerous committees and panels for the American Geophysical Union, the U.S. Geodynamics Committee, and the International Association of Seismology and Physics of the Earth's Interior. Liebermann has also served numerous editorial roles including associate editor of *Geophysical Research Letters* and associate editor, section editor, and senior editor of *Journal of Geophysical Research.*

⊠ Lindsley, Donald H.

(1934–   )
American
Petrologist, Geochemist

Sometimes fate has a way of bringing together the best people into just the right field at just the right time. This is the case with Donald Lindsley. His primary interest through a good part of his career was in the mafic and accessory minerals in a basalt and especially in Fe-Ti oxides. His paper, "Experimental Study of Oxide Minerals," is a classic. He started his career just before the study of basalt was to become critical in several areas of geology. During this time, through experimental and theoretical thermodynamic research, he developed many of the tools that would be used to study basalts. These tools include methods to tell the temperature at which basalts erupted, in addition to how oxidized they were and how much of a role silica played in the chemistry and mineralogy. His work on Fe-Ti oxides was especially popular, but work on pyroxenes and olivine was also timely. He developed unique graphical methods to solve the quantitative results. A computer program and a series of papers to apply these methods are entitled, "Equilibria among Fe-Ti Oxides, Pyroxenes, Olivine and Quartz." This work came just as the deep-ocean drilling project became a regular and popular research area, as well as when the first lunar samples were being brought to Earth. Both of these sources contain overwhelming amounts of basalt. At that time, the State University of New York decided to build a world-class geology department at the Stony Brook, Long Island, campus where only a small one had existed before. This new powerhouse department included many analytical experts, but even more important, many experts on mafic rocks and pyroxenes, in particular. This group developed some of the most fundamental concepts about lunar petrology. Don Lindsley was one of the founding members of this department and participated in the heyday of its success.

After the Apollo missions ended and the furor over lunar processes subsided, Don Lindsley returned to his roots and began a more traditional field petrology research career. He began a study on the Laramie Anorthosite Complex while on a visit to the University of Wyoming in 1979 purely by accident. He was stranded in a snowstorm and just happened to look at the rocks for curiosity's sake. It took a while for that research to reach a self-sustaining level because the profession regarded Lindsley as an experimentalist. He persevered and the research has now lasted 21 years, most of which have been quite successful. Much of this research involves the field-testing and application of the earlier experimental and theoretical work.

Donald Lindsley was born on May 22, 1934, in Princeton, New Jersey. He spent several years in Charlottesville, Virginia, where much of his interest in minerals came from spending time on the University of Virginia campus. He attended Princeton University, New Jersey, where he earned a bachelor of arts degree in geology with high honors in 1956. He earned a doctoral degree in geology with a minor in physical chemistry from the Johns Hopkins University in 1961. His first professional position was as petrologist at the Geophysical Laboratory of the Carnegie Institution of Washington from 1962 to 1970. During this time he was also a visiting associate professor at California Institute of Technology (1969). In 1970, he joined the faculty at the State University of New York at Stony Brook, where he still remains as a professor of petrology. He was awarded the title of distinguished professor in 2001. He was visiting professor several times during his tenure at Stony Brook. Don Lindsley is married and has three children.

Donald Lindsley has been leading a very productive career. He is an author of numerous articles in international journals and professional volumes, many of which are benchmark studies on oxide thermodynamics, among others. He has received several honors and awards for his work

Portrait of Don Lindsley *(Courtesy of Mary Lou Stewart)*

including having a new Ba chromium-titanate mineral named after him as "lindsleyite" in 1983. He received the Roebling Medal from the Mineralogical Society of America in 1996.

Don Lindsley's service to the profession has been outstanding. He was especially active in the Mineralogical Society of America, where he was councilor in 1972 to 1973, vice president in 1981, and president in 1982. He served as associate editor for *American Mineralogist* from 1984 to 1987, and was the editor for the *Reviews in Mineralogy* (volume 25) on oxides. He was equally active in the Geochemical Society, where he served on the program committee from 1969 to 1971, as councilor from 1977 to 1980, as vice president from 1989 to 1991, and as president from 1991 to 1993.

⊠ **Logan, Sir William Edmond**
(1798–1875)
Canadian
Economic Geologist

William E. Logan is generally considered the "father of Canadian geology." His story is straight out of a Horatio Alger novel in that he never received formal training in geology. Instead, he spent more than 20 years in the area of accounting and copper smelter management and just picked up geology as an outside interest. The beginning of his work and research in geology resulted from his interest in understanding economic minerals, such as coal and ores, as an established middle-aged man. He began to gain recognition among Britain's top geologists for his mapping abilities of coal seams and ore deposits. It was for this reason that Logan was selected to establish the Geological Survey of Canada in 1842. Logan headed the survey for nearly 25 years and his geologic achievements and his work as an administrator and financial planner enabled him to gain national and international attention.

William Logan's contributions to geology were significant. He presented a paper to the Geological Society of London in 1840 on Welsh coal seams. He noted the invariable presence of underclays, containing plant remains in the footwall of each seam. With this paper, he established the formation of coal in situ (in place) from the metamorphosis of organic deposits. He confirmed this theory by surveying coal deposits in North America as well.

His other main contribution was to the understanding of the geology of Canada. In his role as director of the Geological Survey of Canada, by 1850, Logan had mapped the Gaspé Peninsula, parts of the Eastern Townships south of the St. Lawrence River, and the area around Lakes Ontario, Erie, Huron, and Superior. He defined three major geological units: folded Paleozoic rocks of Gaspé and the Eastern Townships (Eastern Division), flat-lying Paleozoic rocks extending west from Montreal to Lake Huron (Western Division), and Primitive (Precambrian) rocks to the north (Northern Division). Many of his observations were based on their structure and stratigraphy, on the absence of coal, and on the potential for economic ores in the Lake Superior region. Logan was able to produce maps and reports, and classify the economic minerals and deposits of Canada with all the information he collected. He continued to extend his work throughout his career.

William Edmond Logan was born in Montreal, Canada, on April 20, 1798. At age 16, William was sent to Edinburgh, Scotland, to attend high school. Upon graduation at age 18, he decided to remain and attend Edinburgh University. He studied logic, chemistry, and mathematics. He remained at the university for just one year before joining his uncle in London, England, to work in his accounting house. He remained at this position from 1817–1831, where he excelled at business and management. In 1831, Logan traveled to Wales to work in another of his uncle's businesses, the Forest Copper Works. He worked with practical miners and surveyors in the coalfields. He became joint manager of the Copper Works in 1833 and began establishing quite a reputation as a knowledgeable and field-oriented geologist in Wales. Logan was founder of the Swansea Philosophical and Literary Institution and honorary curator of its geological section. He was elected in 1837 to the Geological Society of London. During this time he also exhibited his geological maps of the Glamorganshire coalfield. Logan's maps caught the attention of Henry De la Beche, who was director of the Ordinance Geological Survey of Great Britain. The maps were so detailed and flawless that they were published without being revised.

In 1838, his uncle passed away and Logan resigned from his job at the Copper Works. He left Swansea in Southern Wales and pursued his geological interests. He visited the coalfields in Pennsylvania and Nova Scotia to continue verify-

ing his observations in the Welsh coals. It was at this time that he applied for a job with the Ordinance Geological Survey in 1841. Canadian governor general Sir Charles Bagot offered Logan the position, and he accepted it on April 14, 1842. This was the birth of the Geological Survey of Canada.

Logan had a hard task ahead of him due to the political conditions of the times. He needed to convince the government and the public of the usefulness of the geological survey. To everyone's surprise, he did so by using practicality and educating the people. Another mammoth task he had ahead of him was the geological mapping of the huge Canadian colony. The geology of Canada was unknown and there were no topographic maps at that time. Logan's home base was located in Montreal, where he put together maps, prepared reports, researched and examined fossil and mineral specimens, and dealt with the government politics that went along with his position. There was never enough funding, so Logan often used his own money to continue his research and to keep his office running. Logan constantly lobbied legislators and submitted a geological survey bill that was passed in 1845. It provided £2,000 annually for the next five years.

By 1863, Logan and his associates had enough mapping completed to release his famous report, "Geological Survey of Canada: Report of Progress from its Commencement to 1863." This report was supposed to be his swan song, but he remained director of the Geological Survey until 1869 when he retired. Even then, he returned as acting director on more than one occasion. In 1874, he returned to Wales to live with his sister, still intent on more geological work. However, his health failed and he died on June 22, 1875.

The contributions of William Logan to the geology of Canada cannot be overemphasized. Against overwhelming odds, he almost single-handedly established the framework for all Canadian geology that was to follow. In recognition of these vast contributions, William Logan received numerous prestigious honors and awards. He was knighted by Queen Victoria in 1856 and named to the French Legion of Honor by Emperor Napoleon III in 1855. He was elected a Fellow of the Royal Society of London in 1851. He received an honorary doctorate from McGill University, Canada, and the University of Lennoxville. Among 22 medals bestowed upon him, he re-ceived the Wollaston Medal of the Geological Society of London 1856. The most prestigious medal of the Geological Association of Canada is named in his honor. Among all of the fossils and a mineral (weloganite), he even has two mountains named after him; one is the highest point in Canada.

M

Mahood, Gail A.
(1951–   )
American
Petrologist, Geochemist

The most common and well-known divergent boundary is a mid-ocean ridge. Indeed, in the life of a margin where plates pull apart, they mark it for probably 99 percent of the time. They produce only basalt volcanism. However, most of these boundaries begin on continental crust. These early stages of rifting are very complex in many ways, including volcanism. Not only is there basalt volcanism, as in the later stages, but there also may be rhyolite or silicic volcanism as well. Gail Mahood is one of the foremost experts on this rare and complex volcanic activity. Rhyolites are blends of magmas resulting from partial crystallization of the basalts and from partial melting of continental crust through which the basalt travels. Because the composition of the basalt and lower crust vary, the composition of the rhyolite is also highly variable. This variability occurs on the major element scale, but even more so on the minor and trace element scale. Because rhyolites are the last bits of liquid in a crystallizing basalt or the first bits of melt from heated crust, they contain only the lowest temperature minerals, as well as all of the elements that do not fit into standard minerals known as incompatible elements. Mahood uses these elements and isotopes, both stable and radioactive, to unravel the processes of formation of rhyolitic magmas. These elements can not only help to determine the source of the magma but also the igneous processes that occur during the ascent and eruption of these rocks. One of her more notable papers is "Synextensional Magmatism in the Basin and Range Province; A Study from the Eastern Great Basin."

Gail Mahood is also interested in the mechanics of the volcanic eruptions. Rhyolite tends to be very sticky and viscous. Coupled with a potentially and commonly high water content that explodes to steam as the eruption occurs, these volcanoes can be very dangerous. They produce enormous amounts of ash in highly explosive eruptions with high eruption columns that result in widespread ash deposits. The biggest volcanic eruptions in North America in recent geologic history were from rhyolite volcanoes. Mahood studies several of these famous deposits like the Bishop Tuff but also several others from the southwestern United States, especially in California and Colorado. Her paper, "Correlation of Ash Flow Tuffs," is seminal reading. She has also done extensive research on rhyolite volcanoes from northwestern Mexico and in Alaska and Italy. Additionally Mahood has investigated the ancient magma chambers that fed the rhyolite volcanoes in the Sierra Nevada, California, and the Andes of Chile.

Mahood has applied her geologic research to Mesoamerican archaeology, where she also has a strong interest. Obsidian (volcanic glass) is commonly produced in these rhyolite volcanoes. Mahood provides constraints on migration and trading routes by determining the source of obsidian, which was used extensively in tools and weapons by the native peoples of the southwestern United States, Mexico, and Central America.

Gail Mahood was born on June 27, 1951, in Oakland, California. She enrolled at the University of California at Berkeley in 1969, but left college after one tumultuous year. She enrolled at the College of Marin in Kentfield, California, in 1971, but returned to the University of California at Berkeley the next year. She earned a bachelor of arts degree in geology in 1974, and remained for graduate studies. She earned a Ph.D in 1980 as an advisee of IAN S. CARMICHAEL. During her graduate career she was a National Science Foundation Fellow as well as a geologist for the Proyecto Arqueologico Copan in Honduras. In 1979, Mahood joined the faculty at Stanford University, California, where she remains today. She served as chair of the department from 1996 to 1999. She was a visiting professor at Pennsylvania State University at State College in 1983 and at the University of Michigan at Ann Arbor in 1989 under a National Science Foundation program. Gail Mahood is married to Wes Hildreth, a well-known volcanologist/petrologist with the U.S. Geological Survey.

Gail Mahood is amid a productive career. She is an author of some 45 articles in international journals and professional volumes. Many of these are seminal papers on igneous processes, especially volcanic, and appear in high-profile journals like *Science* and *Nature*. Gail Mahood has performed outstanding service to the profession. In addition to serving on numerous committees and panels, she served as councilor for the Geological Society of America in 1996 to 1999. She also served on several committees for the American Geophysical Union and panels for the National Science Foundation, National Research Council, and she even testified before the U.S. Congress on the role of the U.S. Geological Survey. Mahood served in numerous editorial roles as well. She was the founding editor of *Proceedings in Volcanology* at the International Association of Volcanology and Chemistry of the Earth's Interior, the executive editor and a member of the editorial board for the *Bulletin of Volcanology,* and an associate editor for the *Geological Society of America Bulletin.* She additionally served as an external reviewer for the geology department at Dalhousie University in Halifax, Canada.

⊠ **Marshak, Stephen**
(1955–)
American
Structural Geologist

After the assembly of continental masses is complete, the stable craton of the continental interior is considered to be an inactive area. Considering that the assembly of most cratons somewhat resembles a war, it is not surprising that the postwar might be overlooked. Stephen Marshak has brought attention to and shown the importance of the tectonic development of continental interiors. He studies folds and faults that form along discrete zones of great length within areas of no other activity. They commonly exert very strong controls on depositional patterns and basin development but they also form the locus of earthquake activity, including current seismicity. Considering that the New Madrid seismic zone, which produced the most powerful earthquakes in the continental United States of Richter magnitude 8.4 and 8.8 in 1811–1812, is one such zone of activity, this study is very important. They also serve as sensitive indicators of changing states of stress inside of continents. The problem is that such zones are barely ever exposed. Their documentation is mostly done using geophysical logs showing the distribution of rock types from oil and gas wells and from seismic reflection profiles, which are like a sonogram of the subsurface. Most of this research has been conducted on the interior United States.

Prior to this new interest, Stephen Marshak was primarily known for his work on fold and thrust belts in the active part of continents. These thick layers of sedimentary rocks are slid like a rug on a floor ahead of an advancing continental collision. They took on great interest in the 1970s because they are areas of significant oil and gas accumulation. Marshak not only performs field studies to identify the detailed processes in these belts, he also attempts analog experiments using sand and clay models and computer simulations. Most of this work concentrates on the three-dimensional structures and processes. He is especially well known for his work on the development of map view curves in these otherwise ruler-straight belts. The field areas for this research include the Appalachians, the Rocky Mountains, and Australia.

To make sure that he leaves no stone unturned in terms of the depth and intensity of features, Stephen Marshak also studies the development of heavily deformed and metamorphosed rocks in Brazil. He has concentrated on

the development of dome and keel structures which apparently result from collapse of the orogen subsequent to collision but his interest extends to many aspects of Precambrian tectonic processes. Much of his Brazilian research is focussed on the São Francisco craton where he has not only studied structural geology, but also the tectonic events and their ages.

Stephen Marshak was born on March 4, 1955, in Rochester, New York. He is the son of the famous physicist Robert E. Marshak, who worked on the Manhattan Project during World War II and later was a faculty member at the University of Rochester, New York. Stephen Marshak attended Cornell University, New York, where he earned a bachelor of arts degree in geology with distinction in 1976. He earned a master of science degree in geology from the University of Arizona in Tucson in 1979. Later, he attended Columbia University, New York, where he earned a Ph.D. in geology in 1983. Upon graduation, Marshak joined the faculty at the University of Illinois at Urbana-Champaign, where he remains today. He has served as department head from 1999 to present. He has also held visiting appointments at the Federal University of Ouro Prêto, Brazil; Lamont-Doherty Geological Observatory, New York; the University of Adelaide, Australia; and the University of Leicester, England, during his tenure at the University of Illinois. Steven Marshak is married to Kathryn Marshak; they have two children.

Steven Marshak is amid a very productive career. He is an author of some 55 scientific articles in international journals, professional volumes, and governmental reports. Many of these papers are seminal reading on the processes of structural geology as well as regional distribution of structural styles. He is also an author or editor of five textbooks and professional volumes. Several of the textbooks, including *Basic Methods of Structural Geology, Earth Structure: An Introduction to Structural Geology and Tectonics,* and *Earth, Portrait of a Planet,* are widely adopted and regarded as of the

Steve Marshak at an exposure of deformed rocks in Brazil *(Courtesy of Steve Marshak)*

highest quality. Marshak has received numerous honors and awards in recognition of his contributions, both in terms of research and teaching. He received the Stilwell Medal from the Australian Journal of Earth Science and several teaching awards from the University of Illinois, including the Luckman Undergraduate Distinguished Teaching Award, the Prokasy Award, the Amoco Foundation Award, and the University of Illinois Course Development Award.

Marshak has also performed extensive service to the profession. He has served on several panels for the National Science Foundation, numerous functions for the Geological Society of America especially in the Division of Structural Geology and Tectonics, and as a board member for both the International Basement Tectonics Association and the Illinois State Surveys. He was an associate editor for *Geology*, a member of the editorial board for *Tectonophysics* and coeditor for the *American Geological Institute Glossary of Geology*. Marshak was also an external evaluator for the University of São Paulo, Brazil.

⊠ Matthews, Drummond H.
(1931–1997)
British
Geophysicist

Drummond Matthews helped revolutionize our understanding of the lithosphere of the Earth. He was trained as a geologist but applied that background to geophysical applications with outstanding success. His most famous research was that done with Fred Vine. They were involved in an expedition to the Indian Ocean, where they performed a detailed geophysical survey over part of the crest of the Carlsberg Ridge in the northwestern part of the ocean. They discovered a large area of reversely magnetized ocean crust. This was the first major discovery of direct evidence of HARRY H. HESS's concept of Sea-Floor Spreading. What ensued was a foot race between an American team

led by ALLAN V. COX and the Vine and Matthews team to fully document the evidence and processes of mid-ocean ridges. Both teams contributed significantly to this, the most important support for the plate tectonic paradigm.

As a reward for this groundbreaking research, Drummond Matthews was put in charge of the marine geophysics group at Cambridge University, United Kingdom. This group participated in some 72 cruises and expeditions. They did research on plate boundaries in several key locations like Azores Gibraltar Ridge, the Gulf of Oman, the eastern Mediterranean and Aegean Seas, and in the North Sea, where they were the first to document crustal thinning. This research program led to many new insights into the development and character of ocean crust.

The third major contribution that Drummond Matthews made to geology resulted from a visiting professorship to Cornell University in New York. There he learned about the U.S. Continental Reflection Profiling Program (COCORP) that was developed by JACK E. OLIVER and colleagues. He modeled and developed funding for a British version of the program, which he called the British Institutions Reflection Profiling Syndicate (BIRPS). In this program, Matthews applied his knowledge of marine geophysical methods to shallow shelf areas with great success. These data revealed lower crustal and upper mantle structures that were previously unknown. The most significant of this work was the Deep Reflections of the Upper Mantle (affectionately called DRUM after Matthews) work in northern Scotland. This research would be instrumental in applications to exploration for North Sea oil reserves as well as its academic implications.

Drummond Matthews was born in 1931 in Porlock, Somerset, England. He attended the Bryanston School in Dorset for his primary and secondary education. He performed his national service with the Royal Navy before attending Cambridge University, where he earned a bachelor of science degree in geology in 1955. For the next two years (1955–1957), Matthews was with the Falkland Islands Dependencies Survey, later to become the British Antarctic Survey, mainly on the South Orkney Islands. He returned to Cambridge University to complete a Ph.D. in marine geophysics in 1960. Upon graduation, he was appointed as a senior assistant in research in the Department of Geodesy and Geophysics at Cambridge University, as well as a research fellow at King's College. From 1961 to 1963, Matthews oversaw the British contribution to the International Indian Ocean Expedition. It was there that he met his first graduate student, Fred Vine. Drummond Matthews married Rachel McMullen in 1963; they would have two children. In 1966, Matthews became assistant director of research at Cambridge University, as well as the head of the marine geophysics group. In 1982, he became the first scientific director of the British Institutions Reflection Profiling Syndicate (BIRPS). Drummond Matthews married his second wife, Sandie Adam, in 1987. Matthews suffered a heart attack in 1989, and took an early retirement in 1990 to professor emeritus as a result of his failing health. Drummond Matthews died in 1997 from complications from diabetes.

Drummond Matthews was an author of more than 200 scientific publications. Several of these papers are true classics on plate tectonics, marine geophysics, and lithospheric structure. In recognition of these contributions to the science, Drummond Matthews received numerous honors and awards. He was a Fellow of the Royal Society of London. He received the Bigsby Medal and the Wollaston Fund from the Geological Society of London, the G.P. Woollard Award from the Geological Society of America, the Chapman Medal from the Royal Astronomical Society, the Arthur L. Day Prize from the U.S. National Academy of Sciences, the Charles Chree Medal from the Institute of Physics (Great Britain), the Hughes Medal from the Royal Society of London, and the Balzan Prize from the Balzan Foundation.

⊠   **McBride, Earle F.**
(1932–   )
American
Sedimentologist

One of the main contributions of geology to society is in providing sources of energy and the main source of energy is petroleum. Petroleum reserves are contained primarily in clastic sedimentary rocks and especially sandstone. The understanding of the processes of sands deposition and the compaction and cementation processes that turn sand into sandstone is of utmost importance to petroleum exploration. Earle McBride has contributed significantly to our understanding of these processes and both indirectly and directly to success in our search for petroleum reserves.

Probably most significant of this work are McBride's contributions to the understanding of the process of diagenesis. His paper "Diagenesis of Sandstone and Shale—Applications to Exploration for Hydrocarbons" is a good example. As sand beds are progressively buried under additional sediments, they mechanically compact, thus reducing the pore space. In addition, groundwater that percolates between the sand grains may become enriched in dissolved minerals, which can then precipitate in the pore spaces. Typical precipitated minerals include quartz and calcite, depending upon the chemistry of the groundwater. The precipitated minerals are called cement and serve to further reduce the pore space and also to isolate the pores from each other thus reducing fluid and gas flow. This process significantly reduces the ability of sandstone to act as a good reservoir for hydrocarbons. His paper "Quartz Cement in Sandstone: A Review" summarizes much of this work. On the other hand, the groundwater chemistry may change with depth and redissolve the cement re-creating porosity and permeability (as described in "Secondary Porosity—Importance in Sandstone Reservoirs in Texas," for example). These ideas are now the mainstay of academic research on diagenesis as well as the constraints on hydrocarbon formation.

Other areas of research for McBride include the origin of horizontal laminations in sandstone and the origin of bedded chert. He is also the foremost expert on the sedimentary geology of the Paleozoic rocks of the Marathon Basin in west Texas. This basin is one of the larger hydrocarbon provinces in the United States and the last area of marine deposition in North America during the building of the supercontinent Pangea.

Earle McBride was born on May 25, 1932, in Moline, Illinois. He grew up in the Quad Cities area of Illinois around the Mississippi River. He attended Augustana College in Rock Island, Illinois, and earned a bachelor of arts degree in chemistry and geology in 1954. He then attended the University of Missouri at Columbia where he earned a master of arts degree in geology in 1956. He earned his doctorate from the Johns Hopkins University in sedimentary geology in 1960 as an advisee of FRANCIS J. PETTIJOHN. While at Johns Hopkins, McBride worked for Shell Oil Company in Texas over two summers as an exploration geologist. In 1959, he joined the faculty at University of Texas at Austin, where he spent his entire career. From 1982 to 1990, McBride occupied the Wilton E. Scott centennial professorship. In 1990, he was named to the J. Nalle Gregory centennial chair in sedimentary geology, which he still holds today. He served as chairman of the department from 1980 to 1985. During that time of high oil prices, there were an amazing 850 undergraduate majors and 250 graduate students. In 1975, McBride served as the Merrill Haas distinguished professor at the University of Kansas. In 1977, he was a NATO visiting professor at the University of Perugia, Italy, and in 1995, he was a Fulbright Fellow to Egypt. McBride wound up in jail on three occasions, once because he needed a place to sleep, once because he forgot his visa while entering Japan, and once because he was carrying a tear gas pen in England.

Earle McBride has had an outstanding career, as demonstrated by his more than 200 articles in international journals and professional volumes, as well as five books and manuals. Several of these papers are seminal works on sedimentology, diagenesis, and petroleum. His research has been well recognized by the profession in terms of honors and awards. While still in graduate school, he was awarded the A.P. Green Fellowship. He received three best paper awards, two from the *Journal of Sedimentary Petrology* and one from the Gulf Coast Chapter of Society of Economic Paleontologists and Mineralogists (SEPM). He was also awarded the Francis Pettijohn Medal from SEPM and two Houston Oil and Minerals Distinguished Faculty Awards from the University of Texas.

Earle McBride has been of great service to the profession. He served on numerous committees and held numerous offices for SEPM, including councilor (1967–1968), secretary-treasurer (1972–1974) and president (1979–1980). He served as vice president of the International Association of Sedimentologists from 1994 to 1998 and numerous functions for the American Association of Petroleum Geologists. He was an associate editor for *Journal of Sedimentary Petrology, Journal of Sedimentary Geology,* and *Journal of Scientific Exploration.* He was also on the editorial board for *Giornale di Geologia* in Italy and the *Egyptian Journal of Petroleum Geology.*

⊠ **McKenzie, Dan P.**
(1942–)
British
Geophysicist

Dan McKenzie has uniquely applied techniques of geophysics and mathematical modeling to a variety of geological problems. Working with SIR EDWARD C. BULLARD, McKenzie first studied the decay of elevated heat flow away from the mid-ocean ridges. The nature of this thermal decay is a direct reflection of the nature of the convection cells in the mantle that drive the lithospheric plates. This work formed the nucleus of the many contributions that Dan McKenzie would make to geology over the years. The first contribution was to take the qualitative description of J. TUZO WILSON on the nature of transform faults and turn it into a quantitative analysis. He showed that the slip vectors of earthquakes along the boundary between the North American and Pacific plates intersect at a point. This became a first step in quantitative modeling of plate movement.

This initial research was expanded in several directions. Dan McKenzie became involved in a major effort to model the generation of mafic magmas both at mid ocean ridges and at mantle plumes. He used both mechanical and geochemical modeling to show the nature of the partial melting of the mantle as well as the melt movement and extraction in these areas (see "Extraction of Magma from the Crust and Mantle," for example). This research traces the degree of partial melting by studying isotopes, the filtering of the magma through the partially melted rock, and the nature of the volcanoes produced in extensional settings as discussed in "Mantle Reservoirs and Ocean Island Basalts." His work on extension was also expanded to the continental crust where he attempted to devise the same sort of simple mechanical models for rifting. He studied the mechanics of normal faulting, the propagation of rift zones and mechanical models for the thinning of the crust and lithosphere (as in his "Geometry of Propagating Rifts," for example). Naturally, he also studied magma genesis in these areas.

In another offshoot of this work on extension, McKenzie developed the basic principles of sedimentary basin modeling. As the surface subsides, the resulting low will fill with sediments. There are a number of factors which control the nature of this sedimentation ranging from the rate of subsidence and active faulting, to compaction and even the heat flow and geochemical evolution of the sediments. His paper, "The Stretching Model for Sedimentary Basins," is a good example

of this research. The work has direct implications for the generation and maturation of hydrocarbon deposits. These methods have been strongly employed by the petroleum industry in exploration worldwide.

All of this modeling of the nature of the outer elastic layer of the Earth and the mantle processes that drive it have been applied to the outer skin of both Venus and Mars in yet another offshoot of Dan McKenzie's research. He modeled the dynamics of movements and magma generation on Venus using gravity and topographic information to show a very active system similar to that on Earth. These are the most comprehensive models for Venusian tectonic activity. The data for Mars has only recently been of high enough quality to perform similar analysis, but even with the primitive data he was able to show that some of the largest convective plumes in the Martian mantle are associated with huge dike swarms and large canyons in many cases. In addition to these extraterrestrial field areas, McKenzie has also studied many specific areas on Earth, including Iceland, Hawaii, the Aegean Sea, the South China Sea, South Africa, and the Zagros Mountains of Iran.

Dan McKenzie was born on February 21, 1942, in London, England, where he spent his youth. He began his college career at Westminster College in London before transferring to Kings College at Cambridge University, England. He earned a bachelor of science degree in physics with a minor in geology in 1963. He remained at Cambridge University for his graduate studies and earned a Ph.D. in geophysics in 1966. McKenzie was an advisee of Sir Edward Bullard. Upon graduation, he accepted a position at Cambridge University, first as assistant in research (1969–1975) and then as assistant director of research (1975–1979), reader in tectonics (1979–1984) and professor in 1984. In 1996, McKenzie was named to a Royal Society research professorship at Cambridge University, where he remains today. Dan McKenzie married Indira Margaret Misra in

1971; they have one child. McKenzie enjoys gardening for recreation.

Dan McKenzie's productive career includes authorship of some 150 scientific articles in international journals and professional volumes. Several of these are seminal papers on the mechanics of rifting, generation of mafic magmas, plate mechanics, and sedimentary basin analysis, and appear in the most prestigious of journals. In recognition of his many contributions to geology and geophysics, Dan McKenzie has been honored with numerous honors and awards. He is a Fellow of the Royal Society of London, a member of the U.S. National Academy of Sciences, and a Fellow of the Indian National Sciences Academy. He received an honorary doctorate from the University of Chicago and an honorary master of arts from Cambridge University. He received the Royal Medal from the Royal Society of London, the Gold Medal from the Royal Astronomical Society, the Arthur L. Day Medal from the Geological Society of America, the Geology and Geophysics Prize from the Balzan Foundation (Switzerland and Italy) and the Japan Prize from the Science and Technology Foundation of Japan.

⊠ McNally, Karen C.
(1940–)
American
Geophysicist

Just like a hunter stalking big game, Karen McNally hunts down earthquakes in the seismically active region of western North America. She was named "The Earthquake Trapper" by the *Los Angeles Times* for her instrumental "capture" of the major 1978 Oaxaca, Mexico, earthquake which had a magnitude of 7.8 on the Richter Scale. She received similar notoriety for her "capture" of the 1989 Loma Prieta, California (World Series), earthquake of magnitude 7.1. In her role as the director of the Charles M. Richter Seismological Observatory at the University of California at Santa Cruz, she serves a dual role as researcher and as monitor and adviser to the people of southern California.

In addition to "capturing" earthquakes, Karen McNally performs research on the source mechanisms of seismic activity. She studies the focal mechanisms of earthquakes as well as foreshocks and aftershocks to create a full picture of the episodic movement on the fault that created a particular seismic event. She also studies swarms of microearthquakes that occur intermittently along active faults. Using these data, McNally models the stress buildup and release within large-scale plate tectonic processes both in California and Mexico as well as within the Central American subduction zone. This work has implications for plate motions in the area, the partitioning of strain and the constantly readjusting manner in which plates move. Papers by McNally on this work include "Non-Uniform Seismic Slip Rates Along the Middle American Trench" and "Seismic Gaps in Time and Space," among others.

McNally evaluates the earthquake potential of California for a number of purposes. She evaluates predictive capabilities for large earthquakes, works to educate the public on earthquake hazards, and acts as an adviser and consultant for zoning and building codes. Her paper, "Terms for Expressing Earthquake Potential, Prediction and Probability," with colleagues is an example of this work. In this regard, she served on the board of directors for the Southern California Earthquake Center, as well as several other committees. McNally has also served in several capacities for the California governor's Office of Emergency Services mainly with regard to earthquake preparedness. She has even been involved with Nuclear Test Ban Treaty verification for the U.S. Congress, among many other groups. Indeed, by virtue of her experience and expertise, Karen McNally is one of the top few seismologists in the evaluation of major earthquakes in California. She is among the first few to be called after an event.

Karen McNally was born in 1940 in Clovis, California. She grew up on a ranch. She married young and had two daughters, but sought a career and was divorced in 1966. She first attended Fresno State College, but soon moved to the University of California in Berkeley, where she earned a bachelor of arts degree in geophysics in 1971. She remained at the University of California at Berkeley for graduate studies and earned a master of arts degree in 1973 and a Ph.D. in 1976, both in geophysics. Upon graduation, McNally became a research fellow and a senior research fellow at California Institute of Technology in Pasadena. During that period, she was also a seismologist and consultant for Woodward-Clyde Environmental Consultants until 1985. In 1982, McNally joined the faculty at the University of California at Santa Cruz, where she remains today.

Karen McNally is amid a productive career. She is an author of some 63 scientific articles in international journals, professional volumes, and governmental reports. Several of these papers are seminal reading for the seismotectonics of California, Mexico, and Central America, as well as stress distributions and earthquake sources at plate margins. In recognition of her contributions to seismology, service to the public, and education, Karen McNally has received several honors and awards. She is a member of the American Academy of Arts and Sciences. She was named an Ernest C. Watson lecturer at California Institute of Technology, a Richtmeyer lecturer by the American Physical Society, and a Sunoco lecturer by the National Science Teachers Association. She was given the first Award of Excellence from the Clovis Unified School District, California, and named a Cientifico Collaborator by the Universidad Nacional of Costa Rica.

The amount of service to the profession and public besides that already mentioned that Karen McNally has performed is extensive. She served on the board of directors for the Seismological Society of America, as well as the Incorporated Re-

search Institutions for Seismology (IRIS), numerous committees for the National Academy of Sciences, NASA, the U.S. Geological Survey, and the American Association for the Advancement of Sciences. She was on the evaluating committee for the Massachusetts Institute of Technology. She also served in several editorial roles including associate editor for *Reviews of Geophysics* and *Geophysical Research Letters.*

⊠ McNutt, Marcia
(1952–   )
American
Geophysicist

Marcia McNutt is one of those people who lets nothing stand in her path toward success. When her undergraduate adviser told her that physics was not a good major for women, she switched advisers. To help her better participate in her chosen field of marine geophysical surveying and analysis, she did not just learn on the job, she completed a U.S. Navy Underwater Demolition Team and SEAL Team training course in addition to scuba diving courses. It is with this determination that McNutt has achieved a meteoric rise to outstanding success and power in the Earth sciences. Her principal research interests involve the use of marine geophysical data to study the physical properties and tectonic processes of the Earth beneath the ocean floor. Some of her more notable studies include the history of volcanism in French Polynesia and how it relates to large-scale convection in the Earth's mantle. Her paper, "The Superswell and Mantle Dynamics beneath the South Pacific," describes the surface features that reflect deep mantle processes that are not otherwise explained by the plate tectonic paradigm. In this same vein, she has been reinvestigating the relation between mantle plumes and hot spots. It appears that old ideas on the source regions, and therefore the processes, may require revising. McNutt has also conducted studies on the mantle processes in the continental breakup in the western United States, and the uplift of the Tibet plateau. This paper, "Mapping the Descent of Indian and Eurasian Plates Beneath the Tibetan Plateau from Gravity Anomalies," shed new light on the Himalayan collision. She has participated in 14 major oceanographic expeditions around the world from Woods Hole Oceanographic Institution, Oregon State University, and Lamont-Doherty Geological Observatory and served as chief scientist on seven expeditions. McNutt's research is not only field based, but also theoretical.

More recently, Marcia McNutt has taken on the role of spokesperson for ocean exploration for the 21st century. She met with President Bill Clinton and has taken on leadership roles in this initiative. She is also president and chief executive officer of the Monterey Bay Aquarium Research Institute, California, a research laboratory that was created to develop and exploit new technology for the exploration of oceans. It is funded by the Packard Foundation. The main objective of the research institute is designing and building innovative underwater vehicles and fixed position sensor packages for increasing the sampling of the ocean and the creatures that inhabit it.

Marcia McNutt was born on February 19, 1952, in Minneapolis, Minnesota, where she spent her youth. In 1970, she graduated high school as valedictorian from the Northrop Collegiate School, Minnesota (later renamed The Blake School), with awards in mathematics, science, and French, and perfect scores on her SAT. She earned a bachelor of arts degree in physics in 1973, summa cum laude and Phi Beta Kappa, from Colorado College in Colorado Springs. She studied geophysics only briefly while at Colorado College, but it piqued her interest enough to spur her on in that direction. She completed her graduate studies at the Scripps Institution of Oceanography in La Jolla, California, where she earned a Ph.D. in Earth sciences in 1978 as a National Science

Foundation Graduate Fellow. Upon graduation, McNutt accepted a position at the University of Minnesota, Twin Cities, where she had a brief appointment as a sabbatical replacement. She then accepted a position as geophysicist at the Tectonophysics branch in the Office of Earthquake Studies of the U.S. Geological Survey in Menlo Park, California, in 1979. McNutt resigned from the U.S. Geological Society to join the faculty at the Massachusetts Institute of Technology, Cambridge, in 1982. She spent the next 15 years at MIT and was appointed the Griswold Professor of geophysics in 1991. From 1995 to 1997, she served as the director of the Joint Program in Oceanography and Applied Ocean Science and Engineering, a cooperative graduate program between the Massachusetts Institute of Technology and the Woods Hole Oceanographic Institute, Massachusetts. In 1997, McNutt made the surprising move to become the president and chief executive officer of the Monterey Bay Aquarium Research Institute in Moss Landing, California, where she remains today. McNutt was a Mary Ingraham Bunting Fellow at Radcliffe College in 1985 and a National Science Foundation visiting woman professor at Lamont-Doherty Geological Observatory in 1989. She has been named to the faculty at both the University of California at Santa Cruz and at Stanford University.

Marcia McNutt's husband died suddenly in the early 1990s, leaving her a widow with three small children. She has since remarried.

Marcia McNutt is amid a very productive career, having been an author of some 80 scientific articles in international journals and professional volumes. Several of these papers are seminal studies on mantle convection and plumes, as well as other areas of marine geophysics. In recognition of her professional contributions, Marcia McNutt has received several honors and awards during her career. She received an honorary doctoral degree from Colorado College. She also received the Macelwane Award from the American Geophysical Union, the Sanctuary Reflections Award from the Monterey Bay National Marine Sanctuary, two Editor's Citations from the *Journal of Geophysical Research,* the Outstanding Alumni Award from The Blake School, and the MIT School of Science Graduate Teaching Prize.

McNutt has performed an outstanding amount of service to the profession and the public. She was president of the American Geophysical Union (2000–2002), chair of the President's Panel on Ocean Exploration and a member of the National Medal of Science Committee. In addition, she has served on numerous committees and panels for the National Science Foundation, the National Research Council, NASA, the Ocean Drilling Program, and the National Academy of Sciences.

⊠   **McSween, Harry (Hap) Y., Jr.**
(1945–)
American
Planetary Geologist, Petrologist

One of the wildest discoveries in the past 20 years in all of science, much less geoscience, is that we may have meteorites here on Earth that originated from Mars. If that is not enough, they may contain evidence of primitive life on Mars. Harry McSween has found himself firmly enmeshed in this controversy. He is one of the main proponents of the Martian origin of the meteorites based upon convincing geochemical and isotopic evidence, which he helped collect. However, he is the voice of caution in the evidence for Martian life within a group of scientists who have been quick to support this high-profile topic. Ironically, his reluctance has resulted in McSween being interviewed in newspapers and on radio and television more often than many of his colleagues. Many in the profession are comforted to have a scientist like McSween who awaits compelling data before accepting new theories.

Harry McSween has been studying meteorites since graduate school and is considered one of

their foremost authorities. Meteorites are leftover material from the formation of the solar system so they have profound implications for the processes of this formation as well as a starting point for its evolution. He has received continuous NASA funding for his research and has written some of the seminal works on meteorites. His interest in extraterrestrial rocks led him to an interest in Mars. He was a member of many NASA teams studying Mars for years that culminated in his pivotal role on the Mars Pathfinder spacecraft mission of 1997. He has continued his role on further missions like the Mars Global Surveyor, which involved mapping the Martian surface from orbit and the Mars Odyssey spacecraft, as well as the ongoing design of Mars Exploration rovers. When any news is released on Martian discoveries, they are sure to have been made at least in part by Harry McSween.

If this extraterrestrial interest is not enough, McSween also conducts petrologic and geochemical research on plutonic igneous rocks and metamorphic rocks of the southeastern United States. Although not as high profile as the extraterrestrial work, it is still well respected in the profession as being of impeccable quality and a contribution to the field.

Harry "Hap" McSween was born on September 29, 1945, in Charlotte, North Carolina. He attended the Citadel in Charleston, South Carolina, and earned a bachelor of science degree in chemistry in 1967 as a Daniel Scholar. He then attended the University of Georgia in Athens as a NASA graduate fellow and earned a master of science degree in geology in 1969. From 1969 to 1974, McSween was a pilot and an officer in the United States Air Force in Vietnam. He earned his Ph.D. in geology from Harvard University, Massachusetts, in 1977. Upon graduation, he joined the faculty at the University of Tennessee at Knoxville where he has remained ever since. During that time, he served as acting associate dean for research and development in 1985 to 1987 and the department head from 1987 to 1997. He was named distinguished professor of science in 1998, a title which he still holds. Also during that time, he was a guest or visiting scientist at the Japanese National Institute of Polar Research, University of Hawaii at Manoa, and the California Institute of Technology. Harry McSween is married to Susan P. McSween, and they have one child.

Hap McSween has been very productive throughout his career. He is an author of more than 100 articles in international journals and professional volumes. More recently, he has begun writing popular books as well. He wrote the three books *Stardust to Planets: A Geological Tour of the Solar System, Fanfare for Earth: The Origin of our Planet and Life,* and *Meteorites and their Parent Planets* to spread his enthusiasm for science to the general public. His research has been well received by the profession, as shown by his numerous honors and awards. He received the Nininger Award for Meteorite Studies (1977), a National Science Foundation Antarctic Service Medal (1982), the Bradley Prize from the Geological Society of Washington (1985) and two NASA Group Achievement Awards (1983 and 1998). From the University of Tennessee, he received the Chancellor's Award for Research and Creative Achievement (1990) and a Senior Research Award (1998), in addition to several teaching awards. The state of South Carolina gave him several awards, including the LeConte Medallion of the South Carolina Science Council (1999), the Order of the Silver Crescent Award from the governor of South Carolina (2001), and he was inducted as the 21st member of the South Carolina Hall of Science and Technology (1999).

Hap McSween has performed extraordinary service to the profession. He served on 14 NASA teams and panels of critical importance, on several of which he was chief. He also served on several committees for the National Research Council. For the Meteoritical Society, he served as president (1995–1996), vice president (1993–1994), secretary, and councilor. For the Geological Society of America he was chair and vice chair of the Planetary Geology Division and chair and vice chair of

Hap McSween at a petrographic microscope with a video monitor attachment showing a microscopic view of a Martian meteorite *(Courtesy of H. McSween Jr.)*

the southeastern section, among many other committees. He was an associate editor for international journals *Icarus, Meteoritics, Geochimica et Cosmochimica Acta,* and the *Proceedings of the 10th Lunar and Planetary Science Conference.* He has also given numerous distinguished lectures and keynote addresses.

⊠ **Means, Winthrop D.**
(1933–)
American
Structural Geologist

Structural geology was largely a descriptive science with only minor quantitative aspects into the

1960s. Then there was a revolution in the field to infuse the theories and applications of engineering and material science. This infusion of quantitative analysis of deformation led to a mass reexamination of previously observed features within this new context. DAVID T. GRIGGS, JOHN G. RAMSAY, and Winthrop Means were the pioneers in this revolution. Means's book, *Stress and Strain: Basic Concepts of Continuum Mechanics for Geologists,* is still a classic even though it was published in 1976 and never revised. This book and the theory it presents form a unique bridge between geology and mechanical (and civil) engineering. Several synchronous and succeeding papers by Means also address this bridging of the fields.

The second main interest of Win Means is the laboratory modeling of microstructures using an analog deformation apparatus that he and Janos Urai invented. The apparatus consists of a standard compression device and a standard shear device, but it is applied to a microscope slide-sized sample of various rock analog materials, such as an organic compound called octochlorophane. This material looks like an aggregate of mineral grains through the microscope. If deformed in a compression vice or press or shear device, the model rock deforms beautifully in a ductile or plastic manner. Grains deform like putty rather than cracking and breaking. Means has simulated many ductile microstructures from real rocks in the device, except they can be observed forming in real time using this method in contrast to the before or after pictures that are the only kind available with real rocks. The processes observed during the simulation have confirmed, modified, and/or revolutionized our understanding of ductile microtextures as well as those from a deforming crystal-melt mixture. His paper, "Synkinematic Microscopy of Transparent Polycrystals," summarizes this concept.

Win Means received continuous National Science Foundation grants since 1976 for this work and great interest from the geologic community. He has participated in setting up the ap-

paratuses in at least 10 other universities. He even taught short courses on the device and observations. The new Earth Science building at the Smithsonian Institution of Washington, D.C., includes some of Means's experimental work.

Winthrop D. Means was born on February 7, 1933, in Brooklyn, New York. He attended Harvard University and earned a bachelor of arts degree in geology in 1955. He then moved to the University of California at Berkeley where he earned a Ph.D. in structural geology in 1960. His first faculty position was at the University of Otago, New Zealand, where he served as a lecturer from 1960 to 1964. He was a postdoctoral fellow at the Australian National University from 1964 to 1965. In 1965, Means joined the faculty at the State University of New York at Albany, where he remained until after his retirement in 1998 as a professor emeritus, his current position. He served as department chairman twice during his tenure at Albany.

Win Means has been very productive throughout his career. He published 46 articles in top international professional journals, 24 of which are single authored. He also was an author of two of the premier textbooks on structural geology, the book mentioned above and *An Outline of Structural Geology* with Bruce Hobbs and Paul Williams, also published in 1976. They are still considered required reading for all students of the field. Means has received numerous honors and awards throughout his career, including a Senior Fulbright Fellowship, CSIRO Geomechanics, in Melbourne, Australia, in 1992. He was also awarded the Career Contribution Award from the Structure and Tectonics Division of the Geological Society of America in 1996. He received the Excellence in Research Award from State University of New York at Albany in 1997, and the Bruce Hobbs Medal from the Geological Society of Australia in 1999.

Win Means served on funding panels for the National Science Foundation in 1983 to 1986 and for Gilbert Fellowships from the U.S. Geological Survey from 1991 to 1994. He was on editorial boards for both *Tectonophysics* (1980 to 1999) and *Journal of Structural Geology* (1983 to 1999). He also served on evaluation committees for Utrecht University, the Netherlands, in 1987 and Northern Arizona University in 1990.

⊠   **Melosh, H. J.**
(1947–   )
American
Planetary Scientist

H. J. Melosh is in the rare position of having started his career as a promising physicist before settling into Earth sciences. He studied elementary particle physics and published studies on quark interactions, which contained the "Melosh Transformation." His work was cited literally thousands of times and sparked a revolution in the field. However, his real passion was in Earth sciences, especially planetary science. He brought his exceptional quantitative capability to the field and added a new dimension to the Earth sciences. Most of his work centers on the mathematical treatment of impact cratering and on the origin and physical interactions of celestial bodies. However, he has also modeled some observable geologic processes that have led to a new understanding.

His most direct contribution to observable processes is a phenomenon that Melosh calls "acoustic fluidization" as explained in his paper, "Acoustic Fluidization: A New Geologic Process," among others. Basically, concentrated sound waves can make a material act like a fluid in terms of physical properties. This unified explanation elucidates the collapse of impact craters, the emplacement of long traveled landslides, and the physics of earthquakes. His interest in earthquakes in this project also allowed him to identify a new phenomenon of long, slow ground motions that occur after major earthquakes. These motions have strong implications for

global positioning system measurements. Other terrestrial work largely involved the application of finite element modeling to subduction zones, mantle and lower lithospheric structure, and the development of normal faults. Many processes and situations are now better understood as a result of these models.

However, Melosh is best known for his study of impact craters. He is the author of the seminal book entitled *Impact Cratering*. In this research, he developed new theories on the deformation caused by meteoroid impacts called "ring tectonics." He developed methods to predict how comets and meteorites break apart in the planetary atmospheres. He modeled ejecta processes and distributions from impacts. He has also been involved with the possibility of meteorites being generated from one planet and impacting another: his paper "Ejection of Rock Fragments from Planetary Bodies" is an example. Because these are such prominent topics, Melosh has achieved a prominent position in the field. Well before it happened, he predicted the breakup of the Shoemaker-Levy 9 comet that struck Jupiter. He has been greatly involved in the Martian meteorite controversy and the possibility of Martian life being preserved in these fragments. He has better defined the mechanics of the Chixulub, Mexico, impact, which is proposed to have caused the extinction of the dinosaurs 65 million years ago. These topics, among many others, are now better understood as the result of Melosh applying his quantitative constraints to geological probabilities.

H. J. Melosh was born on June 23, 1947, in Paterson, New Jersey. He attended Princeton University, New Jersey, and graduated magna cum laude with a bachelor of science degree in physics in 1969. He attended the California Institute of Technology for graduate studies and earned a Ph.D. in physics and geology in 1972. During his last year of graduate school, he was a visiting scientist at CERN in Geneva, Switzerland. In 1972 to 1973, he accepted a postdoctoral position of re-

Jay Melosh examines a rock sample in his laboratory at the University of Arizona (*Courtesy of H. J. Melosh*)

search associate at the Enrico Fermi Institute at the University of Chicago, Illinois. He joined the faculty at the California Institute of Technology in 1973 and worked his way up from instructor to associate professor. In 1979, he accepted a position at the State University of New York at Stony Brook, but moved to the University of Arizona in Tucson in 1982, where he remains today. In 2000, Melosh held the Halbouty Distinguished Visiting Chair at Texas A & M University in College Station.

H. J. Melosh has led a very productive career. He is the author of some 139 articles in international journals and professional volumes. An amazing 22 of these papers appeared in the high-profile journals *Nature* and *Science*. Many of these papers are benchmark studies that have been regularly cited in the literature. He also wrote one book and edited two volumes. He has received several awards from the profession in recognition of his contributions. While still in school, he was

elected a member of Phi Beta Kappa, and received a National Science Foundation Fellowship and the Best Secretary Prize from the International Summer School of Theoretical Physics. He received the American Geophysical Union Editor's Citation for Excellence in Refereeing (1989), a Guggenheim Fellowship (1996–1997), the Barringer Medal of the Meteoritical Society (1999), and the Gilbert Medal of the Geological Society of America (2001).

H. J. Melosh has performed service to the profession. He was a member of the International Lithosphere Program, and a scientific observer for the European Science Foundation. He also served on a NASA working group. He was editor for *Reviews of Geophysics,* associate editor for *Journal of Geophysical Research,* and a member of the editorial board for *Annual Reviews of Earth and Planetary Science.*

⊠   **Menard, H. William**
(1920–1986)
American
Oceanographer, Plate Tectonics

William Menard is one of the pioneers of the plate tectonic revolution. He and a small group of revolutionaries from England and the United States finally put the smoking-gun evidence to ALFRED WEGENER's hypothesis of moving continents. Through a series of research cruises, this group proved seafloor spreading. They looked at the bathymetry of mid-ocean ridges as well as the mirroring of magnetic stripes on the ocean floor on either side of them. These data were assembled to conclusively show that ocean crust was being produced at mid-ocean ridges only to move away in an opposing conveyor beltlike geometry. It was probably the most exciting time in geology and Menard was prominent in the group. It was Menard who recognized the fracture zones that offset these ridges, which would later be called transform faults by J. TUZO WILSON. Several major studies by Menard on these topics include "Marine Geology of the Pacific" and "Topography of the Deep Sea Floor," among others. But Menard was not only interested in one aspect of geology. He was interested in sedimentology and worked on shallow- to deep-ocean sediment flows called turbidites. His ocean voyages would allow him to have firsthand observations of them. Practical application of this work would be to help locate underwater cables to avoid turbidite prone areas. An example of this work is his seminal paper, "Sediment Movement in Relation to Current Velocity." He also discovered manganese nodules on the deep ocean floor and investigated the feasibility of mining them. He was involved in geostatistics of oil drilling, concluding that a good deal of luck went into successful exploration. He calculated that drilling a simple grid pattern in oil-producing areas would yield about as much success as the standard method of picking locations.

One of Menard's real strengths was to meld science with social science. Four of his six books considered the philosophy of a scientific expedition and how the thought processes worked in such an endeavor. He showed the struggles with problems in science as well as all of the outside pressures on a project and how they would alter the outcome. He also described what aspects of a career brought on notoriety in science. Through these works he was considered an expert on the history and development of science as a profession, though the work was all accomplished indirectly. As opposed to being trained as a philosopher, he merely described his own experiences in his research projects. Menard was considered a true scholar with versatility in numerous aspects of geology and philosophy.

William Menard was born on December 10, 1920, in Fresno, California, and attended Los Angeles High School. He attended the California Institute of Technology and earned a bachelor of science degree in geology in 1942. He enlisted in the U.S. Navy soon after the bombing of Pearl

Harbor and served as a photointerpreter and staff intelligence officer in the South Pacific theater. He returned to Cal Tech after the war to earn a master of science degree in geology. He married Gifford Merrill of New York in 1946. They had three children. Menard earned his Ph.D. from Harvard University, Massachusetts, in 1949, though he did his research at Woods Hole Oceanographic Institution. Upon graduation he accepted a position in the Sea Floor Studies Section of the Oceanographic Branch of the U.S. Navy Electronics Laboratory in San Diego, California. In 1955, Menard moved to nearby Scripps Institution of Oceanography of the University of California at San Diego. He remained there until his death on February 9, 1986, except for two leaves of absence. From 1965–1966, he served as technical adviser in the Office of Science and Technology under President Lyndon Johnson. In 1978 to 1981, he served as the 10th director of the U.S. Geological Survey. In addition to being a great scientist, Menard was also a history and English literature buff.

William Menard published six books as well as more than 100 articles in international journals, governmental reports, and professional vol-

William Menard (right) in his administrative role as the chief of the U.S. Geological Survey with Dr. Oskar Adam in 1979 *(Courtesy of the U.S. Geological Survey)*

umes. Several of these papers are definitive studies on plate tectonics as well as geophysics, sedimentology, and geomorphology. Menard was well recognized by the geologic community for his contributions to the science in terms of honors and awards. He was a member of the National Academy of Sciences, the American Academy of Arts and Sciences, and the California Academy of Sciences. He received the Penrose Medal from the Geological Society of America and the Bowie Medal from the American Geophysical Union. Menard also performed significant service to the profession as well as to government as already mentioned. He served on numerous committees and panels for the Geological Society of America and the American Geophysical Union.

⌗   Miller, Kenneth G.
(1956–   )
American
Micropaleontologist

There are minute organisms called foraminifera that float or swim in the near surface waters of our oceans in great abundance. These foraminifera are what many filtering marine animals feed upon. The foraminifera also die in great abundance, sink in the water, and make up a good portion of the sediment on the ocean floor. Foraminifera are rapidly evolving animals that are very sensitive to environmental changes especially with regard to climate changes. Kenneth Miller is one of the foremost experts on the evolution and biostratigraphy of foraminifera. Clearly, because the foraminifera-rich sediments lie deep under the oceans, fieldwork and sampling is no trivial task. Research vessels must travel out to sea and piston-coring devices are driven into the sediment where continuous cores of sample are taken. Miller has made many such cruises to obtain his research material. As early as 1980, he was on a deep-ocean cruise aboard the R/V *Knorr.* He has sampled sediments aboard the *Glomar Challenger,* the *Conrad,* the *Atlantis II,* the *Maurice Ewing,* and the *Cape Hatteras,* among others.

After the cruise, the samples are analyzed in the laboratory. The cores of sediment are painstakingly studied to determine the type and abundance of foraminifera with depth, which is equivalent to time. With colleagues, he determines stable isotope abundances in the foraminifera and sediment. With these data, Miller can then determine the paleoceanography and paleoecology of the ocean basin. He interprets rises and falls of sea level, changes in climate, and catastrophic events like the extraterrestrial impact at the Cretaceous-Tertiary boundary, among others, in addition to evolutionary changes in the foraminifera. Many of the larger questions regarding global changes involve the comparison of the character of certain key strata from core to core and even with strata that can be seen in the Atlantic Coastal Plain onshore in New Jersey. These comparisons permit a more three-dimensional view of a given succession of strata. Such a view allows better environmental interpretations. By performing the same sort of detailed analysis on many sections of the Atlantic Ocean and Coastal Plain stratigraphy, a detailed history is being constructed. Not only will this work better define the stratigraphy in terms of sediment succession, fossil succession, and even isotopic and magnetic succession through the participation of colleagues, it will also lead to a much better understanding of the processes involved in passive margin development. Because Miller's results are so directly reflective of climate changes, those researchers trying to model climate variability and response to predict future changes are especially interested in his findings. Several of Miller's papers on this research include, "Long-Term and Short-Term Cenozoic Sea Level Estimates" and "Control of North Atlantic Deep Water Circulation by the Greenland-Scotland Ridge."

Kenneth Miller was born on June 28, 1956, in Camden, New Jersey. He attended Rutgers University in New Brunswick, New Jersey, where

Portrait of Kenneth Miller *(Courtesy of K. Miller)*

Kenneth Miller has been extremely productive in his still young career. He is an author of some 97 articles in international journals, professional volumes, and governmental reports. He is also an author or editor of four books. Many of his articles are seminal works on the Cenozoic stratigraphy of the Atlantic Ocean and their reflection of climate changes and controls. He is also extremely successful at grant funding, having obtained some $10 million in federal grants.

Miller has performed outstanding service to the profession. He served as vice president and as a member of the board of directors of the Cushman Foundation for Foraminiferal Research. He served on numerous committees and panels for the Geological Society of America, the American Geophysical Union, the Ocean Drilling Program (ODP), and the Joint Oceanographic Institute for Deep Exploration Sampling (JOIDES). His editorial service is also exemplary. He served as editor for *Paleoceanography* and associate editor for *Palaios, Journal of Sedimentary Research, Paleoceanography, Geological Society of America Bulletin,* and *Marine Micropaleontology.*

he earned a bachelor of arts degree in geology with highest honors and distinction in 1978. He completed his graduate studies in the Massachusetts Institute of Technology/Woods Hole Oceanographic Institution Joint Program in Oceanography and earned a Ph.D. in 1982 with a Phillips Petroleum graduate fellowship. He remained as a postdoctoral fellow for one year. He was a postdoctoral research fellow and an associate research scientist at the Lamont-Doherty Geological Observatory of Columbia University, New York, from 1983 to 1988, where he was named an Arco scholar. In 1988, Miller joined the faculty at his alma mater at Rutgers University in New Brunswick, where he remains as of 2002. He was named a prestigious Professor II in 2000 and served as department chair. Kenneth Miller is married to Karen Clark Miller and they have four children.

⊠ Molnar, Peter
(1943–)
American
Geophysicist, Tectonics

The type of continent-continent collision on Earth between northern India and southeastern Asia built not only the Himalayas, the highest mountain range in the world, but also the Tibetan Plateau and mountain ranges to the north, among others. Peter Molnar is one of the foremost experts on this area. His goal has been to bring a simple understanding to the dynamic processes of mountain building and the mechanics of the continental collision using this great example. In 1975, he and Paul Tapponnier showed that the active deformation of Eurasia from the Himalayas to Lake Baikal in Siberia, some 1,800 miles to the north, results from this collision. This work is summarized in the paper, "Active Deformation of Asia from Kinematics to Dynamics," among others. The continuing penetration of the Indian continent into Asia forced large areas of the Asian landmass to move laterally out of the way of this collision. Bulk movement of continental mass out of the way of the colliding plate was accommodated by literally thousands of miles of strike-slip movement. This new class of strike-slip fault rivals the plate bounding transform faults like the San Andreas fault in California. The bulk movement is called "escape" or "extrusion" tectonics and is analogous to smashing a fist into a handful of clay, forcing it to squirt out of the sides. In Asia, the bulk movement was to the east mostly along the Red River and Altyn

Peter Molnar at a rock exposure at the Alpine Fault in New Zealand in 1984 *(Courtesy of Peter Molnar)*

Tagh faults, forcing Indochina to squeeze and rotate into its present configuration.

This groundbreaking research is not the only important contribution to geology that Peter Molnar has made. He also studies the constraints on how high mountains can grow. With several researchers including Philip England (Oxford University), Molnar has placed constraints on crustal thickening in the Tibetan Plateau as the result of the compression of collision. The rheological properties of the crust only allow mountains to grow to a certain height. Even in a highly compressional environment the mountains will collapse by normal faulting and crustal extension.

Since 1990, Peter Molnar has been studying the impact of tectonic processes on climate change. He proposed that mantle processes beneath the Tibetan Plateau might have triggered rapid uplift, which in turn strengthened the Indian monsoon approximately 8 million years ago. Mark Cane of Columbia University and Molnar have proposed that the northward movement of Australia and New Guinea closed the Indonesian Seaway and cooled the waters of the Indian Ocean. This cooling affected the aridification of East Africa at the time that humans evolved. This idea is reported in the paper, "Late Cenozoic Closing of the Indonesian Seaway as the Missing Link between the Pacific and East African Aridification." The blockage of warm water in the western Pacific Ocean ended the perpetual El Niño conditions that prevailed before the sporadic El Niño conditions seen today. They attribute the onset of ice ages roughly 3 million years ago to a decrease in heat transport from the tropics. With MAUREEN E. RAYMO, Molnar has been investigating changes in climate as the result of the massive Himalayan mountain building. The two were featured on a well-received NOVA television special.

Peter Molnar was born on August 25, 1943, in Pittsburgh, Pennsylvania. He moved several times during his childhood, attending both Highland High School, Albuquerque, New Mexico

(1958–60) and Summit High School, Summit, New Jersey (1960–61), where he played forward on Summit's champion basketball team. He attended Oberlin College, Ohio, from 1961 to 1965 and earned a bachelor of arts degree in physics and was cocaptain of the lacrosse team. He completed his graduate studies at Columbia University in New York (1965–1970) and earned a Ph.D. in seismology. From 1970 to 1971, he was a research scientist at the Lamont-Doherty Geological Observatory of Columbia University. From 1971 to 1973, Molnar was an assistant research scientist at Scripps Oceanographic Institute, University of California at San Diego. He was also a National Academy of the Sciences exchange scientist with the USSR for four months in 1973–74. Peter Molnar joined the faculty at Massachusetts Institute of Technology (MIT) in 1974. He served as a visiting scientist at numerous institutions during his tenure at MIT, including Oxford University, England, and Institut de Physique du Globe, Paris, France, among others. In 1986, he quit the faculty at MIT and switched to the research staff. In 2001, he joined the faculty at the University of Colorado at Boulder where he is currently a half-time professor. Molnar has one child from his previous marriage to TANYA ATWATER. Peter Molnar married Sara Neustadt in 1986.

Peter Molnar is an author of numerous scientific articles in international journals and professional volumes. Many of these papers are classic studies on the Himalayas and mountain-building processes. He has received numerous honors and awards for these contributions to the field. Molnar had a Higgins Fellowship in graduate school and a Sloan Fellowship during his first four years at MIT. He was a Guggenheim Fellow in 1980, studying at Cambridge University, England. He was a Harold Jeffreys lecturer, Royal Astronomical Society, London, England, in 1996 and a F.A. Vening Menesz lecturer, Utrecht University, Netherlands, in 1999. Molnar also received an Editor's Citation from American Geophysical Union in 2000.

⊠ Montanez, Isabel Patricia
(1960–   )
American
Sedimentologist, Geochemist

A consequence of modern industrialized society is that the amount of the greenhouse gas carbon dioxide in the atmosphere continues to increase at an alarming rate. The debate now rages as to whether or not this increase is resulting in global warming and erratic weather patterns and what it will mean for the future. Isabel Montanez is a carbonate petrologist and geochemist, a large part of whose research addresses these issues. By studying ancient carbonate deposits, especially during times of rapid environmental change, she seeks to put our current predicament into geologic perspective. By using stable isotope geochemistry, mainly on carbon, coupled with field study of these limestone and dolomite deposits, Montanez documents changes in the composition of seawater and corresponding atmospheric gases through time. This research is designed to determine whether the atmosphere-ocean system can recover from our current catastrophic changes or not. If it can recover, how long will it take for natural processes to do so? How will extinction rates of plants and animals be affected by these changes? By looking at how the Earth has responded to major changes in atmospheric-oceanic chemistry in the past, Isabel Montanez seeks to set a baseline against which our current situation may be compared. Then these important questions may be addressed.

The other main area of research for Isabel Montanez is the basin-wide migration of pore fluids during burial in sedimentary basins. These fluids carry dissolved solids and can either deposit minerals in the pore spaces of the rocks through which they flow or dissolve even more material. The deposited minerals form a cement and inhibit further fluid flow. This research has a direct bearing on whether a rock unit will serve as a good hydrocarbon reservoir or a good aquifer for

Portrait of Isabel Montanez *(Courtesy of I.P. Montanez)*

groundwater or not. This research is therefore of great interest to both the petroleum industry and the environmental industry alike. Examples of papers by Isabel Montanez include, "Recrystallization of Dolomite with Time" and "Evolution of the Strontium and Carbon Isotope Composition of Cambrian Oceans."

Isabel P. Montanez was born on March 17, 1960, in Geneva, Switzerland. She completed her primary education in Manchester, England, and her secondary education in Philadelphia and Altoona, Pennsylvania. She attended Bryn Mawr College, Pennsylvania, where she earned a bachelor of arts degree in geology in 1981. Upon graduation, she worked for Everrett and Associates (environmental consultants), Rockville, Mary-

land, as a research assistant from 1981 to 1983. She also worked as a museum technician for the Smithsonian Institution, National Museum of Natural History, from 1982 to 1983. In 1983, Isabel entered graduate school at Virginia Polytechnic Institute and State University where she worked on carbonate petrology and geochemistry under J. Fred Read. She graduated with a Ph.D. in 1990. She accepted her first faculty position at the University of California at Riverside in 1990 and remained there until 1997. She joined the faculty of the University of California at Davis in 1998 where she is currently a full professor. Isabel Montanez is married to David Osleger, another carbonate sedimentologist at University of California at Davis. They have two sons.

Although Isabel Montanez is still in the early stages of her career, she has already made an impact on the profession. She has published 31 papers in professional journals and volumes. She has earned numerous honors and awards for this research. She received a National Science Foundation–Ford Foundation Dissertation Fellowship and an American Geological Institute Minority Program Scholarship in 1988. She won two awards from the University of California, a Chancellor's Research Fellowship in 1993 and an Acknowledgment of Teaching Excellence in 1994. She was awarded a visiting professorship for women from the National Science Foundation in 1996. She won two best paper awards from the Society of Economic Paleontologists and Mineralogists in 1992 and the J. "Cam" Sproule Memorial Award from the American Association of Petroleum Geologists in 1996. Also in 1996, Isabel won the James Lee Wilson Award for Excellence in Sedimentary Geology.

Isabel Montanez's service to the profession is also outstanding. Between 1990 and 1992, Isabel served as vice chair and chair of the National Carbonate Research Group. She served numerous positions for the Society of Economic Paleontologists and Mineralogists including vice president of the Pacific Section in 1993 and 1994, councilor for research activities in 1996 to 1998. She served in several workshops and as a panelist for the National Science Foundation and she was chosen as a distinguished lecturer for the American Association of Petroleum Geologists for 2000 to 2001. Montanez is currently an associate editor for the *Journal of Sedimentary Research* and coeditor for *Sedimentology.*

⊠   **Moores, Eldridge M.**
(1938–)
American
Structural Geologist, Tectonics

The general public may best know Eldridge Moores as a central character in the John McPhee book *Assembling California.* His work in California, Cyprus, and as a mentor are described in detail in addition to some of his personal life. In the Earth science profession, Eldridge Moores is known as the world's foremost expert on ophiolites. Ophiolites are chunks of ocean crust that are exposed on land. Typically, they are plucked away from the ocean floor during collisional events and then transported a great distance on major faults until they wind up in the mountains. They go from the deepest parts of the Earth's surface to the highest in a single event. Traditionally, they are made up of three parts known as Steinman's Trinity: a base of ultramafic rock (peridotite), sheeted dikes cutting layered pillow lava in the middle, and layered deep-ocean sediments on top. However, all of the parts are not always there. The ultramafic base has become the more common element to be studied or at least used for identification. Eldridge Moores recognized the significance of these rocks as records of lost ocean basins. They commonly mark the suture zone between two ancient tectonic plates and therefore are significant finds for plate tectonic reconstructions. Moores has studied ophiolites in Cyprus (the famous Troodos Complex), in Greece (Vourinos complex), Pakistan, western Nevada, and

those that formed during the Precambrian. More recently, he has been studying processes of seafloor spreading as revealed through these ophiolites. Examples of Moores's publications on this work include *Ophiolites and Oceanic Crust, Geotectonic Significance of Ultramafic Rocks,* and *Ancient Sutures Within Continents.*

Moores is also an expert on California geology. He did research on basement rocks of the Sierra Nevada, the Klamath Mountains, and the surrounding areas. These studies include metamorphic petrology, geochronology, structural geology, and regional tectonics. He worked on the Franciscan subduction melange of the northern coast ranges. He also studied neotectonics of this area as well as that of the Great Valley. He led several field conferences in these areas and published some of the seminal papers and edited volumes on California geology.

Eldridge Moores was born on October 13, 1938, in Phoenix, Arizona, where he grew up. He attended the California Institute of Technology, where he graduated with a bachelor of science degree in geology with honors in 1959. He earned a master of arts degree and a Ph.D. from Princeton University, New Jersey, in geology in 1961 and 1963, respectively. His adviser was HARRY H. HESS, the world's preeminent expert on ultramafic rocks at the time. Eldridge Moores was awarded a postdoctoral fellowship at Princeton University for the years of 1963–66 to study the Vourinos ophiolite complex in northern Greece. He joined the faculty at the University of California at Davis in 1966, where he remains as of 2002. Eldridge Moores is married with three children and enjoys languages, music, reading history, and hiking.

Eldridge Moores has been one of the more productive and prominent members of the geologic community. He published some 105 papers in professional journals and volumes as well as nine books and volumes. Several of these papers are in the most prestigious scientific journals like *Nature* and *Science.* Among these books are popular textbooks including *Structural Geology* and *Tectonics,* both with colleague Robert Twiss. He also assembled an often-cited collection of papers on plate tectonics entitled *Shaping of the Earth, Tectonics of the Continents and Oceans,* published through *Scientific American.* In recognition of these contributions to geology, Moores has received numerous honors and awards. He received an honorary doctor of science from the College of Wooster (Ohio) in 1996, the Distinguished Service Award from the Geological Society of America in 1988, and the Geological Association of Canada Medal in 1994. He was elected a Fellow of the California Academy of Sciences in 1996. He was also elected Honorary Fellow of the Geological Society of London in 1997.

His service to the profession and especially to the Geological Society of America is stellar. Eldridge Moores served as president of that group in 1996, vice president in 1995, council member in 1988–91, and member and chair of numerous committees. He also served as editor of the high-profile journal *Geology* from its inception in 1981 until 1988. He served as science editor for another Geological Society of America publication, *GSA Today,* from its inception in 1990 to 1995. He served on many of the most prominent tectonic geology projects. He served on the advisory committee for the Consortium for Continental Reflection Profiling (COCORP) from 1980 to 1987, the panel for the National Science Foundation Continental Scientific Drilling Committee from 1981 to 1983, and he was the chair of the Ocean Drilling Project Tectonics Panel from 1990 to 1993.

Eldridge Moores on a field trip to the Pinnacles Desert in Western Australia *(Courtesy of Eldridge Moores)*

⊠ Morisawa, Marie
(1919–1994)
American
Geomorphologist

Geomorphology is a subdiscipline of geology that, like others, was historically purely descriptive but which has yet to be fully realized in quantitative terms. There was a revolution in geomorphology in the late 1950s and early 1960s to establish quantitative methods and Marie Morisawa was part of that revolution. This work involves measuring the size of features like watersheds, slopes, and stream channels, and to analyze them both in terms of impact on larger systems and statistical analysis of stability. The problem is that the number of schools with that capability is small and in many places geomorphology remains descriptive. Marie Morisawa, therefore, was free to choose any aspect of geomorphology in which to specialize, as there were really no saturated areas so she chose them all. She worked on talus slopes in the Rocky Mountains, the geomorphology of active fault zones (Wasatch Fault, for example), geomorphology and plate tectonics, coastal geomorphology, geologic hazards (earthquakes, volcanoes, landslides), and environmental geomorphology, of which she was one of the founders. Part of this fame was her pioneering work on the 1959 Hebgen Lake earthquake which caused a massive landslide that dammed a river, producing a huge lake. By 1970, this study of environmental geomorphology had turned into enough of a movement to begin an annual symposium at her home school of State University of New York at Binghamton, which had established itself as a center for geomorphology. Morisawa figured prominently in this reputation.

Marie Morisawa is probably best known for her extensive work on her first love of river systems. Her doctoral work on quantitative geomorphology of streams in Pennsylvania was pioneering and set a new standard for stream studies. She studied channel development and stability as well as channel shifting in addition to watersheds. An example of this work is her paper, "Distribution of Streamflow Direction in Drainage Systems." Her interest crossed from the purely scientific to the aesthetic and what might even be called spiritual. Several of her books are about streams.

Marie Morisawa was born on November 2, 1919, in Toledo, Ohio. Her father was Japanese and her mother was American. She attended Hunter College of the City University of New York system and earned a bachelor of science degree in mathematics in 1941. She then obtained a master of arts degree in theology and held several jobs before returning to school to switch careers to geology. She attended the University of Wyoming at Laramie, where she earned a master of science degree in 1952. Morisawa then moved to Columbia University in New York where she earned a Ph.D. in 1960 as an advisee of Arthur Strahler. She was part of a U.S. Office of Naval Research project to develop methods in quantitative geomorphology. During her graduate career, she also served as an instructor at Bryn Mawr College from 1955 to 1959. In 1959, Morisawa joined the faculty at the University of Montana, but moved to the U.S. Geological Survey in Washington, D.C., in 1961. In 1963, she moved back to academia by accepting a position at Antioch College. She then moved to the State University of New York at Binghamton in 1970, where she spent the rest of her career. She was a Fulbright Scholar in India in 1987–1988 and a geologist in residence at Carleton College in Minnesota in 1990. Morisawa retired in 1990 to professor emeritus. She was still active in the department and it was on the drive from her home to her office that she was in a single-car accident that claimed her life on June 10, 1994.

Marie Morisawa led a varied career. She published many scientific articles in international journals, professional volumes, and governmental reports. She is perhaps best known for her eight books, which include a popular 1975 textbook entitled, *Our Geologic Environment* as well as *Streams: Their Dynamics and Morphology* in 1968, and *Geomorphology Laboratory Manual* in 1977. She also wrote the popular book, *Evaluating Riverscapes,* in 1971. In recognition of her research and teaching contributions to geology, Morisawa received several honors and awards. She received the Distinguished Alumna Award from University of Wyoming, and the Outstanding Educator Award from the Association of Women Geoscientists, among others.

Morisawa was of great service to the profession and the public. She served on numerous committees and working groups, as well as councilor for the Geological Society of America, the American Association for the Advancement of Science, and the American Quaternary Association. She was also chair and board member for the Quaternary Geology and Geomorphology Division of the Geological Society of America numerous times. She also served numerous editorial roles including founder and editor in chief of *Geomorphology,* which was begun in 1986. Morisawa also served in many advisory capacities for town planning both around Binghamton and in Fire Island, New York, where she conducted research.

⊠ Morse, John W.
(1946–)
American
Oceanographer

The oceans can be considered as huge chemical systems with a constant group or series of reactions taking place. These reactions chemically connect the oceans with the solid particles of the Earth, be they sediments or bedrock on the one hand and the gases of the atmosphere on the other. The oceans therefore modulate the entire chemical system of the surface of the Earth. The understanding of these complex chemical systems is, therefore, the key to our understanding of our biosphere. John Morse is one of the foremost experts of this complex system. He studies this chemistry by direct measurements, experimental studies in the laboratory, and theoretical thermodynamic studies. His original work was on carbonates and Morse established himself as one of the foremost experts on their chemistry. One reason that carbonates are so important is that they are directly linked to the greenhouse gas carbon dioxide through the ocean water system. They also control carbon cycling in the atmosphere-hydrosphere. His work is therefore of critical concern to climate change modelers. Morse is especially interested in the surface chemistry of these minerals because that is where the chemical reactions take place. He studied the effect of minor and trace elements on these reactions as well as radioactive isotopes. He also proposed better ways to analyze carbonates and their constituent elements.

As his career has progressed, Morse expanded his research interests into sulfur and sulfur compounds in ocean water. Although not nearly as abundant, sulfides are of major concern because

bacteria in the oceans concentrate sulfur. Sulfide minerals and their production also helps to govern the amount of oxygen in seawater. Morse studied the biogeochemistry of sulfides and sulfates and the trace elements involved in their formation similar to the research he had done on carbonates. In both cases, the interaction of these mineral groups and their interaction with ocean sediments has been a major concern. To predict cycling and circulation the ocean modelers utilize these basic studies. An example of a paper involving this research is "The Chemistry of Transuranic Elements in Natural Waters."

John Morse was born on November 11, 1946, in Fort Dodge, Indiana. He attended the Institute of Technology at the University of Minnesota, Twin Cities, and earned a bachelor of science degree in geology in 1969. He completed his graduate studies at Yale University, Connecticut, and earned a master of philosophy and doctor of philosophy in geology in 1971 and 1973, respectively. His adviser was ROBERT BERNER. He joined the faculty at Florida State University in Oceanography in 1973, but moved to the Rosentiel School of Marine and Atmospheric Sciences at the University of Miami, Florida, in 1976. He served as chair of his division in 1981. Morse accepted a position at Texas A & M University in 1981, and was named the Louis and Elizabeth Scherck Professor of Oceanography in 1998, a position he holds today. He served as chair of the chemical oceanography section of his department in 1985–1990 and 1996–1997. He is married to Sandra Morse and they have one daughter. For recreation, Morse enjoys playing acoustic guitar and fishing.

John Morse has led a very productive career. He is an author of some 118 articles in international scientific journals and professional volumes. Many of these articles are benchmark studies of ocean chemistry and thermodynamics. He also wrote *Geochemistry of Sedimentary Carbonates*, which is regarded as the "bible" on the subject. The research projects that Morse has undertaken have received considerable external funding. He has received several honors and awards for his work including a Fulbright Fellowship (1987), a Sigma Xi Distinguished Scientist Award (1998), and a Distinguished Scientist Award from Texas A&M University (2000). He has performed service to the profession, as well. He served on several panels and working groups for the National Science Foundation and the National Research Council as well as one panel for NASA. He served as the editor in chief for *Aquatic Geochemistry* (1993–present) and associate editor for *Marine Chemistry* (1992–present).

⊠ Muehlberger, William R.
(1923–   )
American
Structural Geologist, Tectonics

William Muehlberger is perhaps most renowned for his work with NASA. He was principal investigator for field geology for the *Apollo 16* and *Apollo 17* lunar landings. His group was involved in landing site selection, detailed geologic analysis of the landing site, sampling traverse design, astronaut training, real-time support during the missions, and post-mission data compilation and analysis. He served this position for three years. Muehlberger was also a co-investigator for the NASA Visual Observations Experiment in Skylab and the Apollo-Soyuz missions. He was responsible for global tectonics, giving lectures to astronauts, debriefing afterward, and offering advice on changes during the mission. This program continued with the space shuttle, where he has been teaching geology to newly assigned astronauts and to crews prior to their flight.

However, the work with NASA is just the tip of the iceberg in a long and distinguished career. Muehlberger is a regional geologist whose scale of observation ranges from outcrop (or even microscope) to satellite images. He is a structural geologist by trade who has studied brittle fault zones and fracture systems worldwide, but especially in Texas, Turkey, Israel, New Zealand, and Guatemala. He also studied basement lineaments and correlated geophysical data with them. Muehlberger studied salt domes and the deformation around them in Texas and Louisiana (for example, the paper, "Internal Structures and Mode of Uplift of Texas and Louisiana Salt Domes"). On the other hand, he studied glacial geomorphology in New England. With all of his extensive observations of the character of the Earth's crust, he was the ideal person to help assemble large-scale maps and to help educate NASA astronauts.

William Muehlberger was born on September 26, 1923, in New York, New York, but grew up in Hollywood, California. He entered college at the

William R. Muehlberger in his office at the University of Texas *(Courtesy of W. Muehlberger)*

California Institute of Technology in 1941, but the U.S. Marine Corps sent him to University of California at Berkeley in civil engineering in 1943. He stayed there until 1944, one semester shy of a degree. He returned to the California Institute of Technology in 1946 and earned his bachelor of science degree and his master of science degree there in 1949 and his Ph.D. in 1954. William Muehlberger married Sally J. Provine in 1949; they have two children. He joined the faculty at the University of Texas at Austin in 1954 and remained there until his retirement in 1992 when he became professor emeritus. He was director of the Crustal Studies Laboratory at the University of Texas from 1962 to 1966. He served as chairman of the department from 1966 to 1970. Muehlberger was on leave from the university from 1970 to 1973 and employed by the U.S. Geological Survey for the NASA Apollo field geology investigations for the Apollo program. He held numerous endowed chairs at the University of Texas, including the Fred M. Bullard Professorship for excellence in teaching (1980–82), the Charles E. Yager Professorship (1982–83), the John E. ('Brick') Elliott Centennial Endowed Professorship in Geological Sciences (1983–85), the William Stamps Farish Chair in Geology (1985–89), and the Peter T. Flawn Centennial Chair in Geology (1989–92).

William Muehlberger has led an extremely productive career publishing more than 200 articles in international journals and collected volumes. He is perhaps better known for producing the *Basement Map of the United States,* published by the U.S. Geological Survey in 1966, and the *Tectonic Map of North America* in plate tectonic format in 1992–1996 and published by the American Association of Petroleum Geologists. He has received many honors and awards from the profession for his contributions to the science. He received the First Award from Ohio State University in 1961, the George C. Mattson Award (best paper) from the American Association of Petroleum Geologists in 1965, and the

Medal for Exceptional Scientific Achievement (1973) and the Public Service Medal (1999) both from NASA. He also received the 1998 Best Paper Award from the Structure/Tectonics Division of the Geological Society of America. In 1978, he was given the Houston Oil and Mineral Corporation Faculty Excellence Award and in 1992 he received the Knebel Distinguished Teaching Award.

William Muehlberger also performed much service to the profession. He served on the U.S. Geodynamics Committee, several committees for the National Research Council, as well as for NASA. He served on many committees for Geological Society of America, American Association of Petroleum Geologists, and the American Geophysical Union. He was an associate editor for *Geological Society of America Bulletin* and for *Geophysical Research Letters.*

⊠ **Mukasa, Samuel B.**
(1955–   )
American
Isotope Geochemist

Samuel Mukasa addresses geologic problems using whatever isotopic system necessary to yield the most definitive results. Perhaps his most famous work is that on the Antarctic. By determining the ages of the exposed (and unexposed) rocks there, he proved that Antarctica was an integral part of the formation of the supercontinent Pangea. He also helped define the tectonics of the area since Pangea broke up. However, describing Mukasa's research in terms of a single geographic area is impossible because he has done work all over the world in as diverse a group of rocks as possible. He started his research career on plutonic rocks in coastal Peru, but branched out into ophiolites (oceanic fragments on land) from the Troodos complex in Cyprus and basalt from Brazil. He studied volcanic rocks from the dangerous Taal volcano in the Philippines and mantle fragments from the French Pyrenees and similar fragments from Italy. He continued his interest in ophiolites from southern Chile and the Philippines. He studied basalts from Thailand and metamorphic rocks from South Georgia Island. He even did research on the Great Dike from Zimbabwe, Africa. His work in the United States includes the study of mantle rocks from Arizona and California.

The research he performs on these rocks is an integrated use of trace elements and Pb, Nd, Sr, Hf, and Os isotopes to model the evolution and dynamics of the Earth's mantle. He accomplishes this by studying materials that either come from the mantle, like alpine peridotite massifs and ophiolites or materials that probe the mantle like mafic volcanic and plutonic rocks from continents and island arcs. On the other hand, he has also looked at the evolution of mountain belts and plate reconstructions especially with regard to the building and breakup of supercontinents (Pangea). This work has direct bearing on the evolution of continents.

He began his work by learning Ar/Ar thermochronology from John Sutter, which he applied to metamorphic rocks in New England. Because the Ar/Ar system measures the age at which the temperature of the rocks cools through a "closure temperature," at which point argon gets locked into the mineral structure, the system typically does not measure the age of formation of the rock or mineral. Instead, the age it records is a cooling age, thus it is called thermochronology rather than geochronology. But Mukasa would not settle for being an expert in just one system. He continued his education with GEORGE R. TILTON where he learned uranium-lead and other systems, which yield formational ages of rocks or geochronology to complement his expertise in thermochronology. In addition to using these systems to find the age of rocks, he also uses them as tracers to understand their origin and evolution. In short, Mukasa studies rocks in diverse settings using diverse techniques.

Sam Mukasa at the South Pole during a research expedition *(Courtesy of Samuel Mukasa)*

Samuel Mukasa was born on September 29, 1955. He attended the University of New Hampshire in Durham, where he earned a bachelor of science degree in geology and chemistry in 1977. He earned a master of science degree in geology from the Ohio State University in Columbus in 1980. He continued his graduate education at the University of California at Santa Barbara, where he earned a Ph.D. in geochemistry in 1984 as an advisee of George Tilton. Mukasa was a postdoctoral fellow in isotope geochemistry at the Lamont-Doherty Geological Observatory of Columbia University, New York, in 1984 and 1985. He joined the faculty at the University of Florida in Gainesville in 1985. In 1989, he moved to the University of Michigan in Ann Arbor, where he remains today. Samuel Mukasa married Claudia McQueen, M.D., in 1984 and together they have one son.

Sam Mukasa has had a productive career. He is an author of 38 articles in international journals and collected volumes. These papers are published in high-quality journals and many are in collaboration with some of the top researchers in geology. Mukasa has also been of service to the profession. He is a member of the American Association for the Advancement of Science. He served as member and chair of the National Science Foundation advisory board for the Office of Polar Programs as well as the panel for postdoctoral fellows. He was an associate editor for the *Geological Society of America Bulletin* from 1995 to 1998. He also hosted a Fulbright Scholar (Ivan Haydoutov) from the Bulgarian Academy of Sciences in 1997–1998.

N

Nance, R. Damian
(1951–   )
British
Structural Geologist, Tectonics

One of the main driving forces that kept the search for the plate tectonic theory alive through the years of opposition was the fact that the continents appear to fit together like a jigsaw puzzle. When there was enough data to prove plate tectonics and reconstructions of ancient worlds began, it was found that indeed the plates had once fit together in such a manner. About 250 million years ago, all of the continents were together and formed a single supercontinent called Pangea surrounded by a single ocean called Panthalassa. As research has continued, we have learned that there was a prior supercontinent to Pangea called Rodinia about 750 million years ago and that it broke apart and dispersed much like Pangea has. Damian Nance took this concept one step further by proposing that there are cycles of supercontinent construction and destruction. His theory is that on a regular 500-million-year cycle, all of the continents will join together through a series of collisions to form a single supercontinent and a single superocean. Because having all of the continental mass in one place on the Earth is gravitationally unstable, that single continent will necessarily split apart and the re-

sulting continental fragments will disperse in all directions. Since the Earth is a sphere, eventually they will all reassemble in another place forming another supercontinent and the cycle begins again. His paper, "The Supercontinent Cycle," summarizes this work.

Damian Nance is a classic regional tectonic geologist and as such he utilizes all types of information to construct regional geologic interpretations. These data include structural geology, stratigraphy, paleontology, Ar/Ar thermochronology, and igneous and metamorphic petrology, among others. The geographic region of expertise for Nance is the Avalon terrane, an exotic volcanic-continental fragment that extends from Rhode Island through coastal Massachusetts and Maine and into maritime Canada. He mostly worked on these rocks in Nova Scotia, Canada, which resulted in the publication of many articles and several books. A summary paper on this research is entitled, "Model for the Evolution of the Avalonian-Camodian Belt." However, he has performed research in many areas from Greece to North Carolina.

Nance also was involved in a rather unique teaching experiment. When the North American Free Trade Agreement (NAFTA) was ratified by Congress, there was a small obscure section on education. Nance masterminded a project that involved comparing and contrasting orogenic belts

Damian Nance on the coast in Nova Scotia, Canada *(Courtesy of R. D. Nance)*

from Canada to Mexico to utilize the funds available in the bill. He collaborated with universities in Canada and Mexico as well as his own and designed a field course in which students from all of the schools would visit the orogenic belts together. It was a very successful project that received multiple years of funding and reportedly was a great benefit to the participants.

Damian Nance was born on October 25, 1951, in Saint Ives in Cornwall, United Kingdom. He attended the University of Leicester, England, where he graduated with a bachelor of science degree in geology with honors in 1972. He completed his graduate studies at Cambridge University, England, where he earned a Ph.D. in geology in 1978. He joined the faculty at St.

Francis Xavier University in Nova Scotia, Canada, in 1976. He moved to Ohio University in Athens in 1980 and has remained there ever since. He served as department chairperson from 1995 to 2000. He was also a senior research geologist at Exxon Production Research Company (1982), a research consultant with Cominco American Inc. (1984), a research adviser at Argonne National Laboratories, and a visiting research scientist at Louisiana State University in Baton Rouge. Damian Nance also has a passion for beam engines and engine houses in mines and canals. He and his wife have published some 15 professional articles on these topics.

Damian Nance has had a very productive career. He is an author of some 67 articles in international journals and professional books and volumes. He is an author or editor of six books and volumes. One such book is entitled *Physical Geology Today.* He is an author of 21 government reports and of three published maps. He has received several acknowledgments for his achievements both in research and teaching. The Atlantic Provinces Intercollegiate Council named him Distinguished Lecturer for the Sciences in 1989. He received both the Distinguished Faculty Award and the Outstanding Teacher Award from Ohio University in 1992. He was named the W. F. James Professor of Pure and Applied Science by St. Francis Xavier University in 1994. Nance's service to the profession includes serving on the editorial board for *Geological Magazine.*

Navrotsky, Alexandra
(1943–)
American
Mineralogist, Material Scientist

The Earth sciences are largely composite sciences that overlap and interact with the other basic sciences. Even the names of the subdisciplines like geophysics and geochemistry reflect that overlap. One of the most impressive straddlers of two fields is Alexandra Navrotsky. She is both a respected mineralogist and a respected material scientist (chemistry-chemical engineering). Her expertise in these two fields has led her to new and exciting research discoveries that would have been otherwise impossible. For this reason she is a true pioneer and one of the foremost experts on the material science of minerals and ceramics.

Navrotsky's research centers on relating the microscopic and submicroscopic features of atomic structure and bonding of minerals, ceramics and other complex materials to their large-scale thermodynamic behavior. She conducts experiments on high temperature and pressure calorimetry of these substances to determine phase changes, thermal expansion and contraction and

the major thermodynamic quantities. These data are then related to the atomic structures of the substances in the study of structure-energy-property systematics especially with regard to order and disorder of the atoms. She has made significant contributions to the understanding of mantle mineralogy and the phase transitions that take place under conditions of elevating pressures and temperatures with depth, thus expanding upon and elucidating the breakthroughs of ALFRED E. RINGWOOD. She has even found that the radically elevated pressures in the subduction of cold ocean crust may be quick enough to form ice in these wet rocks. Navrotsky has also refined the thermodynamics of silicate melts and glass, a complex system of order and disorder as the atoms attempt to form crystalline structures. This order and disorder theme has been applied to the thermodynamics of other minerals like framework silicates (quartz and feldspar), spinels, and several other oxides. Navrotsky also applies this research to more practical problems like ceramic processing, oxide superconductors, nitrides, and natural and synthetic zeolite nanomaterials for a variety of important industrial uses. Alexandra Navrotsky's ability to so elegantly and easily interrelate these two fields makes her unique in the profession and allows her to continue to make numerous important contributions to the Earth sciences. Examples of papers by Alexandra Navrotsky include "Possible Presence of High Pressure Ice in Cold Subducting Slabs" and "Thermochemistry of Pure Silica Zeolites."

Alexandra Navrotsky was born on June 20, 1943, in New York, New York, where she grew up. She graduated from the Bronx High School of Science, New York, in 1960. She attended the University of Chicago, Illinois, where she earned a bachelor of science degree in chemistry in 1963. Navrotsky remained at the University of Chicago for graduate studies and earned a master of science degree in 1964 and a Ph.D. in 1967, both in physical chemistry. Upon graduation, she obtained a position as research associate, first at the

Technische Hochschule Clausthal in Germany and the following year at Pennsylvania State University at University Park. In 1969, Navrotsky joined the faculty at Arizona State University, but first in the department of chemistry before her position became a joint appointment between chemistry and geology. During this time she served as program director for the Chemical Thermodynamics division of the National Science Foundation (1976–1977). She was also named director of the Center for Solid State Sciences at Arizona State University in 1984. In 1985, Navrotsky moved to Princeton University, New Jersey. She served as department chair from 1988 to 1991 and was named the Albert G. Blanke Jr. Professor of Geological and Geophysical Sciences in 1992. In 1997, she again moved to the University of California at Davis as an interdisciplinary professor of ceramic, earth and environmental sciences and remains there today. Over the years, Navrotsky has been a visiting scientist several times including a Kreeger-Wolf visiting scholar at Northwestern University (1999), but also at State University of New York at Stony Brook (1981), University of California at Berkeley (1976), and University of Chicago (1970), among others.

Alexandra Navrotsky is an author of some 200 scientific articles in international journals, professional volumes, and governmental reports. Many of these papers are seminal works on the thermodynamic properties of minerals and their applications as well as ceramics and appear in the most prestigious of journals. In recognition of her vast contributions to mineralogy, geochemistry, and material science, Alexandra Navrotsky has re-

ceived numerous honors and awards. She is a member of the National Academy of Sciences. She was awarded an honorary doctoral degree from Uppsala University in Sweden. She also received the Mineralogical Society of America Award, the Ross Coffin Purdy Award from the American Ceramic Society, the Alexander M. Cruickshank Award from the Gordon Research Conference, the Hugh Huffman Memorial Award from The Calorimetry Conference, and the Ceramic Educational Council Outstanding Educator Award. She was also an Alfred P. Sloan Fellow.

Navrotsky's service to the profession is equally as impressive. In addition to serving on numerous committees and panels, she was the president (1992–1993), vice president (1991–1992) and councilor (1982–1985) of the Mineralogical Society of America. She also served on numerous panels, committees, and advisory boards for the National Science Foundation, the National Research Council, the National Academy of Sciences, NASA, the Geochemical Society, and the American Geophysical Union, among others. She has served on numerous evaluating and advisory committees to universities like the Massachusetts Institute of Technology, California Institute of Technology, Harvard University, and Stanford University, and to national laboratories like Sandia and Los Alamos. Navrotsky has also done editorial work such as serving as editor of *Journal of Materials Research,* associate editor for *American Mineralogist,* North American editor for *Physics and Chemistry of Minerals* and series editor for *Oxford Monographs on Geology and Geophysics,* in addition to several editorial boards.

Oliver, Jack E.
(1923–   )
American
Geophysicist

Jack Oliver is one of the true giants of Earth sciences. First leading the powerful group studying the fundamentals of plate tectonics at Lamont-Doherty Geological Observatory and then leading the powerful group studying the architecture of continents at Cornell University, Oliver has had a profound impact on the science. In the 1950s and 1960s, he was involved with earthquake seismology. With FRANK PRESS, he helped set up a worldwide seismic network installing seismographs on all of the continents, on the deep-ocean floor, down deep mine shafts, and even on the Moon. He studied seismic wave propagation and the use of seismic waves as deep probes of the Earth. With the information from these studies, he became involved in helping to construct the basic plate tectonic paradigm. With Bryan Isacks, Oliver proposed and documented the process of subduction at convergent margins, a fundamental concept of plate tectonics. This research involved the study of island arcs and deep sea trenches all over the Pacific Ocean basin. His 1968 paper, "Seismology and the new Global Tectonics" with Bryan Isacks and LYNN R. SYKES is one of the true classics of plate tectonics. It explains why earth-quakes recur and cluster in specific regions around much of the Earth based on plate tectonic interactions.

The contributions from this early work are enough for two very successful careers in geology but not enough for Jack Oliver. His second effort was to apply his expansive seismic prowess to the study of continental architecture. He established a group including several from Lamont-Doherty to conduct deep seismic profiling of the continents. He used vibroseis, which produces synthetic earthquakes, and then images the crustal and sub-crustal structure much like a sonogram images an unborn baby. This effort produced a series of famous COCORP (Consortium for Continental Reflection Profiling) seismic lines all over the United States. These studies revealed huge deep faults, buried basins, and uppermost mantle structures that were previously unknown. They also showed the fate of major surface features at depth, commonly with surprising results. These data helped many geologists better interpret processes in continental development. The processes of crustal extension were better understood from the work on the Basin and Range Province of the Southwest; transform margins were better understood from the work on the San Andreas fault of California; and plate collisions were better understood from the work on the Appalachian and Rocky Mountains. Many other countries followed

Jack Oliver studying earthquake seismograms in his laboratory at Cornell University *(Courtesy of Jack Oliver)*

his lead and established similar efforts with equally useful results. An example of this work is the paper, "The Southern Appalachians and the Growth of Continents." If all of this work is not enough, Oliver even found time to consider the global distribution of certain fluid-related features in his "spots and stains" theory, which again made a contribution to the Earth sciences.

Jack Oliver was born on September 26, 1923, in Massillon, Ohio, where he spent his youth. He played football on his high school national championship team, which was coached by Paul Brown. His athletic ability earned him a scholarship to Columbia University, New York, in 1941. His college career was interrupted by a three-year tour of duty in the U.S. Naval Construction Battalion in the Pacific theater of World War II. He received a bachelor of science degree in physics in 1947 and a master of science degree in physics in 1950. He remained at Columbia University, where he was

among the founding members of the Lamont-Doherty Geological Observatory with W. MAURICE EWING in 1949. Oliver received his Ph.D. in 1953 in geophysics, whereupon he became a research associate. In 1955, he joined the faculty at Columbia University and soon became the head of the seismology group at Lamont-Doherty Geological Observatory. He was also chairman of the department from 1969 to 1971. Oliver moved to Cornell University, New York, in 1971, where he was named the Irving Porter Church Professor of Engineering. In 1981, he established the Institute for the Study of the Continents at Cornell and served as its first director. He retired to professor emeritus in 1993. Jack Oliver married Gertrude van der Hoeven in 1964; they would have two children and three grandchildren.

Jack Oliver has led an extremely productive career. He is an author of nearly 200 articles in international journals, professional volumes, and governmental reports. Many of these papers are true landmarks in the application of geophysics to plate tectonic and regional tectonic problems. He is also an author of two popular science books entitled, *Incomplete Guide to the Art of Discovery,* and *Shocks and Rocks—Seismology in the Plate Tectonic Revolution.* These research accomplishments have been well received by the geologic community, which in turn has bestowed numerous honors and awards on him. He is a member of the National Academy of Sciences. He received an honorary doctorate from Hamilton College, New York, in 1988. He was also awarded the Walter Bucher Medal from the American Geophysical Union, the Virgil Kauffman Gold Medal from the Society of Exploration Geophysicists, the Eighth Medal from the Seismological Society of America, the Hedberg Award from the Institute for the Study of the Earth and Mantle, and the Woollard Medal from the Geological Society of America.

The impressive number of awards is exceeded only by Oliver's outstanding service to the profession and the public. In addition to numerous committees and panels, Oliver served as both president (1964–1965) and vice president (1962–1964) for the Seismological Society of America and the president (1987) and vice president (1986) for the Geological Society of America. He served as chairman of both the U.S. Geodynamics Committee and the Office of Earth Science of the National Academy of Sciences. He was a member of the President's Advisory Board for seismic monitoring, the U.S. Arms Control and Disarmament Agency, UNESCO earthquake engineering committee, U.S. Air Force Science Advisory Board, and numerous committees for the National Academy of Sciences, the National Research Council, and the National Science Foundation, among others. He also served in several editorial capacities for numerous journals.

⊠ Olsen, Paul E.
(1953–   )
American
Paleobiologist (Climate Change)

Of growing concern in environmental science today is global warming (climate change). Scientists are trying to determine if the addition of massive amounts of greenhouse gases to the atmosphere by humans is causing a radical rise in temperature. Most of the research involves ice coring at the poles and then chemically analyzing that ice to chart the changes. However, can it be verified that what happens at the poles reflects what happens in the mid-latitudes? That is the question that Paul Olsen posed in his quest to establish a baseline for climate variation. After all, normal climate variations must be determined before the current changes can be judged abnormal. His idea was to core the sediment in the base of a large deep lake. The problem is that glaciers made most lakes; they are too young or heavily altered by human activity to yield the fine results required in such an analysis. Older existing large lakes that would meet the criteria, like Lake Tanganyika in eastern Africa, are virtually inaccessible to the type

of ship that would be needed to core sediments to an appropriate depth. Olsen came up with the unique idea of coring an ancient lake system to chart climatic variations over a long period. In a multimillion-dollar project, he drilled a continuous 10,000-foot core of the Mesozoic Newark Basin in New Jersey. The Newark Basin contains the most continuous sequence of lake bed and related sediments of any in the world. It covers literally 30 million years of sedimentation. By studying the variations in lake depths and sedimentation rates, Olsen found multiple cycles of climate change caused by Milankovitch cycles, and other terrestrial and extraterrestrial influences. Even though these sediments are more than 200 million years old, the controlling astronomical processes should not have changed appreciably with time. In this way, he set a baseline against which all other climate change models must be compared.

With all of the attention to the climate change research, it is easy to forget that Paul Olsen is a renowned paleontologist/paleobiologist. His specialty is the systematics of lower vertebrates with emphasis on intrinsic biologic innovations. He and his graduate students have been studying Mesozoic tetrapods and especially their footprints in the rocks of the Newark Basin for several years. Through the combination of the coring (stratigraphy) and studies of animal populations and their evolutionary adaptations, Paul Olsen and his team have established the Newark Basin as the benchmark against which all other multidisciplinary studies must be measured. Even small evolutionary changes can be evaluated in terms of their stimuli. Several important publications that reflect Paul Olsen's research include, "Continental Coring of the Newark Rift Basin," and "Tectonic, Climatic, and Biotic Modulation of Lacustrine Ecosystems: Examples from the Newark Supergroup of Eastern North America." His paper, "The Terrestrial Plant and Herbivore Arms Race—A Major Control of Phanerozoic Atmospheric CO_2," connects all of Olsen's areas of interest.

Paul E. Olsen was born on August 4, 1953, in New York City. He grew up in Newark and Livingston, New Jersey. He attended Yale University, where he earned a bachelor of arts degree in geology in 1978. He continued his graduate studies at Yale University, where he earned a master of philosophy and a Ph.D. in biology in 1984. He was a postdoctoral fellow at the Miller Institute of Basic Research in Science at the University of California at Berkeley in 1983–1984. In 1984, he accepted a faculty position at the Lamont-Doherty Earth Observatory of Columbia University, where he is currently the Storke Memorial Professor of Geological Sciences. He is also a research associate at both the American Museum of Natural History and the Virginia Natural History Museum.

His research accomplishments have attracted much media attention. Interviews with him have appeared in numerous magazines and newspapers including, *Time, Discover, National Geographic, Reader's Digest, American Scientist, Science Digest, The New York Times, Washington Post, Boston Globe, Philadelphia Herald, Los Angeles Times,* and many others. He has also been interviewed on television and radio including, *Good Morning America, NBC News,* and many others both national and international. He also played a prominent role in the acclaimed PBS series *Walking with Dinosaurs.* Primarily as a result of his climate change work, Paul has become one of the most prominent media spokespersons for the geologic profession.

⊠ O'Nions, Sir R. Keith
(1944–)
British
Geochemist

In rare instances, an Earth scientist assumes a position of scientific adviser in an upper level of government. For example, FRANK PRESS was the science adviser to President Carter. Similarly, Sir Keith O'Nions achieved such a stature in the sci-

ences that he was asked to serve as chief science adviser for the Ministry of Defense for the United Kingdom. Now top government officials seek his opinion on geologic issues like uranium resources, oil and gas exploration, but also on biomedical research, astronomy, materials science, and chemical warfare.

The geologic research that brought Keith O'Nions to such a position of distinction involves the large-scale evolution of the Earth from literally a pile of rock at the time of formation to the complex interrelated systems of today. He studies this evolution using geochemical systems. He has used a variety of isotopic systems through novel methods of mass spectrometry. These data have revealed some basic information on fundamental questions like the convective circulation patterns in the mantle as revealed by studying basalts from mantle plumes and mid-ocean ridges. He investigated the origin and growth of continents and continental crust and the construction of mountain ranges using neodymium isotopes. Nd isotopes were also applied to sediment systems and for ocean water mass tracing. Later, O'Nions and colleague Ron Oxburgh correlated He isotope distributions with heat flow from the Earth and documented that there is a slow but constant escape of gases that were trapped deep within the Earth at the time of formation. This research evolved into devising a relationship between groundwater flow and hydrocarbon accumulation.

To address such large-scale questions, O'Nions has traveled the world to find just the right geological feature. He has collected loess from China, rocks and gases from Iceland, samples from the Massif Central, France, and rocks from northwest Scotland, to name a few. He even worked in Africa where he sampled the oldest known gabbro at the time. The Modipe gabbro of Botswana is about 2.5 billion years old. The study was to measure remnant magnetism to determine the magnetic field strength at that time. An accident in a VW microbus at the edge of the Kalahari almost ended O'Nions's career prematurely. Luckily, he survived the accident.

Keith O'Nions was born on September 26, 1944, in Birmingham, England, where he spent his youth. He attended the University of Nottingham, England, where he earned a bachelor of science degree in geology and physics in 1966. He crossed the Atlantic Ocean to complete his graduate studies at the University of Alberta in Edmonton, Canada, where he earned a Ph.D. in geology in 1969. Keith O'Nions married his grammar school sweetheart Rita Margaret Bill in 1967; they have three children. He remained at the University of Alberta as a postdoctoral fellow for one year before accepting a second, Unger Vetlesen Postdoctoral Fellowship at Oslo University in Norway. That position was cut short when O'Nions accepted a position at Oxford University in England. He advanced from demonstrator (now assistant lecturer) in petrology (1971–1972) to lecturer in geochemistry (1972–1975), but left Oxford to join the faculty at the Lamont-Doherty Geological Observatory of Columbia University, New York, in 1975. He returned to England in 1979 as a Royal Society Research Professor at Cambridge University, where he was also named a Fellow of Clare Hall in 1980. O'Nions returned to his alma mater at Oxford University in 1995 to assume the position of professor of physics and chemistry of minerals as well as department head, which he held until 1999. In 2000, he assumed his current position as chief scientific adviser for the Ministry of Defense for the United Kingdom on loan from Oxford University for three years.

Sir Keith O'Nions is amid a very productive career. He is an author of numerous scientific articles in international journals and professional volumes. Many of these papers establish new benchmarks in the geochemical evolution of the Earth and appear in high-profile journals like *Nature*. In recognition of his research contributions to Earth sciences, O'Nions has received several prestigious honors and awards from professional societies. He is a Fellow of the Royal Society of London. He is a fellow or foreign member of the Norwegian Academy of Sciences and the Indian Academy of Sciences. He received the Macelwane Award from the American Geophysical Union, the Bigsby Medal and the Lyell Medal from the Geological Society of London, and the Arthur Holmes Medal from the European Union of Geosciences.

O'Nions has performed significant service to the profession in addition to his obvious service to the public. He served on numerous committees and panels for the Natural Environment Research Council (NERC) of Great Britain. He has been involved in the European Union science committees as well as the Council for Science and Technology of England. He is also involved with the Geological Society of London, as well as the American Geophysical Union though more so while he was in the United States.

⊠ Ostrom, John H.
(1928–   )
American
Paleontologist

The excitement about dinosaurs reemerged several years ago with the release of the motion picture *Jurassic Park*. Although several documentaries produced by the Public Broadcasting System had heralded a new view of dinosaurs, the idea was brought into the public spotlight by this film. No longer were dinosaurs viewed as heavy lumbering monsters but as quick and agile animals that were potent predators and powerful protectors that were similar to mammals in many ways. Later films and documentaries like *Walking with Dinosaurs* furthered this impression. But how did our state of knowledge advance to the point to make this distinction? The answer is that the idea of functional morphology was applied to their study and John Ostrom is perhaps the world's foremost expert. Instead of just adding skin to bones and pushing around models as if they were plastic toys, functional morphology looks at the individual parts of an animal and how they were used. His most famous breakthrough in this application was with the dinosaur Deinonychus (terrible claw). He found that the tendon scars in the long tail made it more of a stiff rudder that would counterbalance the animal in a running position rather than a winding tail like that of a cat as was previously interpreted. With this development, the whole posture of Deinonychus changed to one of an agile, fast-running, fearsome predator with a long slashing vertical talon on each of its hind feet. As a result, the posture of all other bipedal dinosaurs was reexamined and duly changed. Not only did the new movies reflect this change but museums worldwide changed their dinosaur bone displays as well.

The other truly famous work of John Ostrom was on the dinosaur Archaeopteryx, the feathered birdlike dinosaur. Again, he studied the functional morphology of the various features and proposed that it was an active, climbing, running, and gliding dinosaur that acted similar to a bird. Using this intermediate-type dinosaur, he compared features with modern birds as well as with the small upright dinosaurs like Deinonychus. These similarities and changes were used to propose a complete lineage and evolution from dinosaurs to birds as summarized in his paper "Archaeopteryx and the Origin of Birds." This tremendous piece of macroevolutionary work now appears in virtually every textbook on historical geology worldwide.

These breakthroughs may be the most famous of Ostrom's work, but many others are just as important. He studied trackways of small carnivorous dinosaurs to show their social interactions. He studied the eating habits of various dinosaurs to show their diets using both functional morphology of their skulls as in the case of Triceratops and Hadrosaur to contents of their stomachs as in the case of Comsognathus. Papers on several of these topics include, "A Functional Analysis of the Jaw Mechanism of Dinosaurs" and "Functional Morphology and Evolution of Cer-

atopsian Dinosaurs." He carefully investigated the reasoning that at least the bipedal dinosaurs were likely warm-blooded. He looked at the function of the cranial crests on Parasaurolophus. He even trained the outspoken dinosaur enthusiast and researcher, Robert Bakker, who appears in nearly every television documentary on dinosaurs, even more than Ostrom. With these achievements, it may be said that John Ostrom has almost single-handedly pioneered the reevaluation and reemergence of interest in dinosaurs. He is a true giant of paleontology.

John Ostrom was born on February 18, 1928, in New York, New York. He attended Union College, New York, where he earned a bachelor of science degree in geology in 1951. He completed his graduate studies at Columbia University, New York, where he earned a Ph.D. in paleontology in 1960. While still a graduate student, he worked as a research assistant vertebrate paleontologist from 1951–1956. John Ostrom married Janet Hartman in 1952; they have two children. He accepted a position as lecturer at Brooklyn College, New York, in 1955, and joined the faculty at Beloit College, Wisconsin, the next year. In 1961, Ostrom returned to the East Coast to accept a faculty position at Yale University, Connecticut, where he spent the rest of his career, which continues today. At the time he began at Yale University, he was also named the assistant curator for vertebrate paleontology at the Peabody Museum, but he soon became curator in 1971.

John Ostrom is an author of numerous publications in all kinds of international journals from biological to geological as well as museum reports and monographs. Several of these are among the best-known papers on the modern views of dinosaurs and the evolution of dinosaurs to birds. In recognition of his contributions to vertebrate paleontology, John Ostrom has received numerous honors and awards. He is a fellow of the American Academy of Arts and Sciences. He was awarded an honorary doctoral degree from Union College. Other awards include the Romer-Simpson Medal from the Society of Vertebrate Paleontology, the F.V. Hayden Memorial Geological Medal from the Academy of Natural Sciences in Philadelphia, a U.S. Senior Scientist Award from the Alexander von Humboldt Stiftung, Germany, and a J.S. Guggenheim Fellowship.

Ostrom has performed a great amount of services to the profession. Among these, he served as president of the Society of Vertebrate Paleontology (1969–1970) as well as president of Sigma Xi honor society. He also performed several editorial roles including chief editor of *American Journal of Science* and of the *Bulletin of the Society of Vertebrate Paleontology*.

P

Palmer, Allison R. (Pete)
(1927–)
American
Paleontologist

Allison R. (Pete) Palmer has really had three successful careers in one: in government, in academia, and in the premier professional society in geology. The common thread through these various positions is his interest in early Paleozoic invertebrates and especially trilobites as well as Cambrian biostratigraphy. It was on a field trip in his junior year of college that he found his first trilobite fossil and he was hooked. He has been affectionately called "Mr. Trilobite" and "The Trilobite Master" because of this interest. He started out studying Cambrian rocks in the western United States where he was able to subdivide the stratigraphy based upon fossil successions. As a result, the Basin and Range Province went from obscurity to containing the type locale for everything from Cambrian seawater compositions through sequence stratigraphy and rates of animal evolution. Naturally, he also studied the trilobites. Palmer then started a major project to identify trilobites with Laurentian (the name for North America during the Paleozoic) affinities on a worldwide basis. He looked at trilobites in Europe, Russia, Australia, North Africa, Argentina, and China, in addition to more examples in the eastern part of North America. This work was done during the time of the emergence of plate tectonics. His worldwide correlations of trilobites were of great interest to the plate tectonic modelers who used them to prove and disprove their reconstructions. His work on Argentinean trilobites has led to a major revolution in the reconstruction of the ancient supercontinent of Rodinia, which is still being developed today. In all, he is responsible for defining hundreds of new species and genera and has described fossils from Alaska to Antarctica, from Cambrian trilobites to Miocene insects. An example of a paper from this work is "Search for the Cambrian World." As a result of his research career, he has been called "the quintessential American paleontologist."

Palmer continued his studies in academia, especially with regard to the importance of biostratigraphy, which owes much of its development to him. He also helped train a new generation of paleontologists. However, his work with the Geological Society of America earned him even greater fame. He spearheaded a mammoth task of summarizing the state of knowledge on all of North American geology in a project called Decade of North American Geology (DNAG). After that he became a spokesperson for geology. He wrote a regular column entitled "What My Neighbor Should Know About Geology" in the magazine *GSA Today* to extol the virtues of geology and to show how it affects everyday life. In this role, he also appeared in a major role on the "Planet Earth" series from the Public Broadcasting System. It is safe to say that the effort and effectiveness that he put into his outreach efforts were equal in stature to his paleontological achievements. Both stand out as real contributions to the science.

Pete Palmer was born on January 9, 1927, in Bound Brook, New Jersey. He attended the Pennsylvania State University in College Park, where he began by studying meteorology but found his true calling and earned a bachelor of science degree in geology in 1949. For his graduate studies, he attended the University of Minnesota, Twin Cities, and earned a Ph.D. in 1950. It was there that he met and married Patricia Richardson in 1949. They have five children. During his graduate studies he worked as a science aide for the Texas Bureau of Economic Geology but his first permanent position was with the U.S. Geological Survey, which he accepted upon graduation in 1950. He was a Cambrian paleontologist and stratigrapher there until 1966 when he joined the faculty at the State University of New York at Stony Brook. He served as chair of the department from 1974 to 1977. In 1980, Palmer left Stony Brook to become the centennial science program coordinator for the Geological Society of America in Boulder, Colorado. He was also the coordinator of educational programs from 1988 to 1991. He retired from the Geological Society of America in 1993 to become an adjunct professor at the University of Colorado at Boulder where he remains in active research today.

Pete Palmer has led a very productive career authoring some 137 scientific articles in international journals, professional volumes, and governmental reports. He produced nine major monographs. All totaled, he has more than 2,200 printed pages to his credit. Several of these articles are seminal works on Cambrian paleoecology, trilobite morphology, and related studies that appear in top journals like *Science*. In recognition of his contributions to geology, he has received sev-

eral prestigious honors and awards. He received the Charles D. Walcott Medal from the National Academy of Sciences, the Distinguished Service Medal from the Geological Society of America, and the Paleontological Society Medal (United States).

Palmer has been very active in terms of service to the profession. He has served as president for the Institute for Cambrian Studies since 1984 and before that for the Cambrian Subcommittee for the International Stratigraphic Committee (1972–1984). He was president of the Paleontological Society (United States) in 1983 as well. He served on numerous committees for all of these organizations as well as the Geological Society of America.

⊠ Patterson, Clair (Pat) C.
(1922–1995)
American
Isotope Geochemist

Even though he was forever an Iowa farm boy, Clair Patterson made three of the greatest contributions to geology of all time. His first and foremost contribution was to accurately determine the age of the Earth and stony meteorites using isotopic analysis. This research was started during his graduate career under his mentor Harrison Brown and with his friend and colleague, GEORGE R. TILTON, when they were studying meteorites. These radio chemists were investigating the uranium-lead decay series and developed radical new methods for measuring microchemical and precise isotopic ratios using mass spectrometry. Patterson's area of specialization was radiogenic lead. Using these techniques, in 1963, after years of careful and exhaustive research on a variety of terrestrial and extraterrestrial materials, he determined that the age of the Earth is 4.55 billion years. Considering the technological advancement since that time, it is surprising that age has undergone only minor readjustment since then. This benchmark work ranks among the greatest achievements of all time in geochemistry.

Patterson's second great contribution was to establish a fundamental basis and methodology to model the patterns of isotope evolution of terrestrial lead. He collected groups of sediments, rocks, and water samples from the oceans and determined that there were different reservoirs of common lead that were distinctive in each. By analyzing isotopic lead ratios, distinct and common patterns emerged that determined if an area or water body was separate from outside influences or it had mixed sources. This technique has multiple applications for determining origins. It was used to help define ancient plates, which had been amalgamated during plate collisions. In many cases, each plate has a separate and distinct common lead reservoir. This means that even if the rocks look the same and there is no other way to delineate the ancient plates, a geologist can still tell them apart using concentrations of common lead isotopes.

His third great contribution is perhaps the most important to humankind. He provided the first and still the most rigorous analysis of the human induced buildup of lead in our environment. He did this by contrasting current lead concentrations with the natural background. He accomplished this comparison by developing new methods to cleanly extract and analyze minute, nanogram quantities of lead. He exhaustively sampled in remote regions of the Earth, in numerous ocean water environments, and in ancient archaeological sites. He showed that lead concentrations in contemporary humans are elevated 1,000 times greater than that in prehistoric people and just three- to sixfold short of outright poisoning. He even showed how this biologic magnification worked its way up the food chain. He did not just publish these results; he became a spokesperson for the elimination of environmental lead and encountered a great deal of criticism as a result. Industry was especially opposed to his findings and even tried to discredit him. How-

ever, Patterson persevered and the elimination of lead from gasoline, pipes, and solder can be directly attributed to his careful research and refusal to back down in the face of overwhelming odds. As a result, we owe some of our good health to Clair Patterson.

Clair Patterson was born on June 2, 1922, in Des Moines, Iowa. He attended Grinell College, Iowa, and earned a bachelor of arts degree in chemistry in 1943. He earned a master of science degree in chemistry from the University of Iowa in 1944. He served in the armed forces during World War II before returning to graduate school at the University of Chicago, Illinois, where he earned his Ph.D. in 1951 in chemistry. Between 1952 and 1992, Patterson held positions of research fellow, senior research fellow, research associate, senior research associate and finally professor of geochemistry at California Institute of Technology. He retired to a position of professor emeritus in 1993. Clair Patterson died suddenly at his home at The Sea Ranch, California, on December 5, 1995.

Clair Patterson had a highly productive career authoring numerous articles in international journals and professional volumes. Many of these publications are benchmarks in the field of geology, much less their subdiscipline. His research accomplishments have been well recognized by the geologic profession in terms of honors and awards. He was a member of the National Academy of Sciences. He received honorary doctoral degrees from both Grinell College, Iowa, in 1973 and the University of Paris, France, in 1975. Even more impressive was the 1967 dedication of "Patterson Peak" in his honor in the Queen Maude Mountains of Antarctica and the naming of Asteroid 2511 after him. In addition, he received the J. Lawrence Smith Medal from the National Academy of Sciences, the Goldschmidt Medal from the Geochemical Society, the Professional Achievement Award from the University of Chicago, and the Tyler World Prize for Environmental Achievement.

Pettijohn, Francis J.
(1904–1999)
American
Sedimentologist (Stratigraphy)

Francis Pettijohn is considered by many to be the "father of modern sedimentology." He was one of the true leaders in a revolution in sedimentology that occurred after World War II. Classical petrology had advanced to a rigorous science of chemical equilibria and phase relations whereas sedimentary rocks were still considered more through descriptive analyses. Pettijohn almost single-handedly transferred the methods of classical petrology to sedimentary rocks. He used the petrographic microscope extensively on sedimentary rocks and especially sandstones for detailed classification and to determine source areas, distance, and method of transport and lithification processes. These were newly developed methods at the time but are now standard practices thanks to his documentation of their usefulness. The results of these analyses were combined with a new more rigorous analysis of sedimentary structures, which he also pioneered to establish a new subdiscipline of basin analysis. Instead of considering the various aspects of sedimentary rocks separately, he combined them with new statistical methods for paleocurrent analysis to fully analyze the history of sedimentary basins. These methods were immediately used in petroleum exploration, which resulted in great success in locating hydrocarbon reserves.

Francis Pettijohn was a field geologist by training and desire. He found himself in a geology department that was quickly moving away from field research into more high-tech fields. Pettijohn's work was not being supported so he moved to another school. Perhaps as a protest against this new direction, which was not uncommon among geology departments, Pettijohn wrote the book *Memoirs of an Unrepentant Field Geologist.* This book not only expounded upon the joys of field geology but also painted a not-so-complimentary picture of those who were unsympathetic to field research. The book caused quite a stir in the profession.

Francis Pettijohn was born on June 20, 1904, in Waterford, Wisconsin. He graduated from high school in Indianapolis, Indiana, in 1921, and entered the University of Minnesota at St. Paul that year. He graduated with a bachelor of arts degree in geology in 1924 and a master of arts degree in 1925. He was an instructor at Oberlin College, Ohio, for two years before returning to graduate school first at the University of California at Berkeley. He later returned to the University of Minnesota and earned a Ph.D. in geology in 1930. In 1929, Pettijohn joined the faculty at the University of Chicago, Illinois, first as an instructor but later as a professor. In 1952, ERNST CLOOS convinced him to join the faculty at the Johns Hopkins University where he spent the remainder of his career. He served as chair of the department from 1963 to 1968 and acting chair in 1970. Pettijohn died in 1999 and was survived by his three children. His wife predeceased him several years earlier.

Francis Pettijohn was the author of numerous articles in international journals and professional volumes on sedimentary rocks, many of which are seminal studies. He is perhaps best known for his books. His 1949 book, *Sedimentary Rocks,* was widely adopted as a textbook and last reprinted (third edition) an amazing 26 years later. It is still cited in papers today. He was also an author of the widely read and cited books, *Sandstone Petrography* in 1936, *Sands and Sandstones* in 1972, and *Paleocurrents and Basin Analysis* in 1963. In recognition of his research contributions to the profession, Pettijohn received numerous honors and awards. He was a member of the National Academy of Sciences and a fellow of the American Academy of Arts and Sciences. He received an honorary doctoral degree from the University of Minnesota, Twin Cities. In addition, he was awarded the Twenhofel Medal from the Society of Economic Paleontologists and Mineralogists, the Wollaston Medal from the Geological Society of London, the Penrose Medal from the Geological Society of America, the Sorby Medal from the International Association of Sedimentologists and, believe it or not, the Francis J. Pettijohn Medal from the Society for Sedimentary Geology.

Pettijohn performed significant service to the profession. In addition to several committees and panels, he served as president and vice president of the Society of Economic Paleontologists and Mineralogists. He also served in many roles for the Geological Society of America including councilor. His editorial work covered numerous roles in numerous journals, including the *Geological Society of America Bulletin,* and *Journal of Sedimentary Petrology,* among others.

⊠　**Pitcher, Wallace S.**
　　(1919–　)
　　British
　　Petrologist

Wallace Pitcher is likely the foremost authority on the tectonic setting of granites as well as the mechanics of granite pluton (body) emplacement. He has studied all aspects of pluton emplacement from the petrology to structural geology using everything from geophysical techniques to paleontology. As many researchers spent great effort assigning letters (I=igneous, S=sedimentary, etc.) to granite plutons to indicate their heritage based upon their chemistry and mineralogy, Pitcher took a more holistic and less stringent approach. He looked at pluton shape, fabric, deformation, and regional relations to develop an alternative system. He named his plutons by type examples, but related them to a plate tectonic environment. Because the composition of granites can vary so wildly within the same tectonic environment, many researchers found the Pitcher approach more applicable and useful. As a result, his paper "Nature and Origin of Granite" became a true classic.

Most of his research was in the Caledonides of western Europe and notably at Donegal, Ireland, but also in the Peruvian Andes of South America. Two papers describing these two areas are especially notable, "Geology of Donegal: A study of Granite Emplacement and Unroofing" and "Magmatism at a Plate Edge: the Peruvian Andes." In this research, he defined the relationships within zoned and composite plutons based upon textural and chemical differences. The convection of the magma within the hot pluton in the early stages of crystallization can chemically zone plutons, as well as imposing a flow fabric on the rock. He also devised a system to evaluate stress within plutons, but especially within the contact aureole. As the tail of the pluton ascends into the main body, it causes the pluton to "balloon," which causes further deformation of country rock in the contact aureole. It produces triple points of deformation where regional deformation equals that imposed by the ballooning pluton. Different patterns of deformation occur depending upon the tectonic setting. Pitcher even found that country rock units could form a "ghost stratigraphy" both textural and chemical across the pluton, as if the preexisting rock unit was still there and not pushed out of the way as would be expected. This discovery forced the reevaluation of emplacement mechanics. Pluton emplacement appears to be more of an assimilation-type process (eating into the existing rock) rather than a brute pushing country rock aside, at least in some cases.

Surprisingly, Wallace Pitcher also got involved in researching the late Precambrian stratigraphy and sedimentology of the British Isles. He was the first to document that much of the sequence is of glacial origin and includes well developed tillites. This work has been used extensively of late to document the popular "Snowball Earth" hypothesis championed by PAUL HOFFMAN, which has great implications for paleoclimatology. Apparently, the Earth underwent a great cooling event at the end of the Proterozoic.

Pitcher's work was done well before the idea was popular.

Wallace Pitcher was born on March 3, 1919, in England. He attended Acton Technical College and Chelsea College to study chemistry. He obtained a position as an assistant analytical chemist with George T. Holloway and Company in chemical assaying in 1937. In 1939, he enlisted in the Royal Army Medical Corps, where he served for the duration of World War II. By the time he returned to Chelsea College in 1944, he had decided to switch his field to geology. In 1947, Pitcher moved to the Imperial College of London where he was a demonstrator until 1948 when he became an assistant lecturer. It was at this time that he completed his Ph.D. in geology. Wallace Pitcher married Stella Ann Scutt in 1947; they would have four children. Pitcher became a lecturer in 1950 but moved to King's College of London as a reader in 1955. He joined the faculty at the University of Liverpool as the George Herdman Professor of Geology in 1962. He retired from his teaching duties to professor emeritus in 1981. At that time, he also became a Leverhulme Emeritus Research Fellow, a position he held until 1983, when he formally retired. He continues to be active in research today at a slower pace.

Wallace Pitcher led an extremely productive career, having been an author of numerous scientific articles in international journals and professional journals. Several of these papers are definitive studies on granites and especially their emplacement mechanics and tectonic setting. In recognition of his many contributions to the Earth sciences, Wallace Pitcher has received numerous honors and awards. He is an honorary fellow of the Royal Society of London. He has received honorary degrees from the University of Dublin, Ireland, and the University of Paris-Sud. He was also awarded the Bigsby Medal, the Murchison Medal, and the Lyell Fund from the Geological Society of London, the Silver Medal from the Liverpool Geological Society, the Aber-

conway Medal from the Institution of Geologists, England, and the University of Helsinki Medal, among others.

Pitcher served in numerous positions with the Geological Society of London, including president (1976–1977). He was also section president of the British Association.

Porter, Stephen C.
(1934–   )
American
Quaternary Geologist, Glacial Geologist

Mountain climbing in his youth sparked Stephen Porter's interest in alpine glaciation, the study of mountain glaciers during the Earth's ice ages. However, he did not just study mountain glaciers in his home state of California; he studied them all over the world and established himself as one of the foremost authorities on alpine glaciation and the Quaternary glacial ages (the last 2 million years of Earth history). He has studied mountain glaciers in Alaska, the Cascade Range of the northwestern United States, the Argentine and Chilean Andes, the Himalayan and Hindu Kush region, the western Italian Alps, New Zealand, Siberia, the Tibetan Plateau, and Hawaii. Most of these investigations concentrated on the sequence and chronology of glacier advances and retreats as determined by studying the depositional and erosional features of glaciated landscapes. The varying extent, through time, of glaciers on mountain slopes and in adjacent valleys is a measure of local and regional climatic change. Through these studies, glacial and interglacial times can be identified and dated, providing important information for scientists involved in modeling climate change.

Porter also has studied other records of climate change, including investigations of loess deposits in China. Loess is an accumulation of fine windblown dust, representing times of cold, dusty climate. Some of the world's important loess de-

Stephen Porter on a field trip to the Bolshoi Annechag Range in northeastern Siberia *(Courtesy of S.C. Porter)*

posits contain a long and unique geologic record of environmental and climatic change. Porter has carried out his wide-ranging field studies with Chinese colleagues of the National Key Laboratory of Loess and Quaternary Geology in Xi'an. Their research has led them across inner Mongolia, central China, and onto the northeastern Tibetan Plateau. From these deposits, a detailed history of China's monsoon climate can be traced back at least 8 million years. Many of their studies, however, have been concerned with environmental and climatic changes from the last ice age to present, including the interval when the earliest human cultures were succeeded by the early dynastic period of Chinese civilization. An example of this research is the paper, "Correlation between Climate Events in the North Atlantic and China During the Last Glaciation."

Stephen Porter was born on April 18, 1934, in Santa Barbara, California. He attended Yale University in Connecticut and earned a bachelor of science degree in geology in 1955. He then served as an officer in the U.S. Naval Reserve from 1955 to 1957, during which he spent two years aboard a destroyer with the Pacific Fleet. His subsequent graduate studies, also at Yale University, earned him a master of science degree in 1958 and a Ph.D. in geology in 1962. In his final year as a graduate student, he won the Benjamin Silliman Prize from Yale for excellence in defense of his dissertation. While in graduate school, he married Anne M. Higgins, a graduate student in anthropology, in 1959. They had three children, who accompanied their father and mother on field projects. Porter joined the faculty at the University of Washington in Seattle in 1962, and remained there for his entire career. He served as director of the university's Quaternary Research Center from 1982 to 1998. Porter was awarded a Fulbright-Hays Fellowship to the University of Canterbury in New Zealand in 1973–1974 and was a visiting scholar at the Scott Polar Research Institute of the University of Cambridge, England, in 1980–1981.

Stephen Porter is an author of more than 100 articles published in international journals and professional monographs. He also cowrote eight popular introductory textbooks. Several of these textbooks include, *The Blue Planet, An Introduction to Earth Systems Science, The Dynamic Earth, Environmental Geology,* and *An Introduction to Physical Geology.* He has been a guest professor in the Chinese Academy of Sciences since 1987.

Porter has served on numerous national and international professional committees. He was elected president of the American Quaternary Association, and vice president (1991–1995) and president (1995–1999) of the International Union for Quaternary Research. He served on the board of Earth sciences of the National Academy of Sciences/National Research Council and on several panels of the National Science Foundation. He has also held several editorial positions. He was editor of the interdisciplinary journal *Quaternary Research* from 1976 to 2001, has been an associate editor of *Radiocarbon* (1977–1989) and of the *American Journal of Science* (1997–2005). He has served on the editorial board of *Quaternary Science Reviews* (1983–present) and *Quaternary International* (1989–present). He also was a science adviser for the PBS-TV series "The Miracle Planet" and its accompanying book.

⊠ **Press, Frank**
(1924–)
American
Geophysicist

Frank Press is one of the top five most influential Earth scientists of the 20th century. His research area is seismology both in terms of generation (earthquakes) and wave travel but also in terms of the structure of the Earth that it reveals. He was the first to investigate long-period surface waves and free oscillations (two types of seismic waves) as deep probes of the Earth's architecture. During the time he worked with BENO GUTENBERG, he developed new more sensitive instrumentation and recording devices for better resolution of the wave arrivals. The data he collected using these new instruments allowed Press to better define the layers within the Earth from the crust to the deep mantle. He produced detailed profiles that show how seismic velocity changes with depth in the Earth. The basic layers were subdivided and their character better defined at accurate depths. This new level of scientific research defined the beginning of modern geophysics. It also contributed significantly to the understanding of plate tectonics. The new instrumentation allowed him to record events not only on the Earth but also on the Moon and other planets. Through these studies he was able to define the architecture of these extraterrestrial bodies as well.

Not only did his pioneering advances in seismology aid the science of geology, Frank Press was also of great public service. He is especially well known for his international coordination of the exploration of the ocean basins and the continent of Antarctica. The new instrumentation was of great use in monitoring earthquakes. He led several international projects to better monitor earthquakes on a worldwide basis and to formulate plans for better earthquake prediction. This new worldwide network with his new more sensitive instrumentation allowed Press to better monitor nuclear testing on a worldwide basis. He was called into service to help interpret any test that took place. Frank Press was called to the highest level of public service with membership on the science advisory panel to several presidents of the United States and participating sensitive negotiations on nuclear test bans and monitoring. These high-profile positions led Press to be named the "California Scientist of the Year" in 1960 and later as one of the top 100 most important people under the age of 40 in the United States by *Life* magazine in 1962.

Frank Press was born on December 4, 1924, in Brooklyn, New York, where he grew up. He attended the City College of New York and earned a bachelor of science degree in physics in 1944. He completed his graduate studies at Columbia University, New York, where he earned a master of arts in 1946 and a Ph.D. in 1949, both in geo-

Frank Press in his office in Washington, D.C. *(Courtesy of Frank Press)*

physics. Frank Press married Billie Kallick in 1946; they have two children. He joined the faculty at Columbia University upon graduation where he worked with W. MAURICE EWING. In 1955, he accepted a position at the California Institute of Technology in Pasadena where he became director of the Seismological Laboratory in 1957. In 1965, Press moved again to the Massachusetts Institute of Technology in Cambridge, where he assumed the responsibility of department chairman. During this time, Press served as a member of the science advisory committee to both President Kennedy and President Johnson. In 1977, Frank Press was called to Washington, D.C., to serve as science adviser to President Jimmy Carter, as the director of the Office of Science and Technology Policy. He was also the president of the National Academy of the Sciences, a position he held until 1994. He returned to serve as chairman of the department at Massachusetts Institute of Technology from 1980–1982. In 1994, Press became the Cecil and Ida Green Senior Fellow at the Carnegie Institution of Washington, D.C. In 1996, he became a partner in the Washington Advisory Group. In addition to being a great scientist and advocate, Frank Press is a skilled sailor and an authority on baseball and New Orleans-style jazz.

In spite of all of his effort devoted to advisory work, Frank Press has led a very productive scientific career. He is an author of more than 170 articles in international journals and professional volumes. Many of these are benchmark studies that are published in the most prestigious of journals. He is also an author of numerous books including *Earth,* probably the most complete textbook on physical geology, and *Understanding Earth,* probably one of the most popular textbooks on physical geology. The honors and awards that Frank Press has received for both his scientific contributions and advisory work are too numerous to list completely here. He received the National Medal of Science from President Clinton in 1994. He also received the Decorated

Cross of Merit from Germany and the Legion of Honor from France. He was awarded numerous honorary doctoral degrees and numerous society awards, including the Arthur L. Day Medal from the Geological Society of America, the Bowie Medal from the American Geophysical Union, the Ewing Medal from the Society of Exploration Geophysicists, the Gold Medal from the Royal Astronomical Society of England, and public service awards from both NASA and the U.S. Department of the Interior. He even had Mt. Press in Antarctica named after him.

The service that Frank Press has performed to the profession and the public is even more astonishing than his awards. In addition to that described above, he served as an adviser to the U.S. Navy, U.S. Geological Survey, NASA, U.S. Department of Defense, U.S. Arms Control and Disarmament Agency, and the governor of the state of California. He served on the U.S. Nuclear Test Ban Delegation, the UNESCO Technical Assistance Mission, and the U.N. Conference on Science and Technology for Underdeveloped Nations. He served as president of the American Geophysical Union (1974–1976) and president (1962) and vice president (1959–1961) of the Seismological Society of America, among numerous other committees and panels.

Price, Raymond A.
(1933–)
Canadian
Structural Geologist

The Canadian Rockies are famous both for their scenery and because they provide what is probably the best example in the world of a foreland thrust and fold system. They are characterized by conspicuous linear mountain ranges that are formed by overlapping, thick, westward-tilted slabs of sedimentary strata; and they form the eastern side of the North American Cordillera between the Northwest Territories and central

Montana. The sedimentary strata were scraped off the western margin of the North American continent and thrust northeastward in front of a "collage" of overriding oceanic volcanic archipelagos with which North America "collided" as it drifted away from Africa and Europe during the opening of the Atlantic Ocean basin. As the sedimentary strata were slowly shoved northeastward along large, gently inclined thrust faults just like a rug on a floor, they were tilted, folded, and sliced by thrust faults. The resulting structures beautifully illustrate the processes involved in the development of a foreland thrust and fold system. The weight of the resulting gigantic northeastward-tapering wedge of thrust slabs caused the continental lithosphere (the strong outer layer of the solid Earth) of western North America to flex downward, which produced a deeply subsiding sedimentary basin in front of the advancing wedge. Most of the petroleum and coal deposits of western Canada were formed as sediment that was eroded from the advancing wedge of thrust slabs accumulated to a depth of many kilometers in the subsiding basin. Although many geologists have studied this area, Raymond Price has emerged as the foremost expert. He prepared geological maps and cross sections of large areas of this rugged terrain, and he developed models for the movement of the large thrust slabs, the processes of thrusting and folding, and the origin of the foreland basin. He also was the leader of a small group of structural geologists who developed quantitative methods to critically evaluate the evolution of foreland fold and thrust belts. These methods, which are called palinspastic reconstruction, involve the careful analysis of the three-dimensional relationships between the thrust faults and the deformed strata, and the sequential restoration of the strata to their initial undeformed state. One component of the process is the preparation of "retrodeformable" balanced cross-sections. In a "balanced cross section," the configuration of the faulted and folded strata makes it possible to reconstruct

Portrait of Ray Price *(Courtesy of R. Price)*

the initial configuration and location of the undeformed strata without any gaps, overlaps, or other illogical consequences. These procedures have been of great interest to oil companies because the Canadian Rocky Mountains, like the United States Rockies and most other foreland fold and thrust belts worldwide, contain significant petroleum resources.

Raymond Price was born on March 25, 1933, in Winnipeg, Manitoba, Canada, where he grew up. He attended the University of Manitoba, Canada, where he earned a bachelor of science degree in geology with honors and the University Gold Medal in Science in 1955. In 1956, Raymond Price married Mina Geurds; they have three children. He did his graduate studies at Princeton University, New Jersey, and earned a master of arts degree in 1957 and a Ph.D. in 1958. From 1958 to 1968, he worked as a geologist within the petroleum geology section of the Geological Survey of Canada, where he worked in the southern Canadian Rocky Mountains and Yukon. He then joined the faculty of Queen's University, Canada, and served as the head of the department from 1972 to 1977 and a Killam Research Fellow from 1978 to 1980. He moved back to the Geological Survey of Canada in 1981, and served as the director-general from 1982 to 1987 and as the assistant deputy minister for the Division of Energy, Mines and Resources of Canada in 1987–1988. He returned to Queens University as a visiting professor in 1988–1990. Upon his retirement from the Geological Survey of Canada in 1990, he accepted a full-time permanent position at Queens University. Raymond Price retired to a professor emeritus position in 1998.

Raymond Price has led a highly productive career. He is an author of some 175 articles in international journals, chapters in professional books and volumes, and geological maps. Several of these articles are seminal works in the analysis of thrust belts. He was also an author of a popular textbook, *Analysis of Geologic Structures.* He received numerous honors and awards in recognition of his contributions to geology. He is a Fellow of the Royal Society of Canada and a foreign associate for the U.S. National Academy of Sciences. He received honorary doctoral degrees from Memorial University of Newfoundland, Canada, and Carleton University of Ottawa, Canada. He received the Sir William Logan Medal from the Geological Association of Canada, the Major Edward D'Ewes Fitzgerald Coke Medal from the Geological Society of London, England, the Leopold von Buch Medal from the Deutsche Geologische Gesellschaft, the Michael T. Halbouty Award from the American Association of Petroleum Geologists, the R.J.W. Douglas Medal from the Canadian Society of Petroleum Geologists, and the Gold Medal in Sciences from the University of Manitoba. He was named an Officier de l'Ordre des Palmes Académiques, France, in addition to serving numerous named distinguished lectureships from colleges worldwide.

The service that Raymond Price has given to the profession is as impressive as his awards. He has served as member and chair of society and governmental committees and panels too numerous to list more than just the highlights. He served as president of the Geological Society of America in 1989–1990, where he served on numerous committees. He was also president of the Inter-Union Commission on the Lithosphere in 1980–1985. Several other organizations in which he served are the Royal Society of Canada, Canadian Institute for Advanced Research, Ocean Drilling Program, U.S. National Research Council and the International Geosphere-Biosphere Program. His input was sought for issues like energy, nuclear waste disposal, seismic hazards, pure research directions, and others. He has served in editorial roles too numerous to list, but they include such prestigious journals as *Journal of Structural Geology* and *Tectonics.*

Ramberg, Hans
(1917–1998)
Norwegian
Structural Geologist, Tectonics

While most experimental structural and tectonic geologists were building ever more powerful vices and presses to compress and extend rock samples, Hans Ramberg took a different approach. He placed soft ductile materials in a centrifuge and spun them to pressures up to 2,000 times the force of gravity to model deformational processes under metamorphic conditions in the deep crust. These beautiful analogue models illustrated the role of gravity tectonics in deep crustal settings as well as salt tectonics and other ductile substances in shallower settings. The models simulated crustal structures in minutes that would have otherwise required millions of years to form through natural processes. These experiments led to new understanding of the formation and emplacement of diapirs as well as the formation of mantled gneiss domes. These models formed a bit of a revolution in geology when they were first released as many regional geologists attempted to reinterpret their field areas using Ramberg's findings. He continued by constructing scaled models depicting crustal isotasy, rift valleys opening to oceans and growth of continents, and mantle convection. Other models were structures of glaciers and grav-ity gliding of nappes (large horizontally translated sheets or folds of rock) as well as structural patterns observed in orogens and sedimentary basins of all ages worldwide with special emphasis on the Alps of Europe. His book, *Gravity, Deformation and the Earth's Crust* (Academic Press, London, 1967), depicts hundreds of Ramberg's greatest models. On the other hand, there were others who found the whole notion controversial and referred to Ramberg's laboratory as a "baker's shop." In time, his techniques were widely adopted both in academia and in the petroleum industry.

Hans Ramberg began his career and research dealing with the structural and metamorphic geology of real rocks in the Norwegian Caledonides and western Greenland. His main effort involved the chemistry of rocks and minerals and led to his first book entitled, *The Origin of Metamorphic and Metasomatic Rocks*. He then shifted his effort to modeling of processes. He worked first on the formation of pegmatites. Ramberg also used engineering theory to attribute natural and experimental boudinage (regularly spaced bulbous shapes formed in the drawing apart of sheets of rock) with various styles attributed to extension along the thin sheets caused by compression across them. Ramberg's next research project consisted of using fluid dynamics to explain the ratios of wavelength to thickness in ptygmatic folds in terms of buckling of thin sheets.

Toward the end of his career, Ramberg moved into computer modeling of structural and tectonic processes. He developed numerical models for his analogue findings. Later, he modeled simple shear systems (faults).

Hans Ramberg was born on March 15, 1917, in the town of Trondheim, Norway. He attended Oslo University in Norway, receiving his bachelor of science degree in 1943 in chemistry and physics. He continued with his graduate studies at Oslo University and completed a Ph.D. in geology in 1946. He worked as an expedition leader to Greenland during the summers of 1947 to 1951. This part-time position overlapped with his faculty position at the University of Chicago, Illinois, which he obtained in 1948. He took a leave of absence from 1952 to 1955 to be a research associate at the geophysical laboratory at the Carnegie Institution of Washington, D.C. He was also a visiting professor in Brazil in 1959 and 1960. In 1961, Ramberg returned to Scandinavia to join the faculty at Uppsala University, Sweden. It was there that he established the Hans Ramberg Tectonic Laboratory. Between 1970 and 1975, he again took a leave of absence to be a special university professor at the University of Connecticut in Storrs. He retired to professor emeritus in 1982, but continued to be active in research for many years to come. Hans Ramberg succumbed to cancer in May of 1998. His wife Marie Louise (Lillemor) survived him. They had been married since he was an undergraduate in 1942.

Hans Ramberg was an author of more than 100 scientific articles in international journals and professional volumes. He also wrote two highly regarded technical books. Many of his papers are seminal works on analog models of structural and tectonic processes, the thermodynamics of metamorphic rocks, and computer modeling of structural models. In recognition of his many contributions of geology, several honors and awards were bestowed upon him. He received the Arthur L. Day Medal and the Career Contribu-

tion Award from the Geological Society of America, the Asar Haddings Prize, the Celcius Prize and the Bjorkenske Prize from Sweden, the Hans Reusch Medal from Norway, the Arthur Holmes Medal from the European Union of Geoscientists, the Wollaston Medal from the Geological Society of London and the Swedish Royal Academy of Sciences Prize.

⊠ Ramsay, John G.

(1931–)
British
Structural Geologist

John Ramsay is unquestionably the "father of modern structural geology." Although there were several researchers who attempted to integrate quantitative analysis into their studies, structural geology was largely a descriptive discipline into the 1960s. It concentrated on the shapes and associations of folds, faults, and cleavage and devised classifications on these bases. Ramsay assembled all of the quantitative techniques that had been devised by the few structural geologists who had even attempted such exercises. His real contribution, however, was to take the science a step forward. He integrated these studies that attempted to provide a quantitative basis for strain and explained them in terms of continuum mechanics. He also integrated his own studies of deformed passive markers which show the deformation of a rock but which are not formed in the process. These passive markers include features like fossils of all types, certain sedimentary structures (mud cracks, oolites, pebbles in conglomerate, etc.), certain volcanic structures (vesicles, etc.), xenoliths in plutons and others. By knowing the original shape of the feature and comparing it to the deformed state, an equation of strain can be written based upon the geometrical changes. Although these changes are mathematically complex, requiring a tensor solution using matrix algebra, by making certain assumptions and issuing certain requirements to the features, a relatively simple solution can be used in many cases. Ramsay devised a series of trigonometric and statistical solutions to these deformed features that he summarized in a landmark 1967 textbook entitled *Folding and Fracturing of Rocks.*

These new methods, now readily available in a single textbook, sparked a revolution in structural geology that had fallen well behind many of the other subdisciplines of geology in terms of quantitative analysis. Structural geology would go on to utilize many other principles of engineering and material science. Most of Ramsay's work involved the best examples of deformed features rather than field studies. Those field studies that he performed were on single outcrop examples and largely in Great Britain or the Swiss Alps. One of his regional topics of interest was the study of large shear zones especially with regard to their passage from basement to cover rocks. Ramsay was always noted for his ability to find the most beautiful examples of deformed rocks to analyze. Late in his career, he produced a two-volume manual entitled *The Techniques of Modern Structural Geology* with some of the most outstanding photographs of deformed rocks. These volumes also have become classics.

John Ramsay was born on June 17, 1931, in England. He received his primary education at the Edmonton County Grammar School in England before attending Imperial College in London. He earned a bachelor of science degree in geology in 1952. That year he married Sylvia Hiorns but the marriage ended in divorce in 1957. He remained at Imperial College for his graduate studies and earned a Ph.D. in geology in 1955. He then performed military service with the Royal Corps of Engineers until 1957, and he also played in the military band. In 1957, he returned to Imperial College as part of the academic staff and remained until 1973. John Ramsay married Christine Marden in 1960, but

that marriage ended in divorce in 1987. They had four children but one daughter died in her youth. In 1973, Ramsay moved to the University of Leeds, England, where he served as department chair. He joined the faculty at the Swiss Federal Institute (ETH) in Zurich in 1977 and spent the rest of his career there. John Ramsay married Dorothee Dietrich in 1990 and remains married today. He retired to professor emeritus in 1992. Upon retirement he moved to France where he continues to enjoy playing the cello (concert quality) and writing poetry, but devotes less interest to Earth sciences.

John Ramsay led a very productive career having authored numerous scientific articles and reports in international journals and professional volumes. Many of them are groundbreaking studies of the application of continuum mechanics to rocks. He also wrote three textbooks that are regarded by many as the "bibles" of modern structural geology. In recognition of these outstanding contributions to geology, John Ramsay has received numerous honors and awards. He is a Fellow of the Royal Society of London and a member of the U.S. National Academy of Sciences. He received an honorary doctor of science degree from Imperial College. He received both a Best Paper Award and the Career Contribution Award from the Structure and Tectonics Division of the Geological Society of America in addition to the Prestwich Medal from the Geological Society of France. He received most of the awards offered by the Geological Society of London including the Wollaston Award, the most prestigious award.

John Ramsay also performed extensive service to the Earth science profession. He established the first tectonics studies group in the world within the Geological Society of London. He was also the vice president of the Geological Society of France, among other functions. He served on several committees and panels for the National Environmental Research Council (NERC) in England.

⊠   **Rast, Nicholas**
(1927–2001)
Iranian
Tectonics

Although he began his career as a process-oriented structural geologist and stratigrapher, Nicholas Rast became known for being able to utilize any and all geologic information to construct tectonic solutions. He became one of the key researchers in unraveling the complexities of the Caledonian-Appalachian orogen during the period when mountain systems were first being explained in terms of plate tectonics. Because of the extremely complex nature of the mountain system with multiple plate collisions and embedded exotic fragments with complex relations, more geologists performed research here than any other place in the world. This orogen was in the geologic spotlight for many years. With all of the geologists performing research and all of the literature being released on the Caledonian-Appalachian orogen it became very difficult to distinguish oneself in the profession. Yet Nicholas Rast did just that.

One of the main reasons for his insight into regional problems and relations is that Rast worked in so many areas along the orogen. He performed detailed field research in England and Scotland, maritime Canada, New England, and the central and southern Appalachians in the various academic positions he held. This breadth of experience allowed him to recognize regional stratigraphic correlations that few others are capable of. It also allowed him to apply solutions to regional problems that appear sound in one area to other areas. His good memory for stratigraphic re-

Nicholas Rast (center) flanked by his former student Brian Sturt (left) and colleague James Skehan, S.J. (right) on a field trip to northern Newfoundland, Canada *(Courtesy of N. Rast)*

lations, his keen insight, and his ability to synthesize multiple sources of information also contributed to his reputation.

Because Nick Rast was willing to address any aspect of geology in his research, he also participated in process-oriented studies. Early in his career, he studied the mechanics of boudinage, a distinctive structure that forms from the stretching and pulling apart of rock layers into a chain of blocks. Later, he studied the mechanisms involved in the intrusion of magma into preexisting rock layers. He even edited a prominent volume on the topic entitled *Mechanism of Igneous Intrusion.* There are many other examples of his versatility on a worldwide basis. Finally, Rast was a great geological diplomat with his ability to converse with people from any culture reinforced by his air of sophistication.

Nicholas Rast was born on June 20, 1927, in Tehran, Iran, to European parents. He completed his primary education in Nemasi, Iran, and his secondary education was in Shahpour, Shiraz, where he graduated in 1946. He received his technical education at the Technical Institute, Abadan, Iran, where he earned a diploma in industrial chemistry in 1948. He enrolled in University College in London, England, and earned a bachelor of science degree in geology with honors in 1952. Rast completed his Ph.D. at the University of Glasgow, Scotland, in geology in 1956. His first academic position was at University of Wales, England, as a lecturer in 1955. He then accepted a position at University of Liverpool, England, in 1959. In 1971, Rast joined the faculty at the University of New Brunswick, Canada, where he served as chairperson. In 1979, he accepted a position at the University of Kentucky in Lexington as the Hudnall Professor of Geology. He served as chairperson from 1981 to 1989. Rast retired to professor emeritus in 2001. He succumbed to cancer in late August of 2001. Rast was married twice and had two children by the first marriage and three by the second. Rast was multilingual (including Russian).

Nicholas Rast had a very successful career. He is an author of some 110 articles in international journals and professional volumes. Many of these are seminal works on the Caledonian-Appalachian orogen. He was probably best known for his edited volumes, including *Assembly and Dispersal of Supercontinents* and *Profiles of Orogenic Belts,* in addition to his translated volumes like *Geology of the U.S.S.R.* In recognition of this research, he received several honors and awards. He was a Lyell's Fund recipient from the Geological Society of London in 1962, the Geological Society of Liverpool Medal recipient in 1963 and a Royal Society Visiting Professor in the National University of Mexico in 1970. Rast served as editor for *Journal of Geodynamics* and *Geologica Revista Mexicana* as well as on the editorial staff for *Tectonophysics, Earth Science Reviews,* and *Canadian Journal of Earth Sciences.* He also served on numerous committees and panels for the Geological Society of America, the Geological Association of Canada, the Geological Society of Mexico and the Geological Society of London.

⊠ Raup, David M.
(1933–   )
American
Invertebrate Paleontologist

A revolution occurred in the field of paleontology in the mid-1960s. It went from a purely descriptive science to a modern integrative science that utilized mathematical analysis and techniques from other sciences. One of the true leaders of this revolution was David Raup. His main interest was in echinoids. He determined the orientations that the minerals grow to form the shell using advanced optical techniques. He later quantitatively analyzed the coiling geometry of snails. He used computer programs to define a logarithmic spiral for a snail shell and analyze it in three dimensions and in many directions long before anyone else even dreamed of using computers for such applications. The paper, "Theoretical Morphology of the Coiled Shell," is an example of such work. He later developed a computer program to grow an echinoid in a step-by-step progression. Even if this computer model was not exactly the way the echinoid grew, the process forced paleontologists to consider more closely the evolutionary development of animals. For the first time, paleontology was at the same, if not a more powerful, level than evolutionary biology, in sharp contrast to the old system in which a paleontologist became expert by looking at a lot of fossils.

Raup and his students voraciously evaluated the theoretical and functional morphology of every animal from ammonites to brachiopods. He used clever mathematical techniques that he devised or adapted from principles of population biology to uncover patterns of evolution and extinction. This research involves statistical studies of large numbers of fossils to document even small changes and the environmental reasons for them. Some of the biological techniques include survivorship analysis, cohort analysis, and rarefaction. The really perplexing thing about Raup is that he does not know much formal mathematics nor did he attend many such classes in college. He simply has great mathematical insight and readily understands how to apply existing methods to demography and ultimately paleontology without having studied those methods for any length of time. Raup almost single-handedly turned the tables on the evolution biologists who had heretofore claimed evolution as the realm of biology. Now biologists were forced to follow the lead of paleontologists who could evaluate evolution on a macroscopic scale and over long time spans with the same tools as the biologists.

David Raup was born on April 24, 1933, in Boston, Massachusetts. He attended the University of Chicago, Illinois, where he earned a bachelor of science degree in geology in 1953. He completed his graduate studies at Harvard University, Massachusetts, where he earned a master of arts degree in geology in 1955 and a Ph.D. in 1957. In 1956,

Raup was an instructor at the California Institute of Technology in Pasadena before obtaining a permanent faculty position at the Johns Hopkins University in Maryland in 1957. He moved to the University of Rochester, New York, in 1966 and served as department chair from 1969 to 1971. In 1978, Raup moved to his alma mater at the University of Chicago where he was a research associate for two years before becoming a member of the faculty in three programs: geophysical sciences, conceptual foundations of science, and evolutionary biology. He served as department chair from 1982 to 1985, dean of the College of Science from 1980 to 1982, and he was named the Sewell L. Avery distinguished service professor in 1984. He retired to professor emeritus in 1994. Raup was a visiting professor several times during his career at the University of Tübingen, Germany, University of Chicago, and Morgan State College. David Raup was married twice; he has one child.

David Raup led a very productive career. He is an author of numerous articles in international journals and professional volumes. Many of these papers are benchmarks in applying mathematical and other rigorous scientific solutions to paleontological problems. He is also the primary author of the widely adopted textbook *Principles of Paleontology* with STEVEN M. STANLEY. Raup received numerous honors and awards in recognition of his contributions to geology. He is a member of the National Academy of Science and a fellow of the American Academy of Arts and Sciences. He received both the Charles Schuchert Award and the Paleontological Society Medal from the Paleontological Society (United States).

Raup performed significant service to the profession and the public. In addition to numerous committees, he served as president of the Paleontological Society in 1976–1977. He was also vice president of the American Society of Naturalists in 1983. Raup served on several panels and committees for the National Research Council, National Academy of Sciences, the National Science Foundation, NASA, American Association of

Petroleum Geologists, and the American Chemical Society. He was on evaluation committees for Harvard University and the University of Colorado, among others.

⊗ Raymo, Maureen E.
(1959–)
American
Climate Modeling

The question of why there have been so many ice ages in recent geologic history has plagued geologists for many years. Maureen Raymo uncovered a novel possibility that caused scientists to reexamine their ideas. This new "Raymo-Chamberlin Hypothesis" states that the recent cooling of climate was caused in part by enhanced chemical weathering and consumption of atmospheric CO_2 in the mountainous regions of the world and particularly in the Himalayas. This idea means that the growth of the Himalayan Mountains may have caused the onset of ice ages. As with any novel idea, there arose a great interest in supporting or disproving it in a vigorous collection of new data. This interest also includes the popular media, which greatly enhanced Raymo's visibility worldwide. Examples of papers on this research include, "Influence of Late Cenozoic Mountain Building on Ocean Geochemical Cycles" and "The Himalayas, Organic Carbon, Burial and Climate in the Miocene."

In her regular research career, Maureen Raymo examines biogeochemical processes with regard to climate cyclicity. She is especially interested in the Earth's carbon cycle which she studies using carbon isotopes. Much of her work involves studying changes in deep sea cores for geochemical and sedimentological evidence, and fossils and their linkages to ocean water chemistry. Much of this research is conducted in the North Atlantic Ocean. These multidisciplinary approaches to climate modeling show fine scale relations to discern different scales of cyclicity, whether by astronomi-

Maureen Raymo on a field trip to Tibet *(Courtesy of M. Raymo)*

cal Milankovitch-type controls or not. They also set up a series of checks and balances to better constrain the results. An example of this work is the paper, "Late Cenozoic Evolution of Global Climate."

Maureen Raymo was born on December 27, 1959, in Los Angeles, California. Her father was a physics professor who later wrote popular books. Maureen attended Brown University in Providence, Rhode Island, and graduated in 1982 with a bachelor of science degree in geology. She attended graduate school at Lamont-Doherty Earth Observatory of Columbia University, New York, where she earned a master of arts degree in geology in 1985, a master of philosophy degree in 1988, and a Ph.D. in 1989. She accepted simultaneous positions at the University of Melbourne in Australia as visiting research fellow in the meteorology department and associate scientist in the geology department in 1989–1990. In 1991, she accepted a position as assistant professor at the University of California at Berkeley, but departed in 1992 to accept a position at the Massachusetts Institute of Technology. She remained there until

2000, when she accepted a position at Boston University as a research associate professor.

Although still early in her career, Maureen Raymo has already made an impressive impact on the field of Earth science. She has published three books and volumes and some 47 articles in professional journals and collected volumes. She published a popular geology book with her father entitled, *Written in Stone—A Geologic History of the Northeast United States.* It is in its third printing and was greatly revised in 2001. Several of her research papers have appeared in the highly prestigious journals *Science* and *Nature,* in addition to the high-profile journal *Geology.*

Maureen Raymo's work has been well recognized in the field as shown by the number of honors and awards she has received. She received a Presidential Young Investigator Award from the National Science Foundation in 1992. In 1993, she was awarded a Cecil and Ida Green Career Development Chair at Massachusetts Institute of Technology. The Joint Oceanographic Institutions/USSAC in 1994–1995 and the Mountain Research Center in 1996 named her a Distinguished Lecturer at Montana State University. She gave the keynote address at the Chapman Conference on Tectonics and Topography in 1992 and the inaugural lecture at the WISE Public Lecture Series at Syracuse University in 1999. She has been invited to speak at several of the most prestigious international topical conferences worldwide in addition to those at regular geological and oceanographic society conventions. She has also presented 58 departmental seminars over the past 11 years.

Maureen Raymo is probably best known by the public for her science film features. She and her work on climate change and climate modeling with regard to why we have ice ages were featured in four films with general distribution. In 1995, she appeared in the BBC Horizon Series in a film entitled, *Tibet: The Ice Mother.* That same year she appeared in the DSR (German Public Television) film entitled, *Abenteuer Wissenschaft:* *Tibet Teil 1 und 2* (Adventure Science: Tibet part 1 and 2). In 1996, she appeared in a NOVA series production through WGBH public television entitled, *Cracking the Ice Age* with an appearance by Massachusetts Institute of Technology colleague PETER MOLNAR. In 1998, she was featured in the BBC production *Earth Story: Winds of Change.*

⊠ **Revelle, Roger**
(1909–1991)
American
Oceanographer, Science Advocacy

The *New York Times* described Roger Revelle as "one of the world's most articulate spokesman for science" and "an early predictor for global warming." Others have described him as the "grandfather of the greenhouse effect." In any event, Roger Revelle was one of the true giants of Earth science and one of the most influential modern scientists. He is best known for his work on atmospheric carbon dioxide. In 1957, he and HANS E. SUESS were the first to demonstrate that carbon dioxide levels had increased as a result of the burning of fossil fuels. This research led Revelle into his other career as a science adviser. He was named to President Lyndon Johnson's Science Advisory Committee Panel on Environmental Pollution in 1965. This committee published the first U.S. governmental acknowledgment that carbon dioxide from fossil fuels was a problem. In 1977, Revelle served as chair of a National Academy of Sciences Panel on Energy and Climate. They concluded that 40 percent of the anthropogenic carbon dioxide has remained in the atmosphere, two-thirds of which is from fossil fuel and one-third from the clearing of forests. In 1982, Revelle published a widely read article in the magazine *Scientific American* entitled, "Carbon Dioxide and World Climate" that addressed all of the related issues of the greenhouse effect including the rise in global sea level and the relative role played by the melting of glaciers and ice sheets versus the thermal expansion of the warming surface waters.

Carbon dioxide may have provided Roger Revelle with his popular fame but he is just as well known in the Earth sciences for his oceanographic research. He designed and led the famous MidPac Expedition in 1950 in which they mapped the mid-ocean ridges, the 40,000-mile-long, 60-mile-wide seafloor mountain ranges that would later revolutionize the plate tectonic paradigm. The Capricorn Expedition of 1952 involved the dredging of the Tonga Trench in the South Pacific in which the basic ideas for the process of subduction were initiated. These and later expeditions like TransPac (1953), NorPac (1955), Downwind (1957) and NAGA (1959) resulted in important discoveries like the thinness of deep sea sediments, the high heat flow in ocean crust, the young age of sea mounts and the existence of enormous fault-fracture zones, now called transform faults, among others. He even worked with HARRY H. HESS on the Mohole project to drill to the mantle under ocean crust which was never completed but which resulted in the successful Ocean Drilling Program (ODP). These bathymetric, magnetic, and other data collected by Revelle served as the empirical basis for the conception of seafloor spreading and plate tectonic theory.

This pioneering research, however, was not his second career. It was public policy where Roger Revelle was equally effective. In 1961, President Kennedy asked Revelle to become America's first scientific adviser to the secretary of the interior. He was a member for the U.S. National Committee for UNESCO. He was chair of the White House Interior Panel on Waterlogging and Salinity in West Pakistan. He was a member of the International Science Panel of the president's Science Advisory Committee and of the Naval Research Advisory Committee, among many others. This work moved Revelle into population studies and resources versus pollution. He was equally active in that field, working principally in India, Pakistan, and Nepal on

water quality and supply as well as food supply and distribution.

Roger Revelle was born on March 7, 1909, in Seattle, Washington, but his family moved to Pasadena, California, in 1917. He was a gifted student if not a prodigy and entered Pomona College, California, in 1925 at the age of 16. He graduated in 1929 with a bachelor of arts degree in geology after switching his major from journalism. He met Ellen Virginia Clark in 1928, who was attending neighboring Scripps College. She was the grandniece of Ellen Browning Scripps, who was a founder and patron of Scripps College. Roger Revelle and Ellen Clark were married in 1931; they had four children. Revelle began his graduate studies at Pomona College but after one year transferred to the University of California at Berkeley. In 1931, he received a research assistantship at the Scripps Institution of Oceanography in La Jolla, California. He graduated with a Ph.D. in 1936 and was immediately appointed as an instructor at Scripps Institution. However, Revelle spent a year in postdoctoral study at the Geophysical Institute in Norway before beginning his academic career. In 1941, he was called for training duty as a sonar officer five months before being called for active duty at the U.S. Navy Radio and Sound Laboratory in San Diego, California. He was called to Washington, D.C., where he was the commander of the Oceanographic Section of the Bureau of Ships for the duration of World War II. He was intimately involved in the planning of the invasion of Japan. In 1946, he was named to head the geophysics branch for the U.S. Navy. Revelle returned to Scripps Institution in 1948 as associate director and served as director from 1951 to 1963. In 1958, he was named the director of the Institute of Technology and Engineering and in 1960, he became the dean of the School of Science and Engineering and chief administrative officer of the University of California in San Diego, which he helped to establish and which had assumed Scripps Institution. In 1964, Revelle completely changed careers from oceanography to public policy. He founded the Center for Population Studies at Harvard University, Massachusetts, where he was named the Richard Saltonstall Professor of Population Policy. In 1976, he returned to the University of California at San Diego as a professor of science and public policy. Roger Revelle died on July 15, 1991, from heart disease.

In recognition of his contributions to oceanography and public policy, Roger Revelle received numerous honors and awards. Among these, he received the Tyler Ecology Energy Prize, the Balzan Foundation Prize (similar to the Nobel Prize), and the National Medal of Honor from President George H. W. Bush in 1990. He has a research ship named after him at Scripps Institution and a building named after his wife and him at Harvard University. He was a member of the National Academy of Sciences and received 14 honorary degrees from schools like Harvard University, Dartmouth University, Williams College, Pomona College, and Carleton College, among others. Other honors and awards include the Agassiz Medal from the National Academy of Sciences, the Bowie Medal from the American Geophysical Union, the Albatross Medal from the Swedish Royal Society of Science and Letters, the Order of Sitara-I-Imtaz from the government of Pakistan, the Climate Institute Award, and the Vannevar Bush Award from the National Science Board, among many others.

Richter, Charles F.
(1900–1985)
American
Geophysicist

The name of Charles Richter is one of the best known in the Earth sciences by virtue of his Richter scale, used to rank earthquake magnitudes. It is now the scale of choice for reporting by the popular media as well as most of the laboratories. There were other scales for ranking earthquakes as early as the late 19th century. The most popular of these was the 10-point scale of François-Alphonse Forel and Michele Stefano de Rossi. In 1902, Giuseppe Mercalli created a 12-point scale that replaced all of the earlier scales. It measures the intensity of shaking during an earthquake based upon inspection of damage and interviews with survivors. Therefore, the Mercalli number varies with location by proximity to the epicenter as well as the materials through which the earthquake waves pass and even population density. With the advent of more modern advanced seismographs that continuously monitored seismic activity, the Mercalli scale had become outdated. By the 1930s, Richter was recording some 200 earthquakes per year in southern California. He found the Mercalli system so inadequate and misleading when it came to briefing the news organizations that he began investigating alternatives. In 1935, Richter developed a new logarithmic scale that measures the amplitude of the seismic waves from seismograph records and accounts for the material through which they pass. The Richter magnitude determines the amount of energy released by the earthquake rather than the local damage. It relies on measuring the strength of an earthquake at three or more points so that the point of origin can be determined. By comparing the distance with the recorded strength, the strength of the earthquake at the epicenter can be estimated.

Charles Richter and BENO GUTENBERG applied this new system to earthquakes on a worldwide basis. This began a great collaboration for the next decade or so that produced a series of seminal papers with the title "On Seismic Waves." These papers explained how to interpret the seismic wave arrivals that are drawn on a seismogram by the seismograph. These papers provided the basis for modern deep-Earth seismology.

Charles Richter was born on April 26, 1900, on a farm near Hamilton, Ohio. His parents soon divorced and his mother resumed use of her maiden name of Richter as did Charles. In 1909, the family moved to Los Angeles, California, where he spent his later youth. Charles Richter was something of a prodigy and entered the University of California at Los Angeles at 16 years of age. After one year, he transferred to Stanford University, California, where he earned a bachelor of science degree in physics in 1920. he completed his graduate studies at the California Institute of Technology in Pasadena and earned a Ph.D. in theoretical physics in 1928. Richter had planned on a career in astronomy but in 1927 he was invited to become a research assistant at the seismological laboratory of the Carnegie Institution of Washington, D.C., which was also located in Pasadena, California. Charles Richter married his lifelong companion, Lilian Brand, in 1928. The seismological laboratory became part of California Institute of Technology in 1936 and Richter joined the faculty in 1937. He remained a faculty member until his retirement to professor emeritus in 1970. The only time he was absent from Cal Tech was in 1959–1960 when he was a visiting scientist in Japan. Richter remained active after his retirement helping to found the consulting firm of Lindvall, Richter and Associates, which performed seismic evaluations of buildings and other structures. Charles Richter died on April 30, 1985, in Altadena, California.

Charles Richter was an author of more than 200 scientific articles in international journals, professional volumes, and governmental reports. Many of these are true classics on earthquake seismology, seismic wave travel, and the scaling of earthquakes. He was also the author of a widely adopted textbook, *Elementary Seismology* and coauthor of *Seismology of the Earth* with Beno Gutenberg, which was also well received. In recognition of his contributions to geophysics, Charles Richter received a number of honors and awards. He was a fellow of the American Academy of Arts and Sciences. He received an honorary doctorate from California Lutheran College and the Medal of the Seismological Society of America. The Charles F. Richter Seismo-

logical Laboratory at the University of California at Santa Cruz was named in his honor. Richter also served as president of the Seismological Society of America.

⊠ Ringwood, Alfred E. (Ted)
(1930–1993)
Australian
Geochemist

Alfred (Ted) Ringwood is one of the true giants of geology for many reasons. He is best known for his solution of a fundamental problem in geology, the transition between the upper and lower mantle. Seismologists had known for many years that seismic velocities in the mantle increase rapidly between 400- and 900-km depth. This transition zone was speculated to be the result of the crushing down of mineral structures (phase transformations) as the result of the extreme pressure. Such reconfiguring of atoms would be similar to the transformation of graphite to diamond with pressure. The problem was that no laboratory in the world could replicate those conditions. Ringwood overcame that problem by synthesizing an olivine structured mineral (olivine is the common mineral in that part of the mantle) except he used the element germanium instead of silicon (the element in natural olivine) in the structure. Because germanium has a smaller atomic radius than silicon but otherwise fits all of the other requirements, it would transform to the new structure at low enough pressures to be within the experimental range at the time. His experiments showed that the olivine structure would convert to a spinel structure and predicted by extrapolation that it should occur in natural olivine at 400-km depth. Later seismic studies found that indeed there is a seismic discontinuity at 400 km and later experimental work with more sophisticated equipment showed that Ringwood was correct. Continued research by Ringwood showed that pyroxene, the other major mantle mineral, converted to garnet structure between 400 and 650 km and that the spinel structure converted to a "perovskite" structure at still greater depths using his experimental models. Papers on this work include "Phase Transformations in the Mantle," and several other similar titles.

Ringwood then attacked the problem of how basaltic magma was generated in the mantle at the mid-ocean ridge. He bucked traditional wisdom and proposed that there was a strange composition substance in the mantle called "pyrolite" from which basalt was derived. He wrote a classic paper on the topic entitled "The Genesis of Basaltic Magmas" in 1967. The pyrolite model was subsequently disproved but the idea that different compositions of basalt are generated at different depths, which was also incorporated in the model, is still unquestioned and considered a major contribution to the science.

In addition to these gargantuan issues, Ringwood also had several other areas of interest. He gave new insights to the composition of the core of the Earth which are still accepted. He proposed models for the chemical evolution of the Earth, other planets, and meteorites. He proposed models for the composition and origin of the Moon while working on the Apollo lunar missions with NASA. He championed the idea that the Moon may have been spiralled off of the early Earth as the result of a giant impact. Finally, he applied his geochemical expertise to nuclear waste disposal. To each of these highly varied topics he made contributions to the science which still set the standards today.

In recognition of his accomplishments, Ringwood received the Antonio Feltrinelli Prize in 1991 from the Academia Lincei in Rome, the oldest scientific society in the world. Galileo was the head of the society in its early years. The Feltrinelli Prize is similar to the Nobel Prize but awards only one prize per year in all fields. It has been awarded to the likes of Thomas Mann, Igor Stravinsky, Albert Sabin, and Georges Braque. The previous geologist to receive the prize was HARRY H. HESS in 1966, another giant of geology.

Ted Ringwood was born in Kew, near Melbourne, Australia, on April 19, 1930. He attended Hawthorn West State School, Geelong Grammar School, and Melbourne High School as a youth in Melbourne. He enrolled in the University of Melbourne with a Trinity College Resident Scholarship and a Commonwealth Government Scholarship. He graduated with a bachelor of science degree in geology with honors in 1951 and a master of science degree with honors in 1953. He continued at the University of Melbourne to earn a Ph.D. in 1956 at 26 years of age. He became a research fellow at Harvard University in 1957. During this time he made several visits to Sweden to study meteorites and met Gun Carlson, whom he married in 1960. They had two children. He returned to join the faculty at the Australian National University in 1959, where he remained for the rest of his life. He served as director of the Research School of Earth Sciences from 1978 to 1983. Ringwood died of lymphoma on November 12, 1993, at the age of 63.

Ted Ringwood led an extremely productive career. He was an author of more than 300 publications, including articles in international journals and professional volumes as well as books. Many of these papers are benchmark studies on processes, properties, and compositions of rocks in the interior of the Earth. Many appear in prestigious journals like *Science* and *Nature.* He also wrote two widely acclaimed books, *Composition and Petrology of the Earth's Interior* and *Origin of the Earth and Moon,* and even had several patents for high-level nuclear waste disposal. Ringwood received honors and awards too numerous to list completely. He was a fellow of the U.S. National Academy of Sciences and the Australian Academy of Sciences. He received an honorary doctorate from the University of Göttingen, Germany. He received the Mineralogical Society of America Award, the Werner Medaille from the German Mineralogical Society, the Arthur L. Day Medal from the Geological Society of America, the Bowie Medal and the Harry H. Hess Medal from the American Geophysical Union, the Arthur Holmes Medal from the European Union of Geosciences, the Wollaston Medal from the Geological Society of London, the Goldschmidt Award from the Geochemical Society, the Inaugural Rosentiel Award from the American Association for the Advancement of Science, and the Matthew Flinders Lecture and Medal and the J.C. Jaeger Medal from the Australian Academy of Science, among many others. He served numerous endowed lectureships from the most prestigious organizations. He also performed service to the profession like serving as the vice president for the Australian Academy of Science, for example, but it is also too extensive to list here.

Rizzoli, Paola Malanotte
(1946–)
Italian
Oceanographer

The circulation of the water in an ocean basin is highly complex. It depends upon the shape of the basin, temperatures of the air and the water, prevailing winds and storms, in addition to the rotation of the Earth. These processes and interactions of ocean and atmosphere lead to large-scale phenomena like El Niño and North Atlantic oscillations, which in turn can have major effects on our climate. Paola Rizzoli is one of the premier scientists to model such circulation. She applies a strong physics and math background to understand and ultimately to predict these catastrophic changes in ocean circulation. Her first interest was to model the regular and dangerous flooding under varying meteorological conditions in Venice in her native Italy. She expanded these studies to investigate the dynamics of strong oceanographic and meteorological flow structures with long lives, like hurricanes, and their effects on general ocean cir-

culation. These features violate the principles of chaos, the intrinsic unpredictability of ocean and meteorological flow structures. This research was motivated by the work of Edward Lorenz, the developer of chaos theory who helped bring Rizzoli to the Massachusetts Institute of Technology in 1981.

Paola Rizzoli's research then expanded to model general ocean circulation from the global scale to the ocean basin scale. For example, she has been modeling circulation in the Atlantic Ocean, which is a crucial component for modeling the climate system of the Earth. Her research first focused on the Gulf Stream, but more recently it has concentrated on the tropical-subtropical interactions affecting the equatorial Atlantic. She also models marginal seas like the Mediterranean Sea and the Black Sea as a subcomponent of these general circulation models. These local circulation models have major implications for the development of local ecosystems and the flora and fauna that inhabit them. To collect these data, Rizzoli served as chief scientist during oceanographic campaigns in the Adriatic Sea on the research vessels *Adriatic I, II,* and *III* of the Italian National Research Council. Finally, she does research in data collection and assimilation of all available observations for "model data synthesis" for numerical circulation models.

Paola Rizzoli was born on April 18, 1946, in Lonigo, Italy. She attended Lyceum Benedetti in Venice, Italy, where she earned a bachelor of science degree in physics and mathematics with highest honors in 1963. She attended graduate school at the University of Padua, Italy, where she earned a Ph.D. in physics, summa cum laude, in 1968. She completed a one-year postdoctoral fellowship at the University of Padua in 1969 before joining the Istituto Dinamica Grandi Masse, which was created by the Italian National Research Council the next year. She achieved the rank of senior scientist by 1976. In 1971, she became a regular commuter to the Scripps Institution of Oceanography at the University of California at San Diego.

She was a visiting scientist in 1971–1974 before entering the Ph.D. program. Rizzoli earned a second Ph.D. in oceanography in 1978 and accepted the position of Cecil and Ida Green Scholar at the Institute of Geophysics and Planetary Physics at the University of California at San Diego while on leave from the Istituto Dinamica Grandi Masse. In 1981, she joined the faculty at the Massachusetts Institute of Technology in Cambridge, where she remains today. Since 1997, Rizzoli has served as director of the Joint Program in Oceanography and Ocean Engineering between the Massachusetts Institute of Technology and Woods Hole Oceanographic Institution. Paola Rizzoli married Peter Stone in 1987.

Paola Rizzoli is an author of 97 articles in international journals, professional volumes, and governmental reports. She is also an editor of nine professional volumes. Many of these papers are seminal works on the modeling of ocean circulation. She received nine honors and fellowships, including the 1998 Masi Prize from the Italian Ministry of Culture and Education.

Rizzoli has performed significant service to the profession. She served as president (1999–2003) and deputy secretary (1995–1999) for the International Association for the Physical Sciences of the Ocean (IAPSO), in addition to many other committees and panels. She was also president of the Committee on Physical Oceanography of the CIESM (International Exploration of the Mediterranean Sea) 1984–1988) and president of the Italian Commission to Assign University Chairs in the Physics of the Earth (1991–1992). She also served on numerous committees and panels for the National Center for Atmospheric Research, the American Meteorological Society, Institute of Naval Oceanography, National Centers for Environmental Prediction, Goddard Space Flight Center, UNESCO, and the National Science Foundation. She also served as an editor for the *Journal of Geophysical Research,* among other editorial positions.

⊠ Rodgers, John
(1914–)
American
Field-Regional Geologist

John Rodgers collects mountain ranges. This means that he reads the literature, talks to the geologists working in the area, and takes extensive field trips through the mountains. There are very few, if any, mountains that he has not visited. He can be considered the "grandfather of regional tectonics" for this reason. Although he considers himself a synthesizer of information rather than an innovator, few would agree with that evaluation. His experience gives him the unique ability to evaluate the comparative anatomy of mountain belts. He can visit any field area and put those rocks into the perspective of many other similar or contrasting areas. This vision has guided many researchers to better solutions of their work. His work has elucidated the commonalities and variations in the mountain building process especially with regard to fold and thrust belts. Thus there is a baseline from which other observations may be compared. Two of his more recent papers include, "Fold and Thrust Belts in Sedimentary Rocks Part 1: Typical Examples" and "Fold and Thrust Belts in Sedimentary Rocks Part 2: Other Examples, Especially Variants."

John Rodgers's first and primary interest has been the Appalachian Mountains of eastern North America, especially those in New England. He has mostly concentrated on the stratigraphy and structure of the sedimentary rocks, but he is quite comfortable in the metamorphic rocks as well. He wrote a seminal work on the Taconic orogen early in his career but considered the whole Appalachian orogen in his book, *The Tectonics of the Appalachians.* It was at that time that he was in an open controversy over whether the crystalline rocks of the continental basement were involved in the deformation of the sedimentary cover sequence. Rodgers maintained that the basement was not involved, supporting a "thin-skinned" model. He prevailed in the controversy and was proven correct as more and more data were presented. Even through the many years and numerous mountain belts, Rodgers has maintained a strong interest in the building of the Appalachians.

John Rodgers was born on July 11, 1914, in Albany, New York. After graduating cum laude and valedictorian of his class at Albany Academy, New York, he attended Cornell University, New York, where he earned a bachelor of arts degree and a master of science degree in geology in 1936 and 1937, respectively. He earned his doctoral degree from Yale University in 1944 in geology. His graduate studies were interrupted from 1939 to 1946, when he worked for the U.S. Geological Survey as a scientific consultant to the U.S. Army Corps of Engineers during World War II. He helped plan the invasion of Okinawa, Japan, using his geologic knowledge of beaches. In addition, he was instrumental in evaluating the natural resources of Japan. He joined the faculty at Yale University in 1946 and remained there for the rest of his career. He became the Benjamin Silliman Professor of Geology in 1962 and retired to an emeritus professor position in 1985. John Rodgers is an accomplished pianist and an aficionado of classical music. He also enjoys the study of foreign languages and history.

John Rodgers is one of the most influential and respected geologists of all time. Many of the papers and books that he has written are true classics that are studied worldwide. He has also mentored some of the leading structural and tectonic geologists in the world, further extending his influence. This productivity has been well recognized in the profession in terms of honors and awards. He was awarded the Medal of Freedom by the U.S. Army in 1947 for his contributions to the war effort. He is a member of the National Academy of Sciences and the American Academy of Arts and Sciences. The Geological Society of America recognized his achievements by awarding him the Penrose Medal in 1981 and

John Rodgers (third from left) on a field trip for the International Geological Correlation Project (IGCP) in the northern Appalachians in 1979 *(Courtesy of James Skehan, S.J.)*

the Structural Geology and Tectonics Division Career Contribution Award in 1989. He was awarded the Gaudry Prize from the Geological Society of France in 1987 and the Fourmanier Medal from the Royal Academy of Science, Fine Arts and Letters of Belgium in 1987. He was a Guggenheim Fellow in 1973–1974 and a National Academy of the Sciences Exchange Scholar with the USSR in 1967.

John Rodgers has performed more service to the profession than can be listed here. He was the president of the Geological Society of America in 1970 after having served on many committees. He was the secretary of the Commission on Stratigraphy for the International Geological Congress from 1952 to 1960, and the vice president for the Société Géologique de France in 1960. He was an associate editor for the *American Journal of Science* in 1948 to 1954 at which time he became editor, which he has been ever since.

⊠   Roedder, Edwin W.

(1919–   )
American
Geochemist

When a mineral crystallizes, it can trap a small bubble of fluid and/or vapor that is present in the crystallization process. This completely encapsulated bubble is called a fluid inclusion. The fluid within it tells geologists the composition of the fluid that accompanied the crystallization of a pluton or the metamorphism of a rock or terrane, or the mineralization of a vein among other things. By heating or cooling the inclusion until all of the liquids and gases (and even solids) combine on a stage attached to a microscope (to observe the transformation), an experimentally determined "isochore" may be used to determine the pressure and temperature of formation. Before Edwin Roedder did his pioneering research to establish this great use of fluid inclusions, they were merely curiosities to be viewed under a microscope. They were indeed very curious. The fluid in some of the little bubbles would vibrate furiously. It was assumed that this vibrating was an example of "Brownian Motion" and the result of pent-up energy. Roedder showed that the motion was in fact the result of minute thermal gradients across the inclusions. He even patented an ingenious device to sense tiny thermal gradients based upon this observation entitled "Device for Sensing Thermal Gradients." No combination of thermocouples or thermometers has the same delicate sensitivity, nor can they match the speed of response of his device. Edwin Roedder is the true "father of fluid inclusion research," which is now a standard technique for petrologic and ore mineralization research among others. His papers include "Ancient Fluids in Crystals" and "Fluid Inclusions as samples of the Ore Fluids."

Edwin Roedder had another major research direction. He was an experimental petrologist starting at the Geophysical Laboratory of the Carnegie Institution of Washington, D.C., as a graduate student. He found that by adding iron to a relatively common system, two liquids emerged that were immiscible just like oil and water, as reported in the paper "Silicate Liquid Immiscibility in Magmas." The importance of this finding was not fully appreciated until the lunar samples were returned to Earth by the Apollo astronauts. Scientists discovered glass globules in the igneous rocks. When they determined the composition of this glass, they found that it was exactly the same as Roedder found in his experiments. This discovery caused Roedder to renew his work and ultimately to propose that liquid immiscibility is a major process in magmatic differentiation, planetary evolution, and the formation of mineral deposits. As if his fluid inclusion work was not enough of a contribution to the science, this work on immiscibility caused another great impact on the profession.

Edwin Roedder was born on July 30, 1919, in Monsey, New York, but he spent his youth in Philadelphia, Pennsylvania. He attended Lehigh

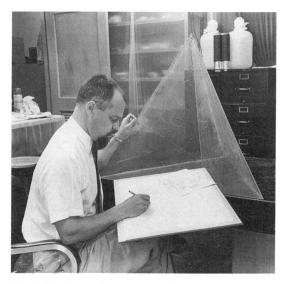

Edwin Roedder performing research on rock compositions on a phase diagram in his laboratory at the U.S. Geological Survey in 1958 *(Courtesy of the U.S. Geological Survey, E.F. Patterson Collection)*

University, Pennsylvania, where he earned a bachelor of arts degree in geology in 1941. He worked as a research engineer at Bethlehem Steel Corporation from 1941 to 1946. Roedder conducted his graduate studies at Columbia University, New York, where he earned a master of arts degree in geology in 1947 and a Ph.D. in 1950. He joined the faculty at the University of Utah upon graduation. He then moved to the U.S. Geological Survey in 1955 as chief of the Solid State Group of the Geochemical and Petrological Branch, but he also held positions of staff geologist and geologist. He remained at the U.S. Geological Survey until his retirement in 1987, whereupon he became an associate of the department at Harvard University, Massachusetts.

Edwin Roedder has led a very productive career with authorship on numerous articles in international journals, professional volumes, and governmental reports. He also edited and wrote several books and volumes. Several of these papers are benchmark works on fluid inclusions (including the definitive book simply entitled, *Fluid Inclusions*) and magma immiscibility. In recognition of his research contributions to the profession of geology, he has received several honors and awards. He is a member of the National Academy of Sciences. He was awarded an honorary doctorate from Lehigh University. He also received the Exceptional Scientific Achievement Medal from NASA, the Werner Medal from the German Mineralogical Association, the Roebling Medal from the Mineralogical Society of America, the Penrose Medal from the Society of Economic Geologists and the H.C. Sorby Medal.

Roedder also performed significant service to the geological profession. He served as president (1982–1983) and vice president (1981–1982) of the Mineralogical Society of America, in addition to numerous panel and committee positions. He also served as president of the Geochemical Society in 1976–1977 as well as committee work. Roedder served on numerous panels and committees for the National Research Council, National Science Foundation, and several governmental advisory committees, especially with regard to nuclear waste disposal.

⊠ Romanowicz, Barbara
(1950–)
French
Geophysicist

How do scientists know what the deeper parts of Earth look like if it cannot be seen? The answer is that seismic waves generated by earthquakes penetrate all of Earth and are received by seismographs worldwide. By analyzing small variations in these waves, seismologists can image the interior parts of the Earth much like a sonogram images a fetus. Barbara Romanowicz is one of the premier experts on imaging the Earth's interior. Her research covers all levels, but some of her most exciting research has been on the core and deep mantle. She found that the inner solid core, rather than simply a solid massive ball as previously considered, has a layering or anisotropy to it. Romanowicz interpreted this anisotropy to reflect convection as reported in her paper "Anisotropy in the Center of the Inner Core," among others. She also performed fine detailed imaging of the lowermost mantle above the outer core. Again, previous assumptions of a uniform character were disproved when she found that there are topographical and density anomalies there. This work is summarized in the paper "Anisotropic Structures at the Base of the Mantle."

Romanowicz also produces seismic tomography images, essentially CAT scans of the upper mantle and lower crust (e.g., "Seismic Tomography of the Earth's Mantle"). She compared this tomography to plate motions, both current and past, and found good correlations. Basically, hot and cold regions develop based upon whether there is mantle upwelling (hot) like at divergent boundaries, or subduction zones where it is cooler because cooler crust is being driven into the asthenosphere. She modeled the deep structure beneath the Tibetan Plateau, the continental United States, and the Atlantic Ocean.

A third area of interest for Romanowicz is earthquake processes. She looks at the release of energy, focal mechanisms, and sources of major earthquakes. This work includes the scaling of events and real-time estimation of earthquake parameters. She also studies the attenuation (loss of energy) of seismic waves as they travel through the Earth. Some of the earthquakes that she has studied include 1999 Izmit, Turkey; 1989 Loma Prieta, California; 1988 Spitak, Armenia; 1985 Chile; 1986 Romania; 1989 Macquarie Ridge, Australia; and earthquakes in Sudan and Syria. This expertise has prompted several earthquake hazard reduction groups in California to solicit her opinion and aid.

Barbara Romanowicz was born on April 5, 1950, in Suresnes, France. She attended the University of Paris, France, where she earned a bachelor of science degree in mathematics with honors in 1974. She attended Harvard University, Massachusetts, for graduate studies and earned a master of science in applied physics in 1975. She returned to the University of Paris and completed doctoral degrees in both astronomy and geophysics in 1975 and 1979, respectively. Romanowicz served as a research associate at the Institute for the Physics of the Earth in Paris, France, in 1978–1979 before becoming a postdoctoral research associate at the Massachusetts Institute of Technology from 1979 to 1981. Romanowicz returned to the Institute for the Physics of the Earth to assume the position of director of research and the director of the geoscope program in 1981. In 1991, she joined the faculty at the University of California at Berkeley where she remains today. She is also director of the Berkeley Seismological Laboratory. Barbara Romanowicz married Mark Jonikas in 1979; they have two children.

Barbara Romanowicz is in the midst of a very productive career. She is an author of more than 110 articles in international journals, professional volumes, and governmental reports. Many of these papers are benchmark studies on the deep architecture of the Earth as well as earthquake processes. Several appear in the high-profile journal *Science.* The contributions to geology and geophysics by Romanowicz have been recognized by the profession as evidenced by her honors and awards. She received the French Academy of Sciences Prize, the Silver Medal from the French National Academy of Sciences, and the Wegener Medal from the European Union of Geosciences.

Romanowicz has also performed significant service to the profession. Among much committee and panel work, she was president of the Seismological Section of the American Geophysical Union. She also served as vice president of the Federation of Digital Seismic Networks and the seismological section of the International Union of Geodesy and Geophysics (France). She served on numerous committees and panels for the National Science Foundation, the National Research Council, and the Seismological Society of America, and on several evaluation committees for departments at Harvard University and University of California at Los Angeles, among others. Her editorial work is also extensive, including serving as the European editor for *Geophysical Research Letters* and editor for *Physics of Earth and Planetary Interiors.*

⊠ Rosendahl, Bruce R.
(1946–)
American
Reflection Seismologist (Tectonics)

Bruce Rosendahl is an Indiana Jones type in the geological community. In a feature article describing his exploits in East Africa, *Esquire* magazine called him "Bwana Boom." The name was ascribed to him by the Turkana tribe of northern Kenya after Bruce blasted a channel into remote Lake Turkana (Lake Rudolf) to launch his research vessel *NYANJA*. Tossing pelican guano hundreds of feet into the air was just one of many adventures that Rosendahl endured as his Project PROBE team conducted the first-ever advanced seismic surveys (like a sonogram of the subsurface) of the African Great Lakes. His work was highlighted in a PBS television documentary in 1986. Facing fierce storms, hurricane-force winds, and the occasional rebel boat over a span of eight years, Rosendahl and his team managed to decipher the faults, folds, and processes of formation for Lakes Tanganyika, Malawi, Victoria, and Turkana. This work led to the realization that continental rift zones are compartmentalized along their lengths into fault-bounded basins (half-grabens), in which the bounding faults tend to alternate on either side of the basins. These flip-flop faults are inclined toward each other. Rosendahl found that the switch of the faults from one side to another form distinct "accommodation zones" marked by characteristic folds, faults, and depositional patterns. The resulting models of linkage of the basins, and their applicability to hydrocarbon exploration, have become a standard in unraveling rift tectonics around the world. The work has also spurred several international research efforts on the African lakes.

Rosendahl also began the PROBE program which uses ultra-deep, multichannel seismic techniques to image the Earth's crust and upper mantle to depths of 40 kilometers along the West African continental margin. The PROBE study resulted in some of the clearest acoustic profiles yet obtained beneath the thick sedimentary sections, which blanket passive oceanic margins. An example is the paper "Nature of the Transition from Continental to Oceanic Crust and the Meaning of Reflection Moho." These images reveal the segmented nature of how continents break up and how the equatorial South Atlantic Ocean was created. Rosendahl is currently extending his work to the matching (conjugate) Brazilian side of the Atlantic to determine the architecture of originally paired margins.

Bruce Rosendahl aboard a research vessel in Florida *(Courtesy of B. Rosendahl)*

Bruce Rosendahl was born December 28, 1946, in Jamestown, New York. He grew up in a small village on the shores of Lake Chautauqua in the western part of the state. He learned to boat, fish, and dive about the time he learned to walk. He undoubtedly became an oceanographer before he could spell the word. Rosendahl earned his bachelor of science degree in geology and his master of science degree in geophysics from the University of Hawaii in 1970 and 1972, respectively. He continued his studies at the Scripps Institution of Oceanography, University of California at San Diego, and earned a doctoral degree in Earth sciences in 1976. His Ph.D. dissertation involved the use of seismic methodology to image a zone beneath the East Pacific Rise. Bruce Rosendahl has been married to Susan E. Rosendahl since 1978. She also has a doctoral degree and is an administrator at the Keys School in Annapolis, Maryland. They have two children.

After a short postdoctoral research position at Scripps Institution, Rosendahl joined the faculty at Duke University, Tennessee, in 1976. During this period he formulated and directed the PROBE project. Bruce moved to the University of Miami, Florida, in 1989 to become dean of the Rosenstiel School of Marine and Atmospheric Science, a position he held until 1995. He also was awarded the Lewis Weeks Endowed Chair in marine geophysics, which he retains today.

Bruce Rosendahl is the author or coauthor of several books, two seismic atlases, and more than 65 papers in scientific journals. His volume *African Rifting* is a classic in the field. He has written numerous articles for technical publications, popular science magazines, trade publications, and newspapers. Bruce's editorial roles have included membership on the board of contributors for the *Miami Herald* newspaper, technical editor of *Sea Frontiers* magazine, and adviser to *Time-Life Books,* PBS's NOVA series, and *National Geographic* magazine. His professional service has included membership on the Joint Oceanographic Institution's board of governors, executive committee of JOIDES, Southern Association of Marine Laboratories, Council on Ocean Affairs, University Corporation for Atmospheric Research, Interunion Commission on the Lithosphere, Marine Geology Committee of the American Association of Petroleum Geologists, and International Lithosphere Program. Rosendahl has been a trustee for the Miami Museum of Science and has served on the board of directors for the Miami Marine Council and the Maritime and Science Technology Academy.

S

⊠ Sagan, Carl E.
(1934–1996)
American
Astronomer, Planetary Scientist

Carl Sagan is among the best-known scientists of the 20th century. He was trained as an astronomer and performed much outstanding research in astronomy. However, relatively early in his career, he turned his efforts to planetary science and contributed greatly in that regard as well. His first planetary research was to study the greenhouse effect on Venus. He showed that the thick caustic clouds that engulf the planet prevented solar radiation from escaping and predicted excessive surface temperatures that were later confirmed by space probes. He studied the atmosphere on Mars, predicting it to be a desert and explaining observed seasonal changes to be a result of windblown dust storms. He completed studies of Saturn's moon Titan, on which he identified organic aerosols in the atmosphere. Because of his expertise on planets and planetary evolution, Sagan was invited to participate on NASA's Apollo missions. He was also one of the leaders in each of the teams for the unmanned Mariner, Viking, Voyager, and Galileo missions to other planets of the solar system.

Carl Sagan later became interested in the origin of life on Earth. He studied the conditions and constraints on how life started but also how it evolved. With his background in these constraints coupled with knowledge of planetary evolution, he helped pioneer a new field of "exobiology" which predicts the form of extraterrestrial life under a typical scenario of development. Sagan was also very interested in mass extinctions, especially involving extraterrestrial impacts. He used the interpreted effects of such impacts to warn against nuclear proliferation to the public as well as in hearings before the U.S. Congress. He asked if such a "nuclear winter" could wipe out the dinosaurs, what could it do to humans?

No description of Carl Sagan would be complete without mentioning his work on popularizing science. Sagan has been called the "world's greatest popularizer of science." His book, *Cosmos,* which accompanied the Emmy Award- and Peabody Award-winning Public Broadcasting Station series, was the best-selling science book ever published in English. It was on the *New York Times* best-seller list for 70 weeks. He has had seven other books on that list. His book, *The Dragons of Eden,* won him a Pulitzer Prize. His novel, *Contact,* was made into a 1997 Warner Brothers movie which Sagan and his wife were coproducing at the time of his death. All of his books are vehicles to share his scientific knowledge with the public. Carl Sagan is a true giant of science.

Carl Edward Sagan was born on November 9, 1934, in Brooklyn, New York. He attended the University of Chicago, Illinois, where he earned a bachelor of arts degree with honors in 1954, a bachelor of science degree in 1955, and a master of science degree in 1956, all of which were in physics. His doctoral degree was in astronomy and astrophysics, which he received in 1960, also from the University of Chicago. He was a Miller Research Fellow at the University of California at Berkeley from 1960 to 1962 before joining the faculty at Harvard University, Massachusetts. In 1968, he moved to Cornell University, New York, where he was director of the Laboratory for Planetary Studies, in addition to his faculty position. Sagan also served as associate director of the Center for Radio Physics and Space Research from 1972 to 1981. In 1976, he was named a David Duncan Professor of Astronomy. He was also president of Carl Sagan Productions beginning in 1977. Carl Sagan's last marriage was to Ann Druyan, who collaborated with him on many of his projects. He was father to five children from previous marriages. Carl Sagan was diagnosed with mylodysplasia, a bone marrow cancer, in 1994 and died of pneumonia resulting from it on December 20, 1996, in Seattle, Washington. His remains were released in space on a subsequent space shuttle mission.

There are not many more productive careers than that led by Carl Sagan. He is an author of more than 600 scientific papers and popular articles in a variety of technical and nontechnical publications. He is also an author or editor of some 20 books and professional volumes. The honors and awards that Sagan received for his contributions both to science and bringing science to the public are equally astounding and too numerous to list here completely. He was a member of the National Academy of Sciences and the American Academy of Arts and Sciences. He received some 22 honorary degrees from American colleges and universities including such schools as Rensselaer Polytechnic Institute, Whittier College,

and the University of Wyoming, among others. He was the recipient of the Public Welfare Medal, the highest award from the National Academy of Sciences. NASA awarded him the Medal for Exceptional Scientific Achievement, the Apollo Achievement Award, and two Distinguished Public Service Awards. He also received the John F. Kennedy Award and the Masursky Award from the American Astronautical Society, the Konstantin Tsiolkovsky Medal of the Soviet Cosmonauts Federation, the Smith Prize from Harvard University, and the Explorers Club 75th Anniversary Award, among at least eight others both scientific and literary. He was named to more than 20 prestigious honorary lectureships at numerous universities and societies.

Sagan was also a great contributor of his time to professional service. He was a cofounder of the Planetary Society, which now boasts more than 100,000 members. He was president of the planetology section of the American Geophysical Union, and chair of the Division of Planetary Sciences of the American Astronomical Society, and the chair of the astronomy section of the American Association for the Advancement of Science, among numerous other committees for these organizations, as well as NASA, National Academy of Sciences, the National Research Council, and the International Society for the Origin of Life, among others. He was also a member of the U.S. Committee for East-West Accord. Sagan served as editor of the planetary journal *Icarus* for more than 12 years.

⊠ Selverstone, Jane
(1956–)
American
Metamorphic Petrologist, Tectonics

Jane Selverstone wears two hats in terms of research. She has done theoretical thermodynamic research with FRANK S. SPEAR, among others, as well as field metamorphic–tectonic research. She has done significant research on the rocks of the

Tauern Window of the eastern Alps. This work involved applying much of the earlier theoretical research to real rocks. She also related deformational processes to metamorphic processes in terms of bulk chemical, isotopic, trace element, and mineralogical changes. The control on these processes largely involves interaction with metamorphic fluids that migrate along deep-seated faults during deformation. The early research concentrated on thrust fault systems, but later work applied these same techniques to normal faults both in the Alps and detachment surfaces in metamorphic core complexes of the Whipple Mountains, California. Fluid inclusions within these fault rocks places pressure-temperature constraints in addition to those from the metamorphic mineral assemblages. This multifaceted, high-quality approach to field metamorphic problems yields intricacies to the processes that are not commonly revealed in the more cursory studies that are commonly performed on such rocks. Selverstone's excellent understanding of the detailed thermodynamics, field metamorphism, and deformational processes also allow her to interrelate these observations in an effective manner. Her ability to form collaborations with the best researchers in each particular discipline also contributes to the quality of her research. Examples of her papers include, "Quantitative Pressure-Temperature Paths from Zoned Minerals: Theory and Tectonic Applications" and "Trace Element Zoning in Metamorphic Garnet."

Jane Selverstone has also taken on a tectonic project to determine the Proterozoic assembly of the crust in the northern Colorado Front Ranges. This project is designed to unravel a complex series of plate collisions that took place about 1.4 billion to 1.7 billion years ago that helped to build the North American continent in the southwestern United States.

Jane Selverstone was born on July 6, 1956, in Cambridge, Massachusetts. She attended Princeton University, New Jersey, where she earned a bachelor of arts degree in geology in 1978 and completed a senior thesis, which she published with her adviser, L. Hollister. She earned a master of science degree in geology at the University of Colorado in Boulder in 1981 with C. R. Stern and J. Munoz as advisers. She moved to Massachusetts Institute of Technology for her doctoral degree, which she earned in 1985 under the advisement of Frank Spear. After a year as an adjunct professor at the University of Colorado in Boulder, her first faculty position was at Harvard University, Massachusetts. From 1990 to 1992, Selverstone was named as a John L. Loeb Associate Professor of Natural Sciences at Harvard. In 1992, she returned to the University of Colorado as a research associate professor. In 1995, she accepted a position at the University of New Mexico in Albuquerque, where she remains as of 2002. She was named regents' lecturer from 1998 to 2001. Jane Selverstone is married to David Gutzler, a professor of climatology, and they have two children.

Jane Selverstone doing fieldwork in the eastern Alps *(Courtesy of Gerhard Franz)*

Jane Selverstone has been very productive throughout these early stages of her career. She is an author of 49 articles in top international journals. Her articles are well received, well cited, and her coauthors are among the top professionals in the field. She has already been recognized for her work with numerous honors and awards. As a student she received the Buddington Award at Princeton University in 1978, the Waldrop Award at University of Colorado in 1981, a National Science Foundation Graduate Fellowship in 1980 to 1983, and a Shell Dissertation Fellowship in 1983 to 1985. As a professional she received a prestigious National Science Foundation Presidential Young Investigator Award from 1987 to 1992. She was named a Mineralogical Society of America Distinguished Lecturer in 1992 to 1993 and awarded an Editor's Citation for Excellence in Refereeing in *Tectonics* in 1993.

Selverstone has also been of service to the profession. She served on the editorial board for *Geology* from 1989 to 1994 and *Journal of Metamorphic Geology* from 1997 to the present, where she also served as coeditor in 1993 to 1997. She served as both vice chair (1998–1999) and chair (2000) for the Structural Geology and Tectonics Division of the Geological Society of America, where she has served on other committees. She also served on committees and panels for the Mineralogical Society of America and the National Science Foundation.

⊠ Sengor, A. M. Çelal
(1955–   )
Turkish
Plate Tectonics

After the plate tectonic paradigm was accepted, there came a group of equally noteworthy Earth scientists to elucidate the details of this coarse model. Premier among that group is Çelal Sengor. One of the reasons for his unparalleled success is his ability to move around the world to find the perfect example of a feature he wishes to study unfettered by differences in language and culture in addition to finding the best possible collaborators. His research covers a large variety of topics; the one he is best known for is the disappearance of ocean basins. When two continents collide, the ocean basin between them is destroyed. The only records of once huge bodies of water are the rocks around the suture zone between the two old continents, which now appear as one. His paper "Classical Theory of Orogenesis," exemplifies this work. Sengor has been especially interested in the destruction of the Tethys Ocean, which included the most spectacular continental collisions of the late Mesozoic to Cenozoic, including the Alps, the Himalayas, and the Zagros of northern Iraq and Iran. Not only do the remnants of this ocean basin contain significant economic deposits, but also the continued convergence is responsible for many of the most destructive earthquakes in history including many in Sengor's home of Turkey. This work constitutes Sengor's regional research.

Çelal Sengor has also performed research on many theoretical aspects of plate tectonics. He discovered that many extensional basins form in continental collision zones at high angles to the suture zones and helped name them "impactogens" as a variation of aulocogens, which are basins, formed in extensional settings. He studied these aulocogens as formed in triple junctions in the initial stages of divergent margins worldwide. He also studied "tectonic escape" of landmasses laterally along strike-slip faults away from continental collision zones. This idea had been previously proposed for the Himalayas but Sengor extended the concept worldwide. He even defined a new type of continental collision he named "Turkic-type orogeny" as described in his paper "Turkic-type Orogeny and its Role in the Making of Continental Crust." Other topics of interest mostly center around plate collisions and the evolution of continental crust. To document these processes, Sengor worked in a phenomenal number of areas from China to the Alps to the Caribbean but mostly in his beloved Turkey.

Çelal Sengor was born on March 24, 1955, in Istanbul, Turkey. Both of his parents' families were immigrants from the Balkan provinces of the Ottoman Empire, which had been ravaged by wars. Both families are among the richest in Turkey. As a result, Sengor grew up in splendor and a highly educated household where he became multilingual. In 1969, he transferred into the Robert Academy in Istanbul, which is reportedly one of the 50 best high schools in the world. He gained his love for geology there and graduated in 1973. Upon graduation, Sengor traveled to Germany for one year where he enrolled in the Goethe Institute in Munich and Berlin. In 1974, he enrolled in the University of Houston, Texas, but transferred to the State University of New York at Albany in 1976. He spent his entire college career there earning a bachelor of science degree in geology, summa cum laude, in 1978, a master of science degree in geology in 1979, and a Ph.D. in geology in 1982. He was an advisee of both JOHN DEWEY and KEVIN BURKE. Upon graduation, Sengor joined the faculty at the Istanbul Technical University in Turkey, where he passed from lecturer to professor over the next several years and where he remains today. He is currently the head of the department and has been since 1998. During this period, Sengor was a visiting scientist at the University of Oxford, England, and the Lunar and Planetary Institute in Texas. Çelal Sengor married Oya Maltepe in 1986; they have one son.

Çelal Sengor is in the midst of an extremely productive career. He is an author of some 165 articles in international journals, professional volumes, and governmental reports. Many of these papers are benchmarks in modern plate tectonics with some of the premier geologists. He is also an author of five books, one of which has been translated into Russian, German, and Chinese. Examples of his books include *Orogeny,* and volumes include *The Cimmeride Orogenic System and the*

Çelal Sengor at the front of the Mont Blanc basement in the Swiss Alps *(Courtesy of C. Sengor)*

Tectonics of Eurasia and *Tectonic Evolution of the Tethyan Region.* In recognition of his many contributions to geology, Sengor has received numerous honors and awards. He is a foreign associate of the U.S. National Academy of Sciences and the Russian Academy of Natural Sciences. He is one of 10 founding members of the Turkish Academy of Sciences (youngest ever) and the first Turkish member of the Academia Europaea (youngest member ever). He received an honorary doctorate from the Université de Neuchatel. He was also awarded the Bigsby Medal and the President's Award from the Geological Society of London, the Medaille du College de France, the Parlar Science, Service and Honor Award from the Middle East Technical University, the Lutaud Prize (Grand Prize) from the French Academy of Sciences, the Director's Plaque from the Turkish Geological Survey, the Rammal Medal from the French Society of Physics, the Social-democrat Populist Party of Turkey Plaque of Recognition and the Science Award from the Turkiye Bilimsel ve Teknik Arasterma Kumuru. He has even had two fossils named after him.

Sengor has also performed outstanding service to the profession and the public. He has served on virtually every major scientific board in Turkey, including serving as the science and technology adviser to the president of Turkey. He represented Turkey in international programs, including the International Lithosphere Project, the Ocean Drilling Program, and even in the NATO Scientific Affairs Division. He has also served numerous editorial roles including associate editor of *Tectonics* and of *Geological Society of America Bulletin,* as well as a member of the editorial board for *Journal of Structural Geology, Tectonophysics, Earth Evolution Sciences, Bulletin of the Turkish Association of Petroleum Geologists, Turkish Journal of Earth Sciences, Earth Sciences History, Geologica Balanica, Terra, Geologische Rundschau, Asian Journal of Earth Sciences, International Geology Review,* and *Eclogae Geologicae Helvetiae.*

⊠   **Shackleton, Sir Nicholas J.**
(1937–   )
British
Climate Modeler

Sir Nicholas Shackleton is one of the true pioneers of climate modeling. He developed a new technique for measuring oxygen isotope concentra-

tions in small samples in the late 1960s. With his new methods, he was able to analyze marine sediments taken from deep sea drilling projects at much higher resolution than was previously possible. These isotopic data provided detailed records showing the history of ice sheet advances and retreats during the Quaternary period. The fluctuations in the size of the polar ice sheets were shown to be much more numerous than was previously thought. These fluctuations were found to be periodic and the periodicity could be modeled. With colleagues JOHN IMBRIE and JOHN M. HAYES, it was shown for the first time that the Earth's orbit was the governing factor in these drastic climate changes and the "Milankovitch Hypothesis" has validity. Examples of his papers on this topic are "Cretaceous Climates and Extraterrestrial Events" and "Constraints on Astronomical Parameters from the Geologic Record for the Last 25 myr." More recently, Shackleton found that these orbital variations actually forced a change in the concentration of carbon dioxide in the atmosphere. The advance and retreat of glacial ice was and is controlled by the carbon dioxide and therefore lags behind the changes in orbital position. This discovery is based upon the detailed study of a 400,000-year-long ice core taken from the Antarctic ice sheet by Soviet drillers at the Vostock station. A summary paper on this work is "Climate Changes Across the Hemispheres."

This original groundbreaking research has been expanded over the years. Shackleton refined the resolution of these climate changes using carbon and oxygen isotopes, sedimentation rates, fossil populations, and magnetic susceptibility of the sediments in deep sea drill cores. His paper, "High Resolution Stable Isotope Stratigraphy from Bulk Sediment," summarizes this work. He found some 20 oscillations between colder and warmer periods over the past 2.5 billion years. Using these data, he developed a new astronomically based timescale for geological sequences for the Quarternary as reported in the paper "Astronomical (Milankovitch) Calibration of the Geologic Time

Scale." This new high-resolution timescale has helped to clarify many of the ocean processes like dissolution rates of minerals and flux of sediments from continents in addition to the behavior of the climate system as a whole. In a recent drilling project on the Ceara Rise in the western equatorial Atlantic Ocean, Shackleton supervised the recovery of a core containing a continuous record of the last 40 million years. This core shows the fluctuations of the sedimentary input of the Amazon River. The flux of sediment directly reflects changes in climate in addition to the isotopic and fossil data. Therefore, this astronomical timescale is being extended back 40 million years as the result of this research.

Nicholas Shackleton was born on June 23, 1937, in London, England. His father is Professor Emeritus Robert Shackleton from the University of Leeds, a prominent exploration geologist who specialized in African and South Pacific geology. Nicholas Shackleton attended the Cranbrook School before attending Clare College of Cambridge University, where he earned a bachelor of science degree in geology and a Ph.D. in 1965 in geophysics. He remained at Cambridge University, where he advanced from senior assistant in research (1965–1972) to assistant director of research (1972–1987) to reader (1987–1991) to professor (1991–present). Shackleton was a research fellow at Clare Hall at Cambridge University from 1974 to 1980 and has been an official fellow ever since. He is also the director of the Goodwin Institute for Quaternary Research (1995–present). He was a visiting scientist at Lamont-Doherty Geological Observatory, New York, in 1974–1975. Nicholas Shackleton has been married to Vivien Anne Law since 1986. He is an accomplished musician, specializing in the clarinet. He is also an expert on the history of the clarinet, about which he has published several articles.

Nicholas Shackleton is amid a very productive career. He is an author of some 250 articles and reports in international journals, professional volumes, and governmental publications. Many of these publications are benchmarks in climate change and marine response to climate change. Many of the articles appear in the high-profile journals *Nature* and *Science.* In addition to having been knighted by Queen Elizabeth in 1998 for his contributions to the Earth sciences, Nicholas Shackleton has received numerous other honors and awards. He is a foreign member of the U.S. National Academy of Sciences and a Fellow of the Royal Society of England. He has received honorary doctorates from Dalhousie University, Canada, and the University of Stockholm, Sweden. He was awarded the Lyell Medal and the Wollaston Medal from the Geological Society of London, the Sheppard Medal from the Society of Economic Paleontologists and Mineralogists, the Huntsman Award from the Bedford Institute of Oceanography, Canada, the Crafoord Prize from the Royal Swedish Academy of Sciences, the Milankovitch Medal from the European Geophysical Society, and the Carus Medal from the Deutsche Akademie der Naturforscher Leopoldina.

Shackleton has performed outstanding service to the profession. He worked extensively with the Ocean Drilling Program in numerous leadership capacities. He was a founding member of Academia Europaea and the chair of the 15th INQUA (Quaternary Research) Congress Program Committee and is currently president of INQUA. He has also served in numerous capacities for the Geological Society of London.

Shoemaker, Eugene M.
(1928–1997)
American
Astrogeologist

In 1994, the proponents of potential extraterrestrial impacts with the Earth having caused mass destruction and many of the great extinction events were given a great piece of support when the Shoemaker-Levy 9 comet was observed break-

Eugene Shoemaker leads a field trip to Meteor Crater, Arizona, in 1967 *(Courtesy of the U.S. Geological Survey)*

quickly recognized as a prodigy. He skipped grades and attended extracurricular classes. He became interested in geology at the Buffalo Museum of Science. The family returned to Los Angeles, where Shoemaker entered California Institute of Technology at 16. He received a B.S. degree in 1947 and an M.S. in 1948. He joined the U.S. Geological Survey as a research assistant that year.

Although he did not start out his career in extraterrestrial studies but rather field mapping and searching for uranium deposits in the Colorado Plateau, he soon became interested in cratering around nuclear explosions. This work led him to question the origin of Meteor Crater in Arizona. In that classic work, he defined the structures and features that would be observable in any impact crater. He earned his Ph.D. from Princeton University, New Jersey, in geology in 1960 based on this work. It also began Shoemaker's fascination with impact craters which drove the rest of his career. This enthusiasm and some prodding convinced NASA to sponsor a lunar geology program that he directed. In this program Shoemaker demonstrated that the surface of a planet could be dated by counting the number of craters and assuming a certain level of influx. This work was carried out using telescopes but in truth, Shoemaker wanted to be an astronaut. Unfortunately, he was diagnosed with Addison's disease in 1962 and was unable to achieve this dream. He had to be satisfied serving as acting director of NASA's Manned Space Sciences Division and helping to design missions and train astronauts.

Shoemaker worked extensively on the Apollo 11, 12, and 13 missions and became quite a celebrity. Any expert information about lunar geology on television usually came from him. He was world famous then as well. However, the high-profile public lifestyle was physically and emotionally taxing and as a result he turned toward teaching. He taught part-time at California Institute of Technology from 1962 to 1985 and even served as chair for the Division of Geological and Planetary Sciences from 1969 to 1972. He also worked on

ing up and crashing to the surface of Jupiter. The damage was obviously intense and visible scars remain today. That event also made the name of Shoemaker world famous. Eugene Shoemaker, his wife Carolyn Shoemaker, and David Levy had been involved in a decade-long study to identify Earth-crossing asteroids and comets when they made the discovery. Although it came a year after his retirement, the discovery and fame served as a culmination of a 45-year career with the U.S. Geological Survey in which he founded a new branch with the title of astrogeology.

Eugene Merle Shoemaker was born in 1928 in Los Angeles, California. The family soon moved to Buffalo, New York, where he was

Project Voyager with Larry Soderblom. In 1982, Shoemaker began a 15-year period of work at the U.S. Geological Survey, Flagstaff office, where he made regular excursions to the Palomar Observatory and occasional trips to Australia to observe comets and asteroids that could collide with Earth. The 1997 *National Geographic* special, "Asteroids' Deadly Impact," prominently featured his work. He was famous yet again.

On one such comet and asteroid observation trip to Australia, Gene Shoemaker was killed in an automobile accident on July 18, 1997. Gene had a wish granted posthumously when on January 6, 1998, a small capsule containing one ounce of his cremated remains was transported to the Moon aboard NASA's *Lunar Prospector* spacecraft. He finally made it to the Moon.

Gene received a number of prestigious awards throughout his distinguished career, including the NASA Medal for Scientific Achievement in 1967, membership in the National Academy of the Sciences in 1980, Geological Society of America's Arthur L. Day Medal in 1982, and Geological Society of America's G. K. Gilbert Award in 1983, as well as being elected a Fellow of the American Academy of Arts and Sciences in 1993. His greatest tribute, however, was being awarded the United States National Medal of Science in 1992. It is the highest scientific honor and is bestowed by the president of the United States.

⊠ Sibson, Richard H.
(1945–)
New Zealander
Structural Geologist

Richard Sibson has looked at faults and earthquake generation in a way that is a bit different from the standard approach. Normally, scientists study the earthquake waves, surface features caused by faults or ancient inactive faults. Sibson considers the actual processes of faulting at the point where the earthquake is generated. This in-

volves a more holistic approach that requires an integration of all of the standard studies to consider the single moment that the earthquake occurs as well as speculation of all of the components that are no longer observable. This work bridges the gap between structural geology and seismology.

Sibson looked at the interaction between irregularities on fault surfaces during earthquakes. If they create gaps, he termed them dilational and if they pressed together, he called them antidilational. Next, he considered how the fluid would behave in the system. He found that in dilational areas, the fault wall rock would implode into an open gap forming a fragmental rock called breccia. The dilation areas would also draw all fluids into them at once, driving the brecciation processes, while the antidilational areas would drive

Richard Sibson in his office in New Zealand *(Courtesy of R. Sibson)*

out the fluids under pressure. These mineral-rich fluids, upon filling an area of relatively very low pressure, would precipitate out their contained minerals due to the drop in solubility. Therefore, this process also explains the genesis of ores within fault zones. Fluids will be drawn and forced from one spot to another through a process called seismic pumping as the fault evolves. In this model, mineralization is induced by pressure gradients as easily as chemical buffering. Sibson is especially interested in gold deposits. The system works like any hydraulic system with certain rock fractures acting like valves. Examples of his papers on this topic are "Crustal Stress, Faulting and Fluid Flow" and "Stopping of Earthquake Ruptures at Dilational Jogs."

Another interest that Sibson pursues is the frictional aspects of faulting. Friction can generate significant heat along fault planes during earthquakes at relatively shallow levels in the crust. The heat can be so intense that it actually melts the rock along the fault walls. The resulting liquid (magma) moves along the fault and freezes into a glass called "pseudotachylite" which looks as if it could be magmatic. Fluids play a role in this melting process. The stresses build up, and releases in faults are directly related to the amount of friction in a system, thus explaining the regularity and severity of earthquakes. Again, higher fluid pressures release the built-up stress at lower levels causing less severe earthquakes. His paper, "Generation of Pseudotachylite by Ancient Seismic Faulting," is an example of this work.

Richard Sibson was born on November 28, 1945, in New Zealand. He attended the University of Auckland, New Zealand, where he earned a bachelor of science degree (first class) in geology with honors in 1968. He completed his graduate studies at the Imperial College of the University of London, England, where he earned a master of science degree and a Ph.D. in 1970 and 1977, respectively. He was a lecturer in structural geology at Imperial College from 1973 to 1981 before accepting a position as a visiting scientist at the U.S.

Geological Survey Office of Earthquake Studies in Menlo Park, California, in 1981. He joined the faculty at the University of California at Santa Barbara in 1982. In 1990, he moved to New Zealand to the University of Otago, where he remains today. He served as department head from 1990 to 1996. Richard Sibson married Francesca Ghisetti in 1999. She is a professor of structural geology at the University of Catania in Italy.

Richard Sibson is leading a very productive career. He is the author of some 74 articles in international journals, professional volumes, and governmental reports. He has also written one book on structural geology. Many of his papers are often-cited seminal studies on active fault processes. He has received several awards in recognition of his work, including the Bertram Memorial Prize at the University of Aukland, the Royal Commission for the Exhibition of 1851 Overseas Scholarship, and the Wollaston Fund from the University of London. He has performed service to the profession including convening several conferences and hosting numerous short courses. He has also served editorial positions including associate editor of the *Geological Society of America Bulletin,* and editorial advisory board for *Journal of Structural Geology* and *Geofluids.*

⊠ Simpson, Carol
(1947–   )
British
Structural Geologist

As faults move, they break rocks in a brittle manner like breaking glass if the rocks are near the surface. Each of these events results in an earthquake. This seismic behavior most commonly occurs at depths shallower than 10 to 15 km for quartz-feldspar-rich rocks. The fault rocks are called gouge if the broken pieces are large and cataclasite for finely ground-up rock. Beneath this depth, rocks stretch like putty and form finely recrystallized, well-layered rocks called mylonite. To tell which way the fault moved, geologists study microscopic to macroscopic features of the mylonite called "kinematic indicators." Carol Simpson is considered the world's foremost expert on kinematic indicators. Her paper with Stefan Schmid in 1983 entitled, "An Evaluation of Criteria to Deduce the Sense of Movement in Sheared Rock," is the seminal work in the field. It has been cited hundreds of times in geologic literature and won her a coveted Best Paper Award from the Structure and Tectonics Division of the Geological Society of America. She spread the word on kinematic indicators to all levels and aspects of the profession, and a respectable furor resulted, with the new terminology finding its way onto field trips worldwide. Simpson further refined and constrained the processes in the development of these features both practically and theoretically by collaborating with experts with complimentary expertise, as in the paper "Porphyroclast Systems as Kinematic Indicators" with Cees Paschier. This work also includes constraining the metamorphic and mineralogical conditions under which kinematic indicators form, mechanics of strain, and Ar/Ar systematics of mylonite systems. Lately, she and her husband, Declan DePoar, another notable structural geologist, have pioneered a new porphyroclast hyperbolic distribution (PHD) technique to study the kinematic history of major mylonite zones. The paper, "Practical Analysis of the Generation of Shear Zones using the Porphyroclast in Hyperbolic Distribution Method: An Example from the Scandinavian Caledonides," is an example of this method. She has applied her studies to mylonite zones worldwide but especially those in the southern Appalachians, Colorado Rockies, Swedish Caledonides, California and Sierra Pampeanas of Argentina.

Carol Simpson was born on September 27, 1947, in Kirkham, England, where she grew up. After experimenting with a career in industrial chemistry, she settled into geology at the University of Wales at Swansea, where she earned a bachelor of science in 1975. She did her senior thesis

Carol Simpson with her husband, Declan DePoar, on a field trip *(Courtesy of C. Simpson)*

under the mentorship of Dr. Rod Graham. She did her graduate studies at the University of Witwatersrand, South Africa, where she received her master of science in geology in 1977. She continued graduate studies under JOHN G. RAMSAY at the prestigious Swiss Federal Institute (ETH) in Zurich, Switzerland, where she received her Ph.D. in 1981. She was a visiting assistant professor at Brown University, Rhode Island, and at Oklahoma State University, before starting her first tenure-track faculty position at Virginia Polytechnic Institute and State University. She taught courses in structural geology at the graduate and undergraduate levels and advised graduate students. She soon left for a position of director of the Structure and Tectonics Division of Earth Sciences at the National Science Foundation, as well as an associate professorship at the Johns Hopkins University, Maryland. Carol Simpson is currently at Boston University, Massachusetts, where from 1995 to 2000 she served as chair of the Department of Earth Sciences. She is now associate provost for research and education.

Carol Simpson is amid a very productive career, serving as an author of more than 50 scientific articles in international journals and professional volumes. Many of these are seminal papers on mylonite zones and kinematic indicators. Simpson has performed extensive service to the profession. She was elected to the Council of the Geological Society of America in 1998. She is chair of the International Union of Geological Sciences Commission on Tectonics—Subcommission on Rheological Behavior of Rocks, she serves on the Solar Observatory Council of the National Solar Observatory, and she has also held various positions with the American Geophysical Union. She has served as editor for *Geology* and associate editor for *Journal of Structural Geology* and she is currently on the editorial boards of *Geologische Rundschau* and the *Journal of Structural Geology*.

⊠   **Skinner, Brian J.**
(1928–   )
Australian
Economic Geologist, Geochemist

Most miners just want to get their ore out of the ground and the public just wants the ore or products made from it. However, there is a whole science of economic geology to study ore genesis and Brian Skinner is one of the true pioneers. Whereas many of the geochemists, petrologists, and mineralogists study the major and accessory rock-forming minerals and their relations, Skinner applied the latest in theories and experimental and analytical techniques to genesis of ore and surrounding gangue minerals. He first worked on volume properties of minerals but also added new information to the science on the behavior of rocks at high temperatures. One of his main interests, however, is sulfide minerals of which he is among the foremost experts. Using innovative experimental techniques, he determined the relations of assemblages of several mineral groups that are common to certain ore deposits. He researched platinum group minerals, copper-silver sulfides, antimony-arsenic minerals, zinc sulfides, and the sulfosalt minerals in general.

He identified, named, and described five new minerals. He was the first to recognize the role of organic sulfur in the formation of low temperature ore deposits. Skinner researched sulfides precipitating from warm brines in the Salton Sea, which provided valuable information about the transport and deposition of metals at low temperature. He investigated the precipitation of sulfides from basaltic lava lakes in Hawaii at high temperature. This work led to a better understanding of sulfide solubility in silicate melts. One such publication on this work is entitled "Mineral Resources of North America."

Later in his career, Skinner's interests began to include more than ore deposits. He evaluated mineral resources and mineral supplies as well as agricultural resources. These interests were directed toward the writing of professional volumes and textbooks, in addition to papers on scientific advocacy and especially resource management. This application of Skinner's extensive scientific knowledge to public issues provided and continues to provide a great benefit to humankind. His book, *Resources of the Earth*, and his paper "Toward a New Iron Age? Quantitative Model of Resource Exhaustion" are examples of this work.

Brian J. Skinner was born in Wallaroo, South Australia, on December 15, 1928. His father was a bank manager and he moved the family from one county town to another. At age 14, Skinner went to Adelaide, the capital of South Australia, where he completed high school at Prince Alfred College. He attended University of Adelaide and earned a bachelor's degree in chemistry and geology and a minor in physics in 1949. After graduation, he worked briefly as a mine geologist for the Aberfoyle Tin Mine in Tasmania. He attended graduate school at Harvard University, Massachusetts, and earned a doctorate in geology in 1954. During graduate school, Skinner worked for the International Nickel Company in Canada and the Reynolds Metals Company in Colorado, New Mexico, and Arizona. He also married fellow geologist H. Catherine Wild at that time. They returned to Australia where Skinner taught at his alma mater, the

Brian Skinner at work in his office in the Department of Geology and Geophysics at Yale University *(Courtesy of B. J. Skinner)*

University of Adelaide, while his wife completed her doctoral degree. In 1958, he accepted a research position with the U.S. Geological Survey in Washington, D.C. By 1961, he had risen to chief of the branch of experimental mineralogy and geochemistry. In 1966, he joined the faculty at Yale University, where he remained for the rest of his career, achieving the position of Eugene Higgins Professor of geology and geophysics in 1972. He served as chair of the department from 1967 to 1973. In his leisure time, he plays tennis and watches birds.

Brian Skinner has been extremely productive throughout his career. In addition to having been an author of more than 80 articles in international journals and professional volumes, he is an author or editor of some 17 books and professional volumes. These books include two books on economic geology and resources and five introductory textbooks. These textbooks include physical, introductory, and environmental geology and are some of the most successful of their kind. Several of these textbooks include *The Blue Planet, An Introduction to Earth Systems Science, The Dynamic Earth, Environmental Geology,* and *An Introduction to Physical Geology.* Literally thousands of students learned all they know about geology from Skinner's books. He has received numerous honors and awards for his research from professional societies, including the first-ever Silver Medal from the Society of Economic Geologists, the Geological Association of Canada Medal, the Neil Miner Award from the National Association of Geology Teachers and the Distinguished Contributions Award from the Association of Earth Science Editors, as well as honorary doctoral degrees from Colorado School of Mines and University of Toronto, Canada. He was also a distinguished lecturer on numerous occasions.

Brian Skinner has performed outstanding service to the profession. He served as president of the Geochemical Society in 1973, the Geological Society of America in 1985 (vice president in 1984), and the Society of Economic Geologists in 1995. He served as chairman of the U.S. National Committee for Geology in 1987 to 1992 and the board of Earth sciences of the National Research Council in 1987 and 1988, among many other positions. Skinner was also coeditor and editor of *Economic Geology,* as well as president of the Economic Geology Publishing Co. He was coeditor of *International Geology Review* and consulting editor for Oxford University Press, among others.

⊠ **Sloss, Laurence L.**
(1913–1996)
American
Stratigrapher

Today the continents are primarily land and the ocean water covers the ocean crust. During the

Paleozoic, however, much of the continental interior of North America was covered by a large shallow sea. Therefore marine sediments of this age span the entire continent. However, the sedimentation reflects several major regressions that exposed most of the continent followed by major transgressions that covered it with ocean again. This periodic rise and fall of sea level had a profound effect on the evolution of life and is the basis for our subdivision of the Paleozoic periods. The main architect of this now classic scheme is Laurence Sloss. He studied the Paleozoic stratigraphy of the United States from coast to coast in a new way. Normally, rock units are subdivided simply on the basis of rock type or lithology. Sloss was the leader in a new way to view sedimentary rocks in packages. Rock units were grouped with their neighbors into sequences to represent related depositional periods. This concept, called "sequence stratigraphy," is now widely accepted and applied. Sloss used this concept to subdivide the sedimentary rocks of North America, which culminated in a landmark 1963 publication entitled *Sequences in the Cratonic Interior of North America.* Each one of these regressive-transgressive cycles was named based upon its best exposure, such as Tippecanoe, Kaskaskia, and others. Considering that most of life on Earth was in the oceans during this time, this draining of the continents, which served as large shelves, resulted in massive extinctions. New groups of animals would dominate during the next transgression, only to be decimated during the subsequent regression. The results of this research now appear in every historical geology textbook in the world.

The natural extension of this revolutionary research was to determine the causes and applications of these cycles. Sloss attempted to correlate the sequences with the plate tectonic events of the continental margins and the entire Earth. After all, to raise and lower sea level to such a degree required a radical change. Certainly the plate collisions with North America had an effect but the

likely culprit appears to be inflation and deflation of the mid-ocean ridges, which displaced the water from the ocean basins onto the land. These inflations and deflations likely reflected the rate of spreading. The application of this work was in oil exploration. Sloss is the author of a classic 1962 paper, "Stratigraphic Models in Exploration," in which these concepts are applied to petroleum deposits. Sloss even performed seminal research on evaporites. Laurence Sloss is one of the true pioneers of stratigraphy.

Laurence Sloss was born on August 26, 1913, in Mountain View, California, on a farm where he spent his youth. He attended Stanford University, California, and earned a bachelor of science degree in geology in 1934. He completed his graduate studies at the University of Chicago, Illinois, where he earned a Ph.D. in 1937. His first professional position was split between the Montana School of Mines in Butte, where he taught paleontology and historical geology, and the Montana State Bureau of Mines and Geology. He remained in Montana until 1947, when he joined the faculty at Northwestern University in Evanston, Illinois. He remained at Northwestern University until his retirement in 1981. Sloss was named the William Deering Professor of geology in 1971, a title he held until his retirement. Laurence Sloss died on November 2, 1996. His wife, whom he married in 1937, predeceased him; they are survived by two sons.

Laurence Sloss led a very productive career. He is an author of numerous scientific articles in international journals, professional volumes, and governmental reports. Many of these papers are benchmarks of sequence stratigraphy and the Paleozoic stratigraphy of North America. He is also an author of a highly regarded and widely adopted textbook entitled, *Stratigraphy and Sedimentation,* with William Krumbein. Sloss received several prestigious honors and awards for his research contributions to geology. He received the William H. Twenhofel Medal from the Society of Economic Paleontologists and Mineralogists, the

Penrose Medal from the Geological Society of America, and the President's Award from the American Association of Petroleum Geologists, among others. In honor of his work, there is now a Laurence L. Sloss Award in stratigraphy issued by the Geological Society of America on an annual basis.

Sloss was also of great service to the profession. He served numerous functions for the Geological Society of America including president and vice president. The same holds true for the Society of Economic Paleontologists and Mineralogists and the American Geological Institute, where he was also very active and served as president and vice president in each. He served on numerous panels and committees for the National Research Council and the National Science Foundation. He served in various editorial roles for the *Geological Society of America Bulletin,* the *American Association of Petroleum Geologists Bulletin,* and the *Journal of Sedimentary Petrology.*

⊠ **Smith, Joseph V.**
(1928–)
British
Mineralogist

There is pure research on minerals to determine their atomic structures, phase relations, compositions, and other properties for scholarly reasons, and applied research to study metallic and industrial minerals. The techniques and much of the outcome are the same, but with industrial minerals, the goal of the mineralogist is to determine an application of a mineral to an industrial process. Joseph Smith has been very successful with both of these directions. Most of his industrial work has been on zeolite minerals. These minerals are formed under the lowest grades of metamorphism. Their atomic structures are therefore very open (loose packing of atoms) in comparison with most minerals, and yet they are solid. They are used in a variety of applications as molecular sieves for both fluid and gas, for anything from reducing pollutants to making the air smell better. Zeolites are also used as catalysts for cracking long hydrocarbon chains to produce the various products in petroleum refining. This research involves the use of both X-ray and neutron diffraction to analyze the materials, in order to model their structures using mathematical analysis of the shape of the atomic framework. This research has been important to the development of many industrial processes that affect our everyday life.

Within the academic realm of geology, Joseph Smith is best known for his multivolume set, *Feldspar Minerals.* This book is the "bible" on the most common mineral group in the Earth's crust. It summarizes his and all other research on feldspars, including that of his close collaborator, JULIAN R. GOLDSMITH. This research continues today, mostly regarding weathering and replacement mechanisms for various feldspar minerals. Smith is also known for his work as a principal investigator on the Apollo program during the late 1960s and early 1970s. He was the first to describe the differentiation of a large body or ocean of magma on the Moon to create the highlands and lowlands. Initially, his conclusions met with skepticism, but it is now the accepted model appearing in all introductory geology textbooks. Smith has also investigated upper mantle processes in the development of certain odd igneous rocks, including carbonitites and kimberlites.

To perform this research involves the use of high-energy analytical equipment. Now the hardware and procedures to obtain quality results are standard. When Joseph Smith began his research they were not. He devoted a large amount of time in the development of the electron microprobe, which is now the accepted technique to analyze mineral compositions thanks to his efforts. He has made similar efforts on other even higher-energy analytical techniques, like the synchotron. Smith has therefore not only added to Earth science with his research results, but also to developing the means to carry it out.

Joseph Smith was born on July 20, 1928, in Derbyshire, England, where he spent his youth on a farm. He attended Cambridge University, England, where he earned a bachelor of arts degree in 1948 and master of science and doctoral degrees in physics in 1951. Upon graduation he was named to a fellowship at the Geophysical Laboratory at the Carnegie Institution of Washington, D.C. Joseph Smith also married his wife, Brenda, in 1951. They have two children. In 1954, he accepted a position of demonstrator at his alma mater of Cambridge University. Smith joined the faculty at Pennsylvania State University at University Park in 1956, and remained until 1960 when he moved to the University of Chicago, Illinois, where he remained for the rest of his career. He was named the Louis Block Professor of Physical Sciences in 1976, a title he retains. In addition to his faculty position, Joseph Smith also held positions of executive director (1989–1993) and coordinator of the science program (1989–1992) for the Consortium for Advanced Radiation Sources as well as a consultant for the Union Carbide Corporation (1956–1987).

Joseph Smith has led a phenomenally productive career with well over 450 scientific articles in international journals and professional volumes to his credit. Many of these papers are benchmarks in the crystallography of feldspars and zeolites ("Topochemistry of Zeolites and Related Minerals," for example), the application of industrial minerals and even the petrology of lunar samples. In recognition of his contributions to geology, Joseph Smith has had numerous honors and awards bestowed upon him. He is a member of the U.S. National Academy of Sciences and the American Academy of Arts and Sciences and a Fellow of the Royal Society of London. He received the Murchison Medal from the Geological Society of London and both the Roebling Medal and the Mineralogical Society of America Award from the society of the same name.

Smith performed service to many societies throughout his career, most notably to the Miner-

alogical Society of America, where he served as president in 1972–1973, in addition to many other positions. He also served on several committees and panels for NASA during his work with the Apollo program. Smith was the editor of the huge *X-Ray Powder Data File* of the American Society of Testing and Materials (ASTM) for about a decade.

⊠   **Spear, Frank S.**
(1949–   )
American
Metamorphic Petrologist

Frank Spear is one of the world's premier metamorphic petrologists/geochemists. He was one of the leaders in a revolution in metamorphic petrology that occurred in the late 1970s and early 1980s. His first paper was perhaps the most cited and important in metamorphic geochemistry. With John Ferry, he did an experimental calibration for the geothermometry of metamorphic rocks containing coexisting garnet and biotite. The two minerals exchange Fe and Mg depending upon the temperature. As a result of this work, by analyzing the Fe and Mg contents of these common minerals in any rock, temperatures of formation can be accurately determined. Nearly everyone who was working in a metamorphic terrane containing aluminous rocks almost immediately determined the geothermometry using these techniques. The impact on the field, which could now determine actual paleotemperatures where previously they had just been estimated, was astounding. That paper heralded a whole series of experimentally and theoretically determined geothermometers and geobarometers based on assemblages of common metamorphic minerals.

Although Spear did some outstanding work on the occurrence and quantitative geochemistry of amphiboles and amphibole-bearing assemblages, his next study that made a big impact on the field was on methods to quantitatively analyze pressure-temperature paths from zoned minerals. The paper, "Quantitative Pressure-Temperature Paths from Zoned Minerals: Theory and Tectonic Applications," was done with his graduate student JANE SELVERSTONE. These analyses allowed researchers to calculate the P-T history of a rock rather than just its final conditions. Based on the shape and position of the calculated path on a graph of pressure versus temperature, regional tectonic models could be interpreted. Zoned minerals acted as recording tape for tectono-metamorphic histories. Characteristic P-T loops of early high pressure followed by high temperature and finally retrogression were found to be common. Spear and his coworkers examined rocks from several areas to perform these analyses, including the Tauern Window, eastern Alps, the Scandinavian Caledonides, and western New England. Spear developed computer programs to perform these complex but exciting analyses and took the unprecedented step of making them openly available to all geologists. Prior to this, quantitative metamorphic geochemistry was somewhat akin to magic.

The next step in this progression was to add a timing component to the path. Frank Spear collaborated with T. MARK HARRISON to analyze Ar/Ar in the metamorphic rocks that were analyzed for P-T paths. The result is an even more powerful Pressure-Temperature-time (P-T-t) path, which he described in his book, *Metamorphic Pressure-Temperature-Time Paths*. Now, not only could geologists tell if the rocks being studied were loaded beneath a thrust nappe, they could tell how fast they were loaded and for how long. It was as if aluminous metamorphic rocks kept a journal of their history. This method was applied to rocks in western New England and the Cordillera of Tierra del Fuego, South America, but it required a lot of cooperation and even then many of the results were questioned and the method was not as vigorously pursued as previous ventures.

Neither rain nor snow stops Frank Spear from visiting rock exposures, as he does here in New Hampshire *(Courtesy of F. Spear)*

Frank Spear then became interested in diffusion in minerals and especially garnet. With these open systems of exchanging elements, how much was really being recorded and under what conditions? He looked at trace element zoning in garnets and developed a new method for performing 3-D imaging of garnets for different elements in contrast to the standard 2-D analysis as described in the paper, "Three Dimensional Patterns of Garnet Nucleation and Growth." The method involved performing a series of 2-D elemental scans at a series of parallel slices through the garnet and then computer stacking the results to form a 3-D image. Mineral zonation analysis was taken to a new level through this method.

Frank S. Spear was born in Connecticut on March 9, 1949. He earned a bachelor of arts degree from Amherst College, Massachusetts, in 1971 with a major in geology. He earned his doctorate from University of California at Los Angeles in 1976, also in geology. He was awarded a postdoctoral fellowship at the Geophysical Laboratory at the Carnegie Institute of Washington, D.C., from 1976 to 1978. In 1978, he joined the faculty at Massachusetts Institute of Technology, where he remained until 1985. He then moved to Rensselaer Polytechnic Institute, New York, where he became a full professor in 1988 and chair of the department in 1999. Frank Spear has two children.

Frank Spear is an author of 79 papers in international journals and professional volumes. He is also an author of two books. He was a Schlumberger Professor at Massachusetts Institute of Technology and a Weeks Visiting Professor at University of Wisconsin at Madison. He was awarded the N.L. Bowen Award from American Geophysical Union. He served as editor for *Geological Materials Research* and associate editor for *American Mineralogist* and *Journal of Metamorphic Petrology*. He has served on countless committees for the American Geophysical Union, the Mineralogical Society of America, and the Geological Society of America. He also presented several short courses worldwide on his P-T-t techniques.

⊠ **Stanley, Steven M.**
    (1941–)
    American
    Paleontologist

Evolution is a concept that is still controversial well over a century after Charles Darwin documented it. Even now, school boards in certain areas hold heated debates about whether it will be taught, and sometimes it is voted down in favor of creationism. Typically, evolution is considered as part of biology rather than geology. However, one of the true scientific leaders on placing the processes of evolution into full and proper context is the geologist Steven Stanley. The reason that he has been so effective in his work is that he

addresses evolution using a holistic approach. Rather than simply considering how an organism or family of organisms is changing over the generations, he considers the stimuli and interactions as well. By understanding the pressures and opportunities that an organism faces, he more fully understands the way in which it adapts. Using this approach, he performed a highly original analysis of how animals go extinct, he clarified the role of species in large-scale (macro) evolution and he analyzed functional shape changes of animals in adaptive evolution.

In his study of extinction, Stanley found that regional climatic cooling during the Plio-Pleistocene ice age caused the disappearance of many species of western Atlantic marine fauna. This and other related work led him to first propose that climate change, whether terrestrial or extraterrestrial, is the main cause of mass extinctions, which he published in a book entitled *Extinction*. Most geologists now accept this idea. He further considered the role of plate tectonics in evolution and extinction in a very successful historical geology textbook entitled *Evolution of Earth and Life Through Time*. This interest in the complex interplay of climate, plate tectonics, and evolution caused him to further consider the dramatic impact of ice ages on human evolution in his popular book *Children of the Ice Age: How a Global Catastrophe Allowed Humans to Evolve*. He continued exploring this complex interplay with Johns Hopkins University colleague Lawrence Hardie. They considered activity of reef-building organisms, seawater chemistry, spreading rates on mid-ocean ridges, and climate changes to model animal and whole-Earth evolution. Such "biocomplexity" is now the direction that most biologic, climatic, and paleontologic research has followed.

Steven M. Stanley was born on November 2, 1941, near Cleveland, Ohio, where he spent his childhood years. He attended Princeton University, New Jersey, and earned a bachelor of arts degree in geology in 1963, graduating summa cum laude. He did a senior thesis on the paleoecology of the Key Largo Limestone, Florida. After one year at the University of Texas at Austin, he entered a doctoral program at Yale University, Connecticut. There he earned a Ph.D. in paleontology in 1968. Before completing his doctorate, he accepted a position at the University of Rochester, New York, in 1967, but only remained until 1969. He joined the faculty at the Johns Hopkins University in 1969, and remains there today. He served as chair of the department in 1987 and 1988 and chair of the Environmental Earth Sciences and Policy Master's Program from 1993 to present. Stanley also spent 1990 to 1991 as chair of the department at Case Western Reserve University, Ohio.

Steven Stanley has been productive throughout his career. He is an author of 57 articles in international journals and professional volumes, many of which are very prestigious including several in the high-profile journal *Science*. However, his real fame in publication lies in his books. He wrote eight books and edited two professional volumes. Several of these books are popular, high-quality textbooks whereas others are scholarly books and more popular science books. His textbooks, *Principles of Paleontology,* with DAVID M. RAUP and his *Earth and Life Through Time* and successors *Exploring Earth and Life Through Time* and *Earth System History* are standard reading for courses in paleontology and Earth history respectively.

Stanley has been well recognized in the profession for his research. He received the Best Paper Award from the *Journal of Paleontology* in 1972. He was awarded the Allan C. Davis Medal from the Maryland Academy of Science in 1973 and the Schuchert Award of the Paleontological Society in 1977. He received a Guggenheim Fellowship in 1980–1981 and an American Book Award Nomination in 1981 for his book, *The New Evolutionary Timetable*. He was elected to the American Academy of Arts and Sciences in 1988 and the National Academy of Sciences in 1994. He re-

ceived the J.A. Brownocker Medal from the Ohio State University in 1998.

Steven Stanley has also performed significant service to the profession. He was a member of several editorial boards including *American Journal of Science* (1975 to present), *Paleobiology* (1975–1982), and the *Proceedings of the National Academy of Sciences* (1999–2000). He served several positions in the Paleontological Society including councilor (1976–1977, 1991–1993) and president (1993–1994), among others. He was president of the American Geological Institute in 2000–2001. He served on the National Research Council, board of Earth sciences, for which he was vice chair in 1987–1988. He also served on several major committees for Geological Society of America, among other societies.

⊠ Stock, Joann M.

(1959–   )
American
Plate Tectonics, Structural Geologist

After the giants of plate tectonics were finished defining the major plate interactions, the next phase of the science was to refine those processes. Now those previously overlooked fine and not-so-fine details and unresolved problems required explaining. Thus began the second phase of the plate tectonic revolution, which continues today. Joann Stock has established herself as one of the up-and-coming leaders in this group. Even though she is still early in her career, she has worked with many of the initial group of pioneers like TANYA ATWATER and PETER MOLNAR, among others, and has already made an impact in her own right.

Joann Stock had an initial interest in major plate motions and especially in quantifying the uncertainties in their positions in recent geologic history. She has been especially interested in the positional problems of Antarctica and Australia and the tectonics of the southern oceans. To these ends, she studies the geometry and processes of the Pacific-Antarctic Ridge in the southern Pacific Ocean basin, among other areas. More recently, she has been interested in the position and movement of the Malvinas Plate of southwest Africa. An example of this research is the paper "The Rotation Group in Plate Tectonics and the Representation of Uncertainties in Plate Reconstructions." In addition to these plate scale projects, she has also studied the state of stress and earthquake potential as well as extensional processes in many other areas. These studies have taken her to Greece, southern Peru, and Yucca Mountain of Nevada, among many other locations.

Stock's primary focus of research, however, has centered on the relations between the Pacific–North American plate boundary deformation (the San Andreas fault system and related deformation in California) and the opening of the Gulf of California in Mexico. See, for example, the paper "Rapid Localization of Pacific–North American Plate Motion in the Gulf of California." The Gulf of California is actively opening in a rather complex divergent boundary whereas the San Andreas fault is a major transform margin. The transition between them occurs near the United States-Mexico border and is marked by rapidly changing and complexly overlapping structural styles. There is opening and closing of basins and rigid rotation of fault-bounded blocks. There is also volcanism that occurs in the transition zone that she has studied. This research includes compositional work as well as documenting the timing, frequency and extent of eruptions, especially in relation to the tectonic development. This research not only has applications to structural geology and plate tectonics but also to earthquake study. Stock has served the public on advisory committees to apply her research to earthquake hazards and prediction.

Joann Stock was born on October 9, 1959, in Boston, Massachusetts. She attended the Massachusetts Institute of Technology in Cambridge, where she earned both bachelor of science and master of science degrees in geophysics in 1981, Phi Beta Kappa, as well as a Ph.D. in geology in 1988. She received a Fannie and John Hertz Foundation Fellowship for her graduate studies. Upon graduation in 1988, she joined the faculty at Harvard University, Massachusetts. She moved to the California Institute of Technology in Pasadena in 1992, and remains there today. During the time between completing her master's and doctoral degrees, she worked concurrently as a geophysicist at the U.S. Geological Survey in Menlo Park, California, from 1982 to 1984. Since 1995, she has been an adjunct investigator with the Centro de Investigacion Cientifica y Educacion Superior de Ensenada, Mexico.

Joann Stock is in the middle of a very productive career. She is an author of some 60 articles in international journals, professional volumes, and governmental reports. Many of these papers are benchmark studies on the new refinement of the plate tectonic paradigm and appear in high-profile journals like *Nature* and *Science*. In recognition of her research potential, Stock received the Presidential Young Investigator Award from 1990 to 1995.

Stock has performed significant service to the profession. She has served on several panels and committees for Geological Society of America, National Science Foundation, American Geophysical Union, Ocean Drilling Program, and the National Earthquake Prediction Evaluation Council, among others. She organized numerous conferences on plate tectonics. She also served several editorial roles including on the editorial board for *Geology* and *Geological Society of America Bulletin*.

⊠ Stolper, Edward M.
(1952–   )
American
Petrologist, Geochemist

When a volcano erupts, a huge amount of gas is released into the atmosphere in addition to the lava and ash. The gas is dominantly water but also carbon dioxide, certain sulfur compounds, and other minor gases. Before the eruption, this gas was contained within the liquid magma similar to carbon dioxide being held in soda. How do these gases interact with the floating unconnected atoms intent on bonding together to form minerals? How do they interact with the newly formed minerals? These are two of the complex questions that Edward Stolper uniquely addresses in his experimental and theoretical research. In simple terms, he provides an unconventional yet brilliant view of the interaction of fluids, melts, and solids in petrogenesis (the making of rock). The experimental results model the processes in a magma chamber and allow him to develop techniques to predict the content and participation of these gases in the crystallization of minerals even after the rock has hardened and all of the gases have escaped. This research allows him to understand the radical geologic processes that occur when a volcano erupts and the complex relations of the solid and liquid components up to that point. Stolper also investigates how different isotopes of elements are divided among the crystallizing minerals and melt in a magma chamber. These experimental and theoretical studies are then applied to real areas for testing. He has investigated magmatic systems in the Marianas, western Pacific, several mid-ocean ridges, Crater Lake in Oregon, Mono Craters in California, and Kilauea in Hawaii. Several of his papers include, "The Speciation of Water in Silicate Melts," and "Theoretical Petrology."

Stolper has also had a long-standing interest in meteorites and lunar samples. He has shown how the different types of achondritic meteorites are formed and even proposed some new types of meteorites. Because meteorites represent the most primitive type of planetary material, Stolper built models for the development of planets with meteorites as a starting point. He developed a chemical model for the differentiation of the Earth into its shells. Not only is there gravitational control on the layering that appears purely based upon density, but also chemical control. This evolving unorthodox model is guided by his experimental research and provides a new look at the chemical evolution of the planet.

Edward Stolper was born on December 16, 1952, in Boston, Massachusetts. He attended Harvard University, Massachusetts, and graduated with a bachelor of arts degree in geological sciences, summa cum laude and Phi Beta Kappa, in 1974. He was married in 1973 and would have two children. He earned a master of philosophy degree from the University of Edinburgh, Scotland, in geology in 1976 before returning to Harvard University for the remainder of his graduate career. He earned a Ph.D. in geological sciences in 1979. Stolper joined the faculty at California Institute of Technology in 1979 and remains there as of 2002. He has been the William E. Leonhard Professor of geology since 1990 and the chair of the department since 1994. During this time, he was a Bateman Visiting Scholar at Yale University, Connecticut (1988), and a Miller Visiting Research Professor at the University of California at Berkeley (1990).

Edward Stolper has published some 133 articles in international journals and professional volumes. The truly impressive part of this productivity is the benchmark nature of the articles and the quality of the journals in which they appear. An amazing 13 papers appear in the prestigious journals *Science* and *Nature.* His collaborators are the top researchers in the world in their respective disciplines. Stolper's research has been well recognized by the profession as demonstrated by his numerous honors and awards. He is a member of the National Academy of Sciences and a Fellow of the American Academy of Arts and Sciences. He received a Marshall Scholarship and a Nininger Meteorite Award while still in graduate school. He was awarded the Newcomb Cleveland Prize by the American Association for the Advancement of Science in 1984, the F.W. Clarke Medal by the

Geochemical Society in 1985, the James B. Macelwane Award by the American Geophysical Union and the Arthur Holmes Medal by the European Union of Geosciences 1997. He was also named a Geochemistry Fellow by the Geochemical Society and the European Association for Geochemistry in 1997.

Edward Stolper has performed significant service to the profession in too great abundance to list here.

⊠ **Stose, Anna I. Jonas**
   (1881–1974)
   American
   Field Geologist

Anna Jonas Stose was a geological pioneer who performed some absolutely incredible geological feats. She was a field geologist who mapped huge areas of the central to southern Appalachian Piedmont during a time when there were few women in the profession much less doing physically taxing work in the field. Most of her fieldwork was done on a reconnaissance rather than a detailed basis, which is intended to produce larger scale regional maps rather than the 5-by-6-mile quadrangle maps. Stose was one of the first to apply the then-advancing petrographic and structural techniques to the Appalachians. She began her mapping in southeastern Pennsylvania and adjacent Maryland with Eleanora Knopf. She continued her mapping southwestward into the crystalline rocks of Virginia throughout the Piedmont and Blue Ridge Provinces with some help from George Stose. She even mapped into North Carolina. Through this work she was a major contributor to both the *Geologic Map of Virginia* (1928) and the *Geologic Map of the United States* (1932), in addition to her numerous state and U.S. Geological Survey reports.

Stose defined many of the major rock units' geologic structures in the central and southern Appalachians and the names are still used today.

Even her interpretations, which have gone in and out of acceptance over the years, are still essentially correct. Considering the adversity that she encountered in the lack of roads, encumbering clothing, and available transportation, in addition to prejudices against women performing such work, these accomplishments become almost unbelievable. She named the Brevard zone, a major structure of North Carolina, and interpreted it as a thrust fault. Unfortunately, she did not live long enough to see the magnitude of this structure as it was imaged on the COCORP seismic reflection profile across the southern Appalachians. It is estimated that this fault may have experienced hundreds of kilometers of thrust and strike-slip movement. She and Eleanora Knopf defined the Martic Line and proposed it to be a major thrust fault. Since then it was interpreted as a major strike-slip fault and a major plate boundary though the thrust fault interpretation has never been abandoned. Stose interpreted the Reading Prong of Pennsylvania to be a series of disconnected thrust fault bounded klippe, an idea that still remains accepted today. Not all of her ideas are still accepted but a surprising number are.

Stose's naming and correlations of rock units from area to area are also surprisingly relevant. She traced the crystalline rocks of the Piedmont Province from Pennsylvania to Georgia and established the fundamental boundaries that still stand today. Many other boundaries were defined since this time but most of them are still debatable. She and Knopf defined and named the Conestoga Limestone a major unit of Pennsylvania. Stose named the major units of the Blue Ridge Province and many of the granite plutons both there and in the Piedmont. She also identified and named the enigmatic but important Mount Rogers volcanic sequence in southern Virginia. These units still retain their names and significance some 60 to 80 years after their identification by Stose. It is rare for interpretations in geology to remain for so long. This longevity is a tribute to the quality of her work. Two of the papers on this research include "Geologic Reconnaissance in the Piedmont of Virginia" and "Stratigraphy of the Crystalline Schists of Pennsylvania and Maryland."

Anna I. Jonas was born on August 17, 1881, in Bridgeton, New Jersey. She was a descendant of one of the pilgrims who came to America on the *Mayflower*. Jonas grew up in Cape May, New Jersey, and attended the Friends Central School of Philadelphia. She received her college education at Bryn Mawr College, Pennsylvania, earning a bachelor of arts degree in 1904, a master of arts degree in 1905, and a Ph.D. in 1912. Her mentor was FLORENCE BASCOM, the grande dame of American geology. The friendship of Jonas and two of her classmates, Eleanora Knopf and Julia Gardner, is legendary. They were pioneers for women in geology. Jonas was an assistant curator at the Bryn Mawr Geology Museum in 1908 and 1909. Jonas worked at the American Museum of Natural History in 1916 and 1917 and as a geologist for the Maryland and Pennsylvania Geological Surveys from 1919 to 1937. She also worked as a contract geologist for the Virginia Geological Survey from 1926 to 1945. Jonas was a geologist with the U.S. Geological Survey from 1930 until her retirement in 1954. Anna Jonas married fellow geologist George Stose in September 1938 and took the name Anna Jonas Stose. George Stose died in 1960. Anna Jonas Stose died on October 27, 1974, of a stroke.

⊠ **Suess, Hans E.**
(1909–1993)
Austrian
Chemistry, Geochemistry

The truly profound research contributions that Hans Suess made to science spanned the range between chemistry and geochemistry. Although his research covered numerous topics, there are four true benchmark contributions. The first was made while Suess was still in Germany. In 1948 and 1949, Suess worked on the nuclear shell model for the architecture of atoms with Hans Jensen and coauthored a study which would later earn Jensen a Nobel Prize. While at the University of Chicago, Illinois, in 1950 and 1951, Suess collaborated with Nobel Prize laureate Harold Urey. Suess had proposed that the relative abundance of each chemical element in the solar system depends in a fairly regular way on the elemental mass. The pattern of abundance is caused by a combination of nuclear properties and the process by which the heavy elements are created in stars. Harold Urey was the founder of modern planetary science and an expert on meteorites. Together they produced a benchmark study on abundances of elements in the solar system based upon meteorite geochemical data. The documentation of this theory was the basis for NASA's Genesis mission. A book by Suess on this topic is entitled *Chemistry of the Solar System* and an example of a paper is "The Cosmic Abundances of the Elements."

These two breakthroughs, however, are not even related to the true reasons for which Suess is famous in the Earth sciences. The first of these reasons is Suess's development and later refinement of the carbon-14 (radiocarbon) method of isotopic dating. He determined experimentally that the relative concentrations of carbon-14 and nitrogen-14 could determine the absolute age of organic matter within the past 5,000 years or so. This method is now used extensively in archaeology as well as recent geologic features and processes. Papers on this research include, "Radiocarbon in Tree Rings," among others. Suess also collaborated with ROGER REVELLE to document the increase of carbon dioxide in the atmosphere and the greenhouse effect. The way that Suess determined the amount of added industrial carbon was by using isotopes. Because industry relies so heavily on fossil fuels, the carbon introduced into the atmosphere comes from old sources (oil, gas, and coal) rather than wood. Old carbon has no radioactive isotopes because it has all decayed away. Therefore the component of radioactive carbon in the atmosphere is continually diluted by the addition of nonradioactive carbon. Wood from 1890 is used as the standard against which the atmospheric carbon is compared. This dilution is referred to as the "industrial effect" or the "Suess effect." This and other work on radioactive elements are included in the paper "Radioactivity of the Atmosphere and Hydrosphere."

Hans Suess was born on December 16, 1909, in Vienna, Austria. He was the son of Franz Suess, a former professor of geology at the University of Vienna, Austria, and the grandson of Eduard Suess, who wrote the book, *The Face of the Earth,* an early work on geochemistry. Even though he had geology in his blood, Hans Suess studied chemistry and physics at the University of Vienna through graduate school. He graduated with a Ph.D. in chemistry in 1935. He was a postdoctoral fellow at the Institute of Chemical Technology in Zurich and the First Chemical University Laboratory in Vienna. In 1938, Suess joined the faculty at the University of Hamburg, Germany, in physical chemistry. He married Ruth Viola Teutenberg in 1940; they would have two children. During World War II, Suess was enlisted into a group of German scientists who were charged with developing atomic weapons. He was also a scientific adviser to the heavy water plant in Vermok, Norway. In 1950, Suess was coaxed to immigrate to the United States where he spent time at the University of Chicago, Illinois, as a research associate working with Nobel laureate Harold Urey. He obtained a position as a physical chemist with the U.S. Geological Survey in 1951 but accepted an offer from Roger Revelle to join the Scripps Institution of Oceanography in La Jolla, California, in 1955. He became one of the first four professors appointed to the faculty at the University of California at San Diego when it was established in 1958 by Roger Revelle. He retired to professor emeritus in 1977, but remained active through the rest of his life including as a visit-

ing scientist at the Geophysical Laboratory at the Carnegie Institution of Washington, D.C. Hans Suess died on September 20, 1993.

Hans Suess was very productive during his career, having been an author of more than 150 scientific articles. Until 1950, nearly all articles were in German and even after that some were. Several of these papers are benchmarks in science on radiocarbon dating, the greenhouse effect, the nuclear shell model, and the origin and synthesis of the elements. Suess was recognized for his contributions to science with several prestigious honors and awards. He was a member of the National Academy of Sciences, the American Academy of Arts and Science, the Heidelberg Academy of Science, and the Austrian Academy of Science. He was awarded an honorary doctoral degree from Queens University in Belfast, Ireland, in 1980. He received the V.M. Goldschmidt Medal from the Geochemical Society, the Leonard Medal from the International Meteoritical Society, the Alexander von Humboldt Prize from the Humboldt Society, and a Guggenheim Fellowship.

⊠ **Suppe, John E.**
(1942–)
American
Structural Geologist

John Suppe is among the top few active structural geologists in the world. Although he is best known for his work on foreland fold and thrust belts, his interests and expertise are vast. One of the concepts that he is known for is fault-bend folding introduced in the paper "Geometry and Kinematics of Fault Bend Folds." They are a whole class of folds that are formed as a result of movement on faults. Strata are bent into folds because they are forced to rotate as they move around a bend in a fault. These folds are generally broad and extensive but small fault bend folds are possible also. Fault bend folds are an integral process in the formation of foreland fold and thrust

John Suppe on a field trip to the Himalayas *(Courtesy of J. Suppe)*

belts. They are also important in evaluating seismic hazards because the folds indicate a nearby fault.

The second major concept for which Suppe is recognized is critical taper wedge as introduced in the paper "Mechanics of Fold and Thrust Belts and Accretionary Wedges." Working in the Taiwan subduction complex led Suppe with colleagues Dan Davis and Anthony Dahlen to produce a dynamic analog model of the process. They found that foreland fold and thrust belts form similar to the plowing of snow. A wedge of snow develops in front of the plow that has consistent slope from the plow downward to the unaffected snow in front. To increase the height of the wedge of snow, the length of the wedge will be proportionately increased to maintain the same slope. The same critical taper (slope) is maintained in subduction wedges and foreland fold and thrust belts. In order to increase the height of the mountains in a foreland fold and thrust belt, it must be widened. A best paper award was received by the three researchers for this study. Other areas of foreland fold and thrust belts that Suppe studied and mastered include new methods for balancing cross sections. The theory of cross-section balancing or retrodeformation involves the pulling apart of these highly deformed strata to model what they looked like prior to deformation. Although he did not invent the method, he is now one of the foremost experts.

If there is active sedimentation taking place during faulting, there is thickening of the strata as a direct result. In curved faults, the sedimentary layers form wedges which are folded in fault-bend folds. Suppe and associates Chou and Hook found that the rates of folding and faulting could be determined by studying these relations. The folds are called rollover in normal faulted areas and can contain significant amounts of petroleum. A paper on this topic is entitled, "Origin of Rollover."

Suppe was also involved in measuring the state of stress across the San Andreas Fault of California. Because it is a strike-slip fault, the expected orientation of the maximum stress (force) direction is about 45° from the fault, according to the theory. However, when the stresses were resolved, it was found that the maximum stress direction was perpendicular to the fault. This discovery caused an uproar in the seismic hazards community and a reevaluation of previous theory. This work placed Suppe into the middle of the efforts to evaluate and predict earthquakes in California.

Suppe also became a major contributor to our understanding of the tectonics of Venus. Using remote sensing data (SAR) from the NASA Magellan mission, he applied his knowledge of Earth processes to the development of deformational features on the Venusian surface. He evaluated the age of the features on Venus by using impact crater density assuming a constant fallout rate. This work appears in his paper, "Mean Age of Rifting and Volcanism on Venus Deduced from Impact Crater Densities." This work was a major contribution to how Venus is studied.

John E. Suppe was born on November 30, 1942, in Los Angeles, California. He grew up in South Gate, California, and found an interest in the outdoors and especially mountaineering from trips with the YMCA summer camp and the Sierra Club. He attended University of California at Riverside and earned a bachelor of arts degree in geology with honors in 1965. He met his wife, Barbara, in college and they were married soon after graduation. He attended Yale University for graduate school and earned a Ph.D. in 1969 in structural geology. His adviser was JOHN RODGERS. He was a National Science Foundation postdoctoral fellow at the University of California at Los Angeles from 1969 to 1971. He joined the faculty at Princeton University, New Jersey, in 1971, and has remained there ever since. Suppe served as department chair from 1991 to 1994 and was named Blair Professor of geology in 1998, a title he still holds. During his time at Princeton, Suppe was a visiting professor at National Taiwan University, California Institute of Technology, University of Barcelona, Spain, and Nanjing University, China.

John Suppe has had a very productive career. He has been an author of 74 articles in international journals and professional volumes. Many of these papers are true classics of structural geology. He is an author or editor of five books and professional volumes. One of these books is a very successful textbook entitled, *Principles of Structural Geology.* Suppe's work has been recognized by the profession through numerous honors and awards. He is a member of the National Academy of Sciences. He received an unprecedented two Best Paper Awards from the Structural Geology and Tectonics Division of Geological Society of America. He was a Guggenheim Fellow and a Guest Investigator for the NASA Magellan mission to Venus.

⊠ Sykes, Lynn R.
(1937–   )
American
Geophysicist

Lynn Sykes has had two major interests in his career and he has established himself as one of the true leaders in each. His major area of research is earthquake seismology both in terms of seismic sources and wave travel. Sykes is a pioneer in terms of explaining earthquakes in terms of plate tectonics. His 1968 paper "Seismology and the new Global Tectonics" with Bryan Isacks and

JACK E. OLIVER is one of the true classics of plate tectonics. This work demonstrated why earthquakes are concentrated in certain geographic regions because they are at plate margins. He proved the importance of transform faults that offset mid-ocean ridges in accommodating plate motion on a spherical Earth. His research laid the groundwork for deciphering plate motions from the focal mechanisms of earthquakes as well as the precise locating of earthquake epicenters at mid-ocean ridges, transform faults and deep-sea trenches. The research that the team from Lamont-Doherty conducted showing that earthquake foci get progressively deeper from a trench to beneath an island arc proved the geometry of a subduction zone, where ocean crust is consumed in the mantle. The earthquake foci form a Benioff Zone that images the top of the ocean crust as it is driven progressively deeper into the Earth. Much of this research was carried out on the Fiji-Tonga region but Sykes also studied earthquakes in Alaska, the Puerto Rican-Virgin Islands region as well as earthquakes produced by extension in Iceland and Nevada. Other classic studies on this work include *Earthquake Swarms and Sea Floor Spreading* and *Seismicity and the Deep Structure of Island Arcs*.

With this new understanding of the control of earthquakes, Sykes applied his knowledge to societal needs. He spent a good deal of his career investigating earthquake prediction and prevention. The paper, "Earthquake Prediction: A Physical Basis," is an example of this work. In addition to participating in projects in the United States, mainly in New York and California, he was also a leading participant in several international efforts. He worked collaboratively with the former Soviet Union as well as the People's Republic of China, among others. This work involved establishing a worldwide network to monitor earthquakes but it also served as a monitor for nuclear testing. As a result, Sykes became one of the scientific leaders in the establishing of thresholds for nuclear testing and later for the banning of underground nuclear testing. He testified six times before the U.S. Congress as an expert witness on nuclear test verification and served on the presidential advisory board for the same reason. He is still consulted by the press on issues of nuclear test monitoring and the relaxation of treaties.

Lynn Sykes was born on April 16, 1937, in Pittsburgh, Pennsylvania. He attended the Massachusetts Institute of Technology in Cambridge and earned both bachelor of science and master of science degrees in geophysics in 1960 on a Procter and Gamble scholarship. He attended Columbia University, New York, where he earned a Ph.D. in geophysics in 1964 as an advisee of Jack Oliver and on an Edward John Noble Leadership Award. He remained at Lamont-Doherty Geological Observatory of Columbia University as a research associate before becoming a member of the faculty in 1968. In 1972, Sykes was appointed as the head of the seismology group. In 1978, he was chosen as the Higgins Professor of geology. He retired to professor emeritus in 1998. Sykes was a visiting professor several times including the Earthquake Research Institute at the University of Tokyo, Japan, in 1974. Lynn Sykes married Katherine Flanz in 1986.

Lynn Sykes has led a very productive career. He is an author of more than 100 articles in international journals, professional volumes, and governmental reports. Several of these papers establish new benchmarks for the science of geology. Sykes's research contributions have been well received by the profession and recognized in terms of honors and awards. He is a member of both the National Academy of Sciences and the American Academy of Arts and Sciences. He was awarded an honorary doctorate from State University of New York at Potsdam in 1988. He has held both Sloan and Guggenheim Fellowships. He received the Macelwane Award and Walter H. Bucher Medal from the American Geophysical Union, the Medal of the Seismological Society of America, the Public Service Award from the Federation of American Scientists, the H.O. Wood Award from the Carnegie Institution of Washington, D.C., and the Vetlesen Medal from the Vetlesen Foundation.

Sykes has also performed outstanding service both to the profession and the public. He was president of the American Geophysical Union and the Geological Section of the New York Academy of Sciences. He served on numerous panels and committees for National Academy of Sciences, National Research Council, National Science Foundation, American Geophysical Union, Geological Society of America, U.S. Geological Survey, Seismological Society of America, NASA, and the New York State Geological Survey. In addition to the public service already described, Sykes served as a consultant to the U.S. Air Force, the U.S. Nuclear Regulatory Commission, and the State of New York. Sykes also served numerous editorial roles including associate editor of *Journal of Geophysical Research.*

⊠ Sylvester, Arthur G.
(1938–)
American
Structural Geologist

Arthur Sylvester has two main areas of interest in research, deformation associated with pluton emplacement and strike-slip deformation. When a magma intrudes preexisting rock units, it imposes a complex deformation sequence on them. Granite and granitic plutons best display these sequences. The magma intrudes in an elongate inverted teardrop shape that rises through the crust to a level where its buoyancy is balanced by the pressure of the surrounding rock. At that point, the bulbous top of the tear drop stops moving but the tail continues to rise and adds its volume to the bulbous part causing it to balloon. The initial intrusion causes some deformation, and the ballooning adds more deformation locally. All of this deformation of the surrounding rock

happens while it is being quickly heated to great temperatures. The result is a unique series of combined deformation-metamorphic features. Sylvester has studied such features in granitic plutons in eastern California and Norway, establishing himself as one of the foremost experts in the field.

As a Californian, Sylvester has experienced firsthand the effects of one of the most famous strike-slip faults on Earth, the San Andreas Fault. A whole slew of geologists both from California and elsewhere have conducted extensive research on the San Andreas Fault. Even with all of this research activity by many prominent geologists, Sylvester managed to distinguish himself as one of the premier experts on the deformation associated with it. He was the first to show that significant vertical movements and associated deformation could be synchronous with the dominantly lateral movements of the fault. His identification of "keystone structures" led the way to the identification of a new type of deformation called trans-

Arthur Sylvester sitting in a field of flowers in California *(Courtesy of Arthur Sylvester)*

pression. His paper on this work is entitled "Tectonic Transpression and Basement Controlled Deformation in the San Andreas Fault Zone, Salton Trough, California." Because of his contributions to the study, Sylvester agreed to write an elegant summary paper entitled "Strike-Slip Faults" for the centennial volume of Geological Society of America, which is now standard reading in structural geology classes.

In addition to his research accomplishments, Arthur Sylvester is an inspired teacher. He uses the field as his classroom to convey the processes of deformation as well as regional geology. He runs summer field camps for undergraduate and graduate students as well as field structure courses for a number of petroleum companies, geological societies, geologic surveys, and other universities. He has established a reputation for his prowess in the field as well as his ability to convey complex ideas in an understandable manner.

Arthur Sylvester was born on February 16, 1938, in Altadena, California. He attended Pomona College in Claremont, California, and earned a bachelor of arts degree in liberal arts and geology in 1959. He attended graduate school at University of California at Los Angeles and earned a master of arts degree in 1963 and a Ph.D. in 1966. During this time, he was a Fulbright scholar at Oslo University in Norway in 1961 and 1962. Arthur Sylvester was married in 1961 and has two children. He worked as a research geologist for Shell Development Company in California from 1966 to 1968. He joined the faculty of the University of California at Santa Barbara in 1968 and has remained there throughout his career. From 1972 to 1974, he served as associate director of the overseas program at University of

Bergen, Norway. He served as department chair from 1980 to 1986 and directed the department field camp numerous times. He returned to Oslo University, Norway, as a Fulbright scholar in 1995–1996 and he was a visiting professor at University of Svalbard, Norway, in 2001. Sylvester is fluent in Norwegian, German, and Italian.

Arthur Sylvester has led a very productive career both as a researcher and a teacher-mentor. He is the author of numerous articles in international journals and professional volumes. Many of these articles are often-cited research papers and review articles. He has been well recognized in the profession both for his research and teaching. He was named a Fellow of the Norwegian Research Council in 1996. He received the Distinguished Service Award from the Geological Society of America in 1995. For his teaching and mentoring, he received the Distinguished Teaching Award (1996–97) and the President's Award for Mentoring (1994) from the University of California and the Distinguished Teaching Award from the Pacific Section of the American Association of Petroleum Geologists in 1994. He has also been named a distinguished lecturer by several organizations.

Sylvester has performed much service to the profession. He served on numerous committees for the American Association of Petroleum Geologists including serving as the director of the Structural Geology School in 1984–1986. He was associate editor of *American Association of Petroleum Geologists Bulletin* in 1984 to 1988. He served on numerous committees for the Geological Society of America and was the chief editor for the *Geological Society of America Bulletin* from 1989 to 1994.

T

Talwani, Manik
(1933–   )
Indian
Geophysicist

There are two parts to the measured gravitational field, the main field of the Earth and the anomalous field. The anomalous field reflects any body of rock in the crust that does not have a density of 2.65 gm/ml, the average density. Before Manik Talwani, the anomalous field was separated from the main field and geometrically evaluated to estimate the type of body that might be producing the anomaly. Talwani developed mathematical methods to model the shape and density of the subsurface body or structure that produced the anomaly. He first developed methods to model it in two dimensions but later methods modeled bodies in three dimensions, both using integral calculus. He later extended the same methods to magnetic data. These elegant methods sparked a revolution in field geophysics. Much old data were reevaluated with the new methods and astonishing new structures and bodies were revealed. The methods were quickly adopted by the petroleum industry greatly increasing their exploration success and allowing them to find new targets that previously were not even imagined. The methods were written as interactive computer programs that are in use today in both research

labs and classrooms with virtually no alteration to the now more than 35-year-old equations. The papers introducing these breakthroughs are "Rapid Computation of Gravitational Attraction of Three-Dimensional Bodies of Arbitrary Shape" and "Computation with the Help of a Digital Computer of Magnetic Anomalies Caused by Bodies of Arbitrary Shape."

Manik Talwani applied most of his geophysical ability to studies of the oceans. To accurately measure gravity on the research ships, he invented a cross-coupling computer to compensate for the roll of the waves. He went aboard submarines to make highly accurate pendulum measurements underwater. With these tools in hand, Talwani traveled the four corners of the world to do geophysical surveys over every type of plate tectonic margin in as many different variations as possible. He was even on a ship (*R/V Vema*) that took him to 81 degrees north latitude. Most of his efforts were to study the gravity of the various features but later in his career, he became interested in seismic reflection profiling of margins. Through the famous EDGE project, Talwani studied the U.S. East Coast, the conjugate South Atlantic margins off of Brazil, and Namibia, as well as the southwest margin off of India. A book resulting from this work is entitled *Atlantic Rifts and Continental Margins*. Much of his later research was tied to petroleum exploration during and after he was employed in the petroleum industry. In these studies, he uses both 3-D seismic reflection processing techniques and a new highly sensitive technique called gravity gradiometry.

Manik Talwani even designed and oversaw the "Traverse Gravimeter" experiment on the Moon. This project, which involved instrumentation design as well as surveying, resulted in the only gravity measurements ever to have been made on the Moon.

Manik Talwani was born on August 22, 1933, in India. He attended University of Delhi, India, and earned a bachelor of science degree in 1951 and a master of science degree in 1953. He earned his Ph.D. in geophysics at Lamont-Doherty Earth Observatory of Columbia University, New York, in 1959. He accepted a research position at Lamont-Doherty in 1957 and a faculty position in 1970. He was the director of the Lamont-Doherty Geological Observatory from 1973 to 1981. He moved to Gulf Research and Development Company in 1981 as director of the Center for Crustal Studies and he became chief scientist in 1983. In 1985, he joined the faculty at Rice University, Houston, Texas, as the Schlumberger Professor of geophysics and the director of the Geotechnology Research Institute of the Houston Advanced Research Center. Talwani was a Sackler Distinguished Lecturer at the University of Tel Aviv, Israel, in 1988. Manik Talwani married Anni Fittler in 1958 and they have three children.

Manik Talwani has had an extremely productive career authoring some 150 articles in international journals and professional volumes. Many of his papers set benchmarks for geophysics that still stand today. He also edited five volumes. His research has been well recognized in the profession through numerous honors and awards. He is a member of the Norwegian Academy of Arts and Sciences, and a foreign member of the Russian Academy of Natural Sciences. He was awarded an honorary doctoral degree from the University of Oslo in Norway. He received the First Krishnan Medal from the Indian Geophysical Union in 1965, the James B. Macelwane Award from the American Geophysical Union in 1967, and the NASA Exceptional Scientific Achievement Award in 1973. He was given the George P. Woollard Award by the Geological Society of America in 1983, the UNESCO Toklen Award by the National Institute of Oceanography of India in 1990, and the Alfred Wegener Medal by the European Union of Geosciences in 1993. Talwani was a Hays-Fulbright Fellow in 1973 and a Guggenheim Fellow in 1974.

Talwani has performed service to the profession that is too extensive to report here individually. He served as a member or an official of virtually every committee involving ocean studies, and geophysics for that matter, including the Joint Oceanographic Institute for Deep Ocean Sampling (JOIDES), the Joint Oceanographic Institute, and several committees in the National Research Council and the National Academy of Sciences. He also served on several committees for the American Geophysical Union and the Geological Society of America. He was even called upon to help resolve a boundary dispute between Iceland and Norway because he had intimate knowledge of the Norwegian Sea. In another case, he helped negotiate the first United States–China cooperative scientific project since World War II.

⊠ Taylor, Hugh P., Jr.
(1932–   )
American
Isotope Geochemist

In addition to the more commonly known radioactive isotopes, there are also stable isotopes, which do not decay. These isotopes act as tracers for geologic processes that involve fluids or melts. There are characteristic signatures of stable isotope ratios that depend upon the source of the fluid or melt. The reason for the variation in the

Hugh Taylor on a field trip in California *(Courtesy of Arthur Sylvester)*

ratios is because the different isotopes will tend to enrich or partition by different geologic processes. Hugh Taylor is likely the foremost expert on stable isotopes with special interest in oxygen, hydrogen, carbon, and silicon. He has analyzed stable isotopes on rocks and soils from all over the Earth and the Moon as well as meteorites. He analyzed the lunar rock and soil samples retrieved during the Apollo missions and produced the definitive works on stable isotopes on the Moon (for example, the paper, "Oxygen and Silicon Stable Isotope Ratios of the Luna 20 Soil"). He also did some groundbreaking work on stable isotopes in tektites (meteorites) and impact generated glass.

Taylor's most notable research, however, is on terrestrial geology, which is much more varied because the systems are more complex. His research is primarily on plutonic rocks, both granitoid and

layered mafic intrusions, but he has also done work on volcanic rocks and metamorphic rocks. The two geographic areas where he has concentrated his work on granitoid plutonic rocks are in western North America where he studies Mesozoic and Cenozoic plutonism, and western Europe where he studies late Paleozoic Hercynian plutonism. This research involves the determination of the source of the magma that formed the plutons as well as the interaction of the fluids expelled during the crystallization of the magma with the country rock. Examples of publications include *The Oxygen Isotope Geochemistry of Igneous Rocks* and *Stable Isotope Geochemistry.* This aspect of his research has shed light on the processes of contact metamorphism around these plutons as well as the hydrothermal activity generated by dewatering plutons. This activity has produced significant deposits of ore minerals, an example of which is the Comstock Lode, Nevada, which Taylor studied. An example of this work is the paper "Hydrogen and Oxygen Isotope Ratios in Minerals from Porphyry Copper Deposits."

His interest in ultramafic rocks and layered mafic intrusions has taken Taylor all over the world. He studied layered intrusions from Africa (Skaergaard) to Alaska and ophiolites from Oman (Samail) to Cyprus (Troodos) to California, in addition to mantle fragments (xenoliths), komatiites from Australia and ocean fragments. This research has focused on both the alteration of these bodies after they were emplaced, and also their origins. The mantle processes involved in their generation is further elucidated in studies of volcanic rocks. An example of this work is the paper "Stable Isotope Studies of Ultramafic Rocks and Meteorites." He studied potassic volcanic rocks from all over the world including Australia, East Africa, Antarctica, Italy, and the central United States. This work led Taylor to propose that there are even fluid-rock interactions in the upper mantle. Taylor has shown through stable isotopes that this interaction of fluids and solid rock, with or without melt as a chemical system, explains many of the features and processes we see in plutons, volcanoes and ore deposits.

Hugh Taylor was born on December 27, 1932, in Holbrook, Arizona. He attended the California Institute of Technology in Pasadena, where he earned a bachelor of science degree in geochemistry in 1954. He entered Harvard University, Massachusetts, for graduate studies and earned a master of arts degree in geology in 1955. Upon graduation he accepted a position as chief scientist aboard a cruise vessel for the U.S. Steel Corporation, where he explored for iron ore in southeast Alaska in 1955 and 1956. He returned to graduate school at California Institute of Technology and earned a Ph.D. in geochemistry in 1959. Taylor was also married that year. He then joined the faculty at California Institute of Technology, where he remains today. He was a member of the faculty at the Pennsylvania State University at University Park in 1961–1962, a visiting professor at Stanford University, California, in 1981–1982, a Crosby Visiting Professor at Massachusetts Institute of Technology and a geologist for the U.S. Geological Survey in Saudi Arabia in 1980–1981. Taylor was named Robert P. Sharp Professor of geology in 1981, a title which he still holds. He was also the executive officer for geology at Cal Tech from 1987 to 1995.

Hugh Taylor has led an extremely productive career. He is an author of 159 articles in international journals and professional volumes. Many of these papers are the often-cited definitive studies on stable isotopes and appear in the top journals in the profession. Taylor has received several honors and awards for his contributions to the science. He is a member of the National Academy of Sciences and a Fellow of the American Academy of Arts and Sciences. He was awarded the Urey Medal by the European Association of Geochemistry and the Arthur C. Day Medal by the Geological Society of America. He received a Best Paper Award from the U.S. Geological Survey. He has been named to several prestigious endowed lectureships including a Cloos Memorial Scholar

at the Johns Hopkins University, a Turner Lecturer at University of Michigan, the First Hoffman Lecturer at Harvard University, the 30th William Smith Lecturer for the Geological Society of London, and an Invited Lecturer at the Italian Academy of Sciences.

Taylor has also performed service to the geological profession. He has served on numerous committees and panels and even held offices for the Geological Society of America, the Geochemical Society, and the Mineralogical Society of America. He has also served in editorial positions, including editor for *Chemical Geology* and associate editor for *Geochimica et Cosmochimica Acta* and *Geological Society of America Bulletin*.

⊠ Teichert, Curt
(1905–1996)
German
Paleontologist

Curt Teichert is an example of the quintessential international geologist. He held faculty positions with seven universities on three continents and governmental positions in Denmark, Australia, and the United States. His expertise was in the study of cephalopods but he extended his areas of interest into biostratigraphy, plate tectonics, and even energy and mining. To these ends, he literally traveled the world (every continent except Antarctica) to conduct research.

His early work on the morphology and evolution of cephalopods was done in conjunction with research on biostratigraphy and paleoenvironmental analysis of the rocks in which the fossils were found. His paper, "Main Features of Cephalopod Evolution," is a summary of his paleontologic work. This research was done in such diverse areas (Canada, Greenland, Australia, etc.) that it began to have implications for plate correlations. His publication, *Stratigraphy of Western Australia,* exemplifies his long-distance stratigraphic correlations. Of course, his early work was before plate tectonics and he was instrumental in the geosynclinal theory. However, his career spanned the acceptance of the plate tectonic paradigm and he contributed there as well. Teichert's vast travels and observations made him invaluable to regional stratigraphic projects, many of which were through his efforts. Perhaps the best known of these efforts was through a project sponsored by the U.S. Agency for International Development and carried out by the U.S. Geological Survey in Quetta, Pakistan. He trained numerous Pakistani geologists in stratigraphy, helped establish a National Stratigraphic Committee and helped establish a program of stratigraphic correlation among Pakistan, Iran, and Turkey as part of the Central Treaty Organization (CENTO). He contributed directly to this effort by performing an inch-by-inch analysis of the Permian-Triassic sequence in the Salt Range, a fundamental boundary in Earth history by virtue of the greatest extinction event ever.

Even this description of his varied history does not fully portray his vast experiences. For example, he studied the Great Barrier Reef in Australia and other coral reefs in the Indian Ocean as described in his paper, "Cold and Deep-Water Coral Banks." He studied the fauna across the Cambrian-Ordovician boundary in Northern China. He studied the Devonian stratigraphy and biostratigraphy of Arizona. He studied the Cambrian to Holocene stratigraphy of Australia. Even these research experiences do not cover his work on fuels and energy. It is through this vast experience that Teichert was able to help guide some of the fundamental theories on evolution, Earth history, and plate reconstructions.

Curt Teichert was born on May 8, 1905, in Konigsberg, East Prussia (Germany). He studied at universities in Munich, Freiburg, and Konigsberg, ultimately receiving a Ph.D. degree from Albertus University in Konigsberg in 1928. That year, he married Gertrud Kaufman, the daughter of a physics professor at Konigsberg. He accepted a postdoctoral fellowship at the University of Freiburg. In 1930, he received a Rockefeller Foundation award for paleontologic studies in Washington, D.C., New York City, and Albany, New York. This international exposure led to a position as geologist on a Danish expedition to Greenland in 1931–1932. When Teichert returned to Germany, he found that the political conditions had so deteriorated that he moved to Copenhagen, Denmark, where he received a small stipend as a research paleontologist at the university there. By 1937, the situation in Europe was far worse. Teichert applied for and received a grant from the Carnegie Foundation that made it possible for him to obtain a position as research lecturer at the University of Western Australia in Perth. In 1945, he accepted a position as assistant chief geologist in the Department of Mines for Victoria, Australia, but moved to the University of Melbourne as a senior lecturer in 1947. In 1952, Teichert began his North American career as a professor at the New Mexico School of Mines in Socorro. By 1954, he moved yet again to the U.S. Geological Survey in Denver, Colorado, to organize and direct a Fuels Geology Laboratory. Teichert left the U.S. Geological Survey in 1964 to return to academia as a Regents Distinguished Professor at the University of Kansas, where he remained until his retirement in 1977. He then moved to New York where he was an adjunct professor at the University of Rochester, where he remained until 1995. His wife Gertrude died in 1993. Teichert moved to Arlington, Virginia, in 1995 and died on May 10, 1996.

Curt Teichert led a very productive career. He was an author of some 325 scientific articles in international journals, professional volumes, and governmental reports. Several of these are seminal papers on cephalopods and biostratigraphy. He was also an editor of 13 professional volumes, including seven volumes of the *Treatise on Invertebrate Paleontology,* symposium volumes for the International Geological Congress and a Geological Society of America volume on the paleontology of China. In recognition of his research contributions to geology, Teichert received several honors and awards. He received the David Syme Prize from the University of Melbourne, the Raymond C. Moore Medal from the Society of Economic Paleontologists and Mineralogists, and the Paleontological Society Medal.

Teichert performed excellent service to the profession and the public. In addition to numerous committees, he served as president of the Paleontological Society (1971–1972). He served on many committees and as president of the International Paleontologic Association (1976–1980). He served on numerous committees and panels for the Geological Society of America and the International Geological Congress. He was also a founder of the Geological Society of Australia. He served in numerous editorial roles with the *Geological Society of America Bulletin* and the *Journal of Paleontology*, among others.

⊠ Thompson, James B., Jr.
(1921–)
American
Metamorphic Petrologist, Field Geologist

Soon after World War II, there was a revolution in petrology and geochemistry to better formulate geologic problems by classical thermodynamic methods. This heralded a major change in metamorphic petrology from classic mostly descriptive efforts to a modern geochemical approach. James B. Thompson can be considered the "father of modern metamorphic petrology" because he emerged as the leader in this revolution. It was his work in developing a physical framework that guided experimental petrologists to choose critical petrologic systems in which to conduct their experiments. This physical framework was based on observations of rocks in the field or through a microscope. Throughout his theoretical research, he was always sure to return to the real rock systems to make sure that he remained solidly based. His

famous statement exemplifies this attitude: "It would be embarrassing indeed if we were to construct an internally consistent geology, chemically and physically sound, perfect in fact but for one flaw—the lack of a planet to fit it."

Thompson's research dealt with the thermodynamics of individual minerals as part of larger chemical (metamorphic) systems. He employed the Gibbs method in new ways to explain metamorphic facies (conditions). Standard triangular diagrams used in plotting the minerals found in aluminous metamorphic rocks (schist to gneiss) are referred to as "Thompson Diagrams." Famous papers on this work are "The Graphical Analysis of Mineral Assemblages in Pelitic Schists" and "A Model System for Mineral Facies in Pelitic Rocks." He later wrote a treatise on a new idea of "reaction space," a kind of thermodynamic virtual space in which metamorphic reactions could be displayed. The major publication on this work is entitled *Reaction Space: An Algebraic and Geometric Approach.* Finally, he also worked on the properties of certain mineral systems including amphiboles, white mica, and feldspars. He even has minerals named after him, jimthompsonite and clinojimthompsonite.

Jim Thompson was able to base his thermodynamic work on real rocks so well because he was also a talented field geologist. Much of the stratigraphy and structural geology of New England, and especially that of Vermont, is understood as the result of his work. He mapped from low-grade to high-grade metamorphic rocks and looked at the characteristic structures of New England like domes and large flat folds called nappes. The astounding discovery of fossils in high grade metamorphic rocks, where they should have been destroyed, allowed Thompson to connect the stratigraphy of the crystalline part of New England to the sedimentary part. This alone is a major contribution to the understanding of the geology of New England.

James B. Thompson Jr. was born on November 20, 1921, in Calais, Maine. He attended

Dartmouth College, New Hampshire, and earned a bachelor of arts degree in geology in 1942 and served as an instructor that same year. He entered the U.S. Army Air Force in 1942 and served as a first lieutenant for the duration of World War II. After his discharge, he entered Massachusetts Institute of Technology and earned a doctorate in geology in 1950. He joined the faculty at Harvard University, Massachusetts, in 1950 and remained there for the rest of his career. He was named the Sturgis Hooper Professor of Geology in 1977 and retired as an emeritus professor in 1992. During his tenure at Harvard he was a visiting professor at University of Bern, Switzerland (1963), Dartmouth College (1988–1992, part-time) and Arizona State University (1991), a distinguished visitor at University of Cincinnati, Ohio (1974), a guest professor at Swiss Federal Institute of Technology (1977–1978) and a visiting research geologist for the U.S. Geological Survey (1985–1986), among several others. James Thompson married Eleonora Mairs in 1957; they have one child.

James Thompson has had a very productive career. He was an author of some 41 articles in international journals and professional volumes as well as 12 field guides and four geologic maps. He was coeditor of one professional volume. The absolute numbers may not be as impressive as others in this book but publishing was much less convenient and emphasized in the 1950s and 1960s than it is today. In addition, many of his papers set new benchmarks in petrology and geochemistry. Thompson has been well recognized by the geologic profession for his contributions in terms of honors and awards. He is a member of the National Academy of Sciences, and the American Academy of Arts and Sciences. He received the Arthur L. Day Medal from the Geological Society of America (1964), the Roebling Medal from the Mineralogical Society of America (1977), and the Victor M. Goldschmidt Medal from the Geochemical Society (1985). He received a Ford Faculty Fellowship (1952–1953), a Guggenheim Fellowship (1963), and he was an Ernst Cloos

Memorial Scholar at the Johns Hopkins University (1983) and a Fairchild Distinguished Scholar at California Institute of Technology (1976), among numerous distinguished lectureships.

Thompson has also performed significant service to the profession serving on numerous committees and panels for the National Science Foundation, National Research Council, the Geochemical Society, Geological Society of America, and the Mineralogical Society of America. He was president of the Mineralogical Society of America in 1967 and 1968, as well as the Geochemical Society in 1968 and 1969.

⊠   **Tilton, George R.**
(1923–   )
American
Isotope Geochemist

Geologists now determine the ages of rocks using isotope geochemistry on a routine basis. But it was not that long ago that such determinations were impossible. It was only after World War II that the understanding and capability to analyze radioactive isotopes became available. Probably the most important system to determine formational ages of rocks was and still is that of the decay of parent uranium to daughter lead. One of the true pioneers in applying U-Pb isotopic methods to rocks is George Tilton. With colleague CLAIR C. PATTERSON, Tilton developed new techniques to estimate the age of granites and granitic rocks using the accessory mineral zircon. The first rock ever to be analyzed for absolute age was a 1-billion-year-old granite from the Canadian Shield. From that starting point, Tilton joined a group of pioneering scientists who were charged with refining the methods and applications for all isotopic systems. Tilton was to lead the research on U-Pb and devise and confirm the basic procedures for analysis, which are still in practice today. His paper "Uranium-Lead Ages" is a classic. He is most certainly the "father of uranium-lead geochronology" for both granitic rocks and others. Zircon still provides the most reliable ages and it is still the mineral of choice for isotopic analysis.

After establishing these techniques, Tilton still determined the age of rocks but his main research efforts moved in other directions. He devised methods to use isotopes to gain information on the sources of volcanic magmas. The information he gathered allowed him to better understand the geochemical processes in the Earth's mantle. By looking at the isotope ratios of given whole volcanic rocks rather than of single minerals, Tilton could predict the character of the source region in the mantle for these volcanic rocks. He could tell whether the source material for magma was recycled (melted) crustal material or directly from the mantle. By more detailed studies, he could even glean some information on the mantle from which the magma was derived. He could tell if there were other melting episodes from that mantle or if the magma studied was the first. Several papers on this work include "Evolution of Depleted Mantle: The Lead Perspective" and "Isotopic Evidence for Crust-Mantle Evolution with Emphasis on the Canadian Shield." With all of this research under his belt, Tilton is still working on a timescale for the production of continental crust. He is attempting to determine when the first granitic crustal rocks appeared and what their rate of production has been throughout geologic time.

George Tilton was born on June 3, 1923, in central Illinois, where he spent his youth. He became interested in chemistry in high school and continued that interest in Blackburn College, a small two-year school near Saint Louis, Missouri. After three and one-half semesters, he was drafted into the army in February 1943 to serve in World War II. In September 1945, he was discharged and resumed his college career at the University of Illinois at Urbana-Champaign, where he graduated with a bachelor of science degree with high honors in chemistry in 1948. There he met a colleague, Elizabeth Foster, whom he later married.

George Tilton with his solid source mass spectrometer at the University of California at Santa Barbara *(Courtesy of G. R. Tilton)*

They would have four children. He enrolled at the University of Chicago, Illinois, for graduate studies intent on radiochemistry. He graduated in 1951 with a Ph.D. degree in geochemistry and accepted a position as a geochemist at the Geophysical Laboratory of the Carnegie Institution of Washington, D.C. In 1965, Tilton joined the faculty at the University of California at Santa Barbara, where he remained for the rest of his career. He retired to a professor emeritus position in 1991, but he still remains active in research.

George Tilton has had a very productive career. He is an author of more than 100 articles in international journals and professional volumes. Many of these papers are landmark studies in the field of isotope geochemistry both in terms of techniques and applications. His work has been well received by the geologic profession as evidenced by his numerous honors and awards. He is a member of the National Academy of Sciences. He was awarded an honorary doctor of science degree from the Swiss Federal Institute of Technology (ETH) in Zurich in 1984. He received the

Alexander von Humboldt Foundation Senior Scientist Award in 1989 and the Alumni Achievement Award from Blackburn College, Illinois, in 1978.

Tilton was also very active in service to the profession, serving on numerous committees and panels for both government and society functions. He was the president of the Geochemical Society in 1980. He also served in editorial positions for several prominent journals.

⊠ **Tullis, Julia A. (Jan)**
(1943–)
American
Structural Geologist (Rock Mechanics)

Under shallow crustal to surface conditions, rocks crack and break in a brittle manner when stressed to their breaking point. This cracking releases energy as earthquakes. The fault rocks under these conditions crush into rocks called breccias and cataclasites. Deeper in the crust, the temperature and pressure are elevated and the rocks flow like gum in plastic or ductile behavior when stressed to their breaking point. Few if any earthquakes are produced under these conditions. The rocks produced are called mylonites. Jan Tullis conducts experiments to determine the conditions under which this transition from brittle to plastic behavior occurs. These experiments involve squeezing or compressing rocks in a press capable of exerting many tons of force under any temperature condition until the rock cracks or flows. She then studies the deformed rock both with an optical microscope and a transmission electron microscope (TEM) to determine the processes of failure. Because she has determined this transition for the most common minerals, quartz and feldspar, her work is commonly used as the constraints for structural studies on faults and fault processes. Researchers of natural fault zones in deformed orogens have physical constraints for interpreting the thermo-mechanical history from preserved fea-

tures in the rocks, thanks to Tullis's research. Two examples of this work are papers entitled "Experimental Faults at High Temperature and Pressure" and "High Temperature Deformation of Rocks and Minerals."

Tullis determined that the transition from brittle to plastic behavior for quartz occurs at about 250°C and for feldspar it occurs at about 450°C. These temperatures correspond to depths in the Earth of about 10 km and 15 km, respectively, assuming a normal geothermal gradient. Papers on this work include "Experimental Deformation of Dry Westerly Granite" and "Dynamic Recrystallization of Feldspar: A Mechanism for Ductile Shear Zone Formation." Plastic deformation involves the sequential breaking, shifting, and reattaching of the chemical bonds between atoms in contrast to brittle deformation, which breaks the rock like glass. Tullis discovered a transition state between the two types of behavior, which appears to be plastic under a microscope but using a transmission electron microscope (TEM), it is clearly brittle, just at a much smaller scale. It appears that the scale of breakage progressively decreases at the transition. Further experimentation involves the evaluation of other factors in this transition including presence and amount of water, melted rock (magma) as well as grainsize. Water in the rock weakens it significantly so that it deforms under much lower pressures as described in "Pressure Dependence of Rock Strength: Implications for Hydrolytic Weakening." Fine-grained aggregates of minerals tend to slip on each other to absorb much of the deformation.

Jan Tullis is also interested in the mechanisms that produce layering or preferred orientations in deformed rocks. These mechanisms operate on the submicroscopic scale and involve the movement of chemical species, irregular chemical bonds, and holes in the mineral structures where atoms are missing. The movement induces shape changes in the minerals and ultimately recrystallizes them into shapes that are better suited to the high strain environment. This shape is relatively flattened and even the chemical bonds become aligned through this process. It is what produces the highly strained and flattened mylonites in deep fault zones.

Julia Ann Tullis was born in Swedesboro, New Jersey, on February 21, 1943. She attended Carleton College, Minnesota, and earned a bachelor of arts degree in 1965. She attended graduate school at the University of California at Los Angeles, and earned a Ph.D. in structural geology and rock mechanics in 1971. Her advisers were DAVID T. GRIGGS and J. Christie. Tullis accepted a position as research associate at Brown University, Rhode Island, in 1970 and became a research professor the following year. She became a full member of the faculty in 1979 and has remained ever since. Jan Tullis was married to Terry Tullis, a fellow structural geologist and Brown University geology professor, in 1965.

Jan Tullis is in the middle of a very productive career. She is an author of more than 50 articles in international journals and professional volumes. Several of these papers are definitive studies on the mechanical response of minerals to stress under various pressure-temperature conditions. They are very commonly cited in geological literature. Tullis has received several honors in recognition of her contributions to the science in both teaching and research including the Phil Bray Award for Teaching Excellence from Brown University and the Woodford-Eckis Distinguished Lecturer at Pomona College, California.

Tullis has also performed service to the profession. She served on several committees and panels for the National Research Council, the American Geophysical Union, the National Science Foundation, the U.S. Geological Survey, the American Geological Institute (for which she helped found the Women Geoscientists Committee), and the Geological Society of America. She was on the evaluation committee for Massachusetts Institute of Technology and Carleton College. Tullis also served on the editorial boards for *Tectonophysics* and *Journal of Structural Geology.*

⊠ **Turcotte, Donald L.**
(1932–)
American
Geophysicist (Fluid Dynamics)

Donald Turcotte has truly led two successful careers. He began as an aerospace engineer successfully studying combustion, magnetohydrodynamics, plasma physics, and lasers. He was a National Science Foundation postdoctoral fellow in 1965 at Oxford University in England when he learned about plate tectonic processes and had a reawakening. He proceeded to apply his considerable expertise in engineering to geology in a thoroughly unique manner. He developed a new direction in geology by constructing sophisticated mathematical models of the processes especially with regard to plate motions. In one citation, a colleague said, "It is fair to say that if it moves, Don will model it." Turcotte proposed a boundary-layer theory of convection for the circulation of the mantle. This modeling has been designed to address the question of how the plates are driven around the Earth. This extensive work culminated in a book entitled *Geodynamics: Application of Continuum Mechanics to Geological Problems* and really defined a new field. He applied this modeling to the development and nature of other planetary interiors. His "membrane theory" accounts for the correlation of gravity and topography on the Moon and Mars. However, Turcotte's modeling did not end there. He worked on such diverse topics as sediment compaction and lithification, petroleum maturation, the thermal evolution of basins, hydrothermal flow patterns around heat sources, strain accumulation and release in earthquakes on the San Andreas fault and even on the geometrical forms of volcanoes. He also modeled global geochemical cycles (oxygen, carbon, and so forth) to better understand the distribution of major elements, trace elements and isotopes of elements.

More recently, Turcotte became a pioneer in applying fractal and chaos solutions to geological problems. Fractals are basically a mathematical analysis that shows how small-scale relations reflect the larger-scale relations in what is described as a self-similar manner. Field geologists have qualitatively used fractals for years, studying outcrop scale structures to help with interpreting the map scale structures but it was never quantified. Turcotte applied fractals to the topography of Arizona, to seismic hazard assessment, crustal deformation and mineral deposits among many others and again began a new method to analyze geological features. This research is summarized in his publication *Fractals in Geology and Geophysics.* He applied chaos theory to mantle convection as well as stress distributions around faults, also setting new standards for geological analysis. This pioneering research of Donald Turcotte has truly opened a whole new aspect to the Earth sciences.

Donald L. Turcotte was born on April 22, 1932, in Bellingham, Washington. He attended the California Institute of Technology and graduated with a bachelor of science degree in mechanical engineering in 1954. He earned a master of science degree in aeronautical engineering from Cornell University, New York, in 1955. He then returned to California Institute of Technology to complete a Ph.D. in aeronautical engineering in 1958. In 1958–1959, he was a research engineer at the Jet Propulsion Laboratory in Pasadena, California, and an assistant professor at the U.S. Naval Postgraduate School in Monterey, California, before joining the faculty at Cornell University where he has remained ever since. He began his tenure at Cornell University in the Graduate School of Aerospace Engineering before moving to the Department of Geological Sciences in 1973. He served as chair of the department from 1981 to 1990. His current title is the Maxwell M. Upson Professor of engineering, which he has held since 1985. During his years at Cornell University, he has been a visiting professor at such schools as Oxford University and he has consulted for such firms as

TRW, Monsanto Inc., Corning Glass Inc., and the U.S. Department of Defense. Donald Turcotte has been married since 1957 and is the father of two children.

Donald Turcotte has led an impressively productive career with hundreds of articles published in international journals, professional volumes, and governmental reports in both aeronautical engineering and geology. The papers in geology are seminal works on unique mathematical treatments of a whole range of geological problems like mantle convection. These contributions to the science have been well received and recognized by the profession as shown in the numerous honors and awards that he received. Turcotte is a member of the National Academy of Sciences. He was awarded the Arthur L. Day Medal from the Geological Society of America, the Regents Medal of Excellence from the State of New York, the Alfred Wegener Medal from the European Union of Geosciences, and the Charles A. Whitten Medal from the American Geophysical Union. He was also the recipient of a Guggenheim Fellowship and named a William Smith Lecturer at the Geological Society of London.

Turcotte was president of the Tectonophysics Section of the American Geophysical Union for which he served on numerous committees and panels. He also served on committees for the International Union of Geodesy and Geophysics, the Seismological Society of America, and the American Physical Society among others.

⊠ Turekian, Karl K.
(1927–   )
American
Geochemist, Atmospheric Scientist

There is a complex exchange of chemicals among rock, ocean, and air. If the transmission of chemicals from rocks to air and water occurs during weathering, imagine the complexity of the interaction of the swirling mass of liquid in the oceans with the swirling mass of gas in the atmosphere. These interactions and cycling of elements that results is the mainstay of the climate change movement. Karl Turekian began his interest in these complex interactions well before it was fashionable. He was involved in the initial research to determine the fundamental distribution of chemical properties of the ocean in the early 1960s. Once this distribution was determined, the next step was to determine the static and dynamic processes that control it, and Karl Turekian was the leader in this research.

Because estuaries serve as the chemical reactors that control the passage of elements from land to sea, that is where Turekian focused his efforts. In his work, he turned Long Island Sound into the world model for coastal studies. He used short-lived radioactive elements (isotopes) as tracers to track ocean circulation and processes. Movement of water masses, zones of ascending or descending ocean currents, and chemical interactions could also be studied using these tracers. The tracking of isotopes additionally helped determine scavenging, sediment accumulation rates, bioturbation, residence times, and atmospheric deposition. An example of this work is "The Osmium Isotopic Composition Change of Cenozoic Sea Water." Turekian even looked at extraterrestrial input into the ocean. Because the ocean directly exchanges elements with the atmosphere, it constrains the chemistry. Turekian also studies the atmosphere in terms of residence times of certain elements and origin of ozone in the troposphere as well as interaction with rocks. His results led him to consider the origin and evolution of the oceans and atmosphere.

Karl Turekian was born in New York, New York, on October 25, 1927. He attended Wheaton College, Illinois, and earned a bachelor of arts degree in chemistry in 1949. He also served in the U.S. Navy as an aviation electronic technician's mate third class. He completed his graduate studies at Columbia University, New York, where he earned a master of arts and a

Ph.D. in geochemistry in 1951 and 1955 respectively. He was a research associate at the newly established Lamont-Doherty Geological Observatory in 1954 to 1956. He joined the faculty at Yale University, Connecticut, in 1956, and remained there for the rest of his career. Turekian was named a Henry Barnard Davis Professor of geology and geophysics from 1972 to 1985. He was then named the Benjamin Silliman Professor of geology and geophysics in 1985, and he retains that title today. He served as department chair from 1982 to 1988. He is also currently director of the Yale Institute for Biospheric Studies. Turekian married Roxanne Hagopian in 1962 and they have two children. His son, Vaughan, followed in his father's footsteps and the two recently published several articles together on atmospheric-oceanic interactions.

Karl Turekian has enjoyed an extremely productive career. He is an author of more than 210 articles in international journals and professional volumes. He also published several books. These articles appear in some of the best journals in the profession and many set new benchmarks for the science. His popular textbook is entitled *Global Environmental Change: Past, Present, and Future*. He has been richly recognized for his contributions to the science in terms of honors and awards. He is a member of the National Academy of Sciences and a fellow of the American Academy of Arts and Sciences. He received honorary doctoral degrees from Yale University and State University of New York at Stony Brook. He was awarded the V. M. Goldschmidt Medal from the Geochemical Society (1989) and the Maurice Ewing Medal from the American Geophysical Union (1987). He was a Guggenheim Fellow at Cambridge University (1962–1963) and a Sherman Fairchild Distinguished Scholar at California Institute of Technology (1988).

Turekian has performed exemplary service to the profession. He served on the U.S. National Committee on Geochemistry (1970–1973), the Climate Research Board (1977–1980), the Ocean

Science Board (1979–1982), the U.N. Council on the Scientific Aspects of Marine Pollution, and several committees and panels for the National Academy of Sciences, the National Research Council, and the National Science Foundation. He was president of the Geochemical Society in 1975 and 1976 and served on many committees for that society as well as the Geological Society of America and American Geophysical Union. Turekian served as editor for some of the top journals including *Journal of Geophysical Research, Geochimica et Cosmochimica Acta, Earth and Planetary Science Letters* and *Global Biogeochemical Cycles,* in addition to the *Proceedings of the National Academy of Sciences.*

⊠   **Tuttle, O. Frank**
(1916–1983)
American
Geochemist, Petrologist

Frank Tuttle was one of the greatest experimental petrologists to grace the field. He invented the "Tuttle press" and the "Tuttle bomb," which allowed him for the first time to adjust the temperature and pressure of his experiments at will to simulate virtually any conditions in the Earth's crust. These inventions revolutionized the entire field of experimental petrology. The current equipment for experimental studies is really just a modified version of that which he invented in the late 1940s. Not only did he perform his own experiments, the data from which has withstood the test of time, but he also set the foundations for all work to follow. His experiments centered on multivariate chemical systems in the felsic range of compositions. Most of this work was done with his close colleague NORMAN L. BOWEN. They did experimental studies on quartz, defining the stability fields for its many polymorphs, feldspars, and feldspathoids. They conducted experiments on synthetic systems of MgO-SiO_2-H_2O and K_2O-Al_2O_3-SiO_2-H_2O and would provide the

basis for standard petrogenetic grids. In addition, they conducted melting relations in natural and synthetic granite and defined the entire granite system. Tuttle visited many locations worldwide to collect samples of classic and odd granites for this work. He sampled the Harker Collection at Cambridge University in England and visited the French Pyrénées, the Isle of Skye, Scotland, Finland, and Norway among others. His publication "Origin of Granite in the Light of Experimental Studies in the System $NaAlSi_3O_8$-$KAlSi_3O_8$-SiO_2-H_2O" with N. L. Bowen in 1958 is still considered a classic work on granites. Other publications include "Chemistry of the Igneous Rocks: I. Differentiation Index" and "The Granite Problem: Evidence from the Quartz and Feldspar of a Tertiary Granite."

Tuttle investigated other systems as well. Notably, he and PETER J. WYLLIE investigated the system CaO-CO_2-H_2O and defined the origin and processes in the genesis of the odd carbonitite magmas. This work resulted in a book entitled *Carbonatites.* They also investigated the hydrothermal melting of shales in a publication of the same name and the effect of volatile components with sulfur, phosphorus, lithium, and chlorine on granite magma. He worked with RICHARD H. JAHNS on pegmatites, among others. Each of these projects established a new benchmark in petrology.

Frank Tuttle was born on June 25, 1916, in Olean, New York. He grew up in Smethport, Pennsylvania, and graduated from Smethport High School in 1933. He worked in the Bradford, Pennsylvania, oil fields for several years and enrolled in the Bradford Campus of the Pennsylvania State University on a part-time basis. He enrolled at the main campus in State College in 1936, and earned a bachelor of science degree in geology in 1939 and a master of science degree in 1940. He enrolled at Massachusetts Institute of Technology and completed his coursework by 1942 before his graduate work was put on hold because of World War II. He and his lifelong partner, Dawn Hardes, were married in 1941. They had two daughters. Tuttle's war effort was in research and it took him from Massachusetts Institute of Technology to the Geophysical Laboratory of the Carnegie Institution in Washington, D.C., to the U.S. Naval Research Laboratory in Maryland. He was involved in the synthesis and characterization of crystals for defense applications. It was during this time that he met N. L. Bowen. Before even completing his Ph.D. in 1948, Tuttle joined Bowen at the Geophysical Laboratory in 1947. He joined the faculty at Pennsylvania State University at State College in 1953, and served as dean of the College of Mineral Industries in 1959 and 1960. In 1960, he was diagnosed with Parkinson's disease and resigned his position as dean. The symptoms would recur throughout the remainder of his life. In 1965, Tuttle moved to Stanford University in California, where he spent the remainder of his career. In 1967, he requested a medical leave from Stanford University as a result of his declining health and tendered a formal resignation in 1971. In 1977, he was diagnosed with Alzheimer's disease and moved to a nursing home. He died on December 13, 1983, one year after his wife. Tuttle was an avid golfer.

Frank Tuttle had a very productive career despite his health problems. He was an author of numerous articles in international journals and professional volumes in collaboration with several of the top geologists ever. These papers are true classic works on experimental geochemistry and petrology especially with regard to granite and have been cited in other articles countless times. His contributions to geology were well received by the profession as evidenced in his numerous honors and awards. Tuttle was a member of the National Academy of Sciences. He received the first ever Mineralogical Society of America Award (1952). He also received the Roebling Medal from the Mineralogical Society of America (1975) as well as the Arthur L. Day Medal from the Geo-

logical Society of America (1967). He received other honors too numerous to fully list here.

Twenhofel, William H.
(1875–1957)
American
Sedimentologist

William Twenhofel has been called the "patriarch of sedimentary geology." He was originally trained as a paleontologist and he practiced paleontology, but he soon observed that the sediments in which they occurred were a key element to the interpretation of the paleoecology. He slowly became one of the true pioneers in the up and coming field of sedimentology. For his graduate research, Twenhofel studied the fossils of the Ordovician-Silurian boundary on Anticosti Island, Quebec, under the great paleontologist Charles Schuchert. He walked some 700 miles around the island and even though it was "ram-jammed full of beautiful fossils," it was the sedimentary sequences that induced him to return to the depositional sequence in that or nearby areas during numerous field seasons. He studied Ordovician and Silurian strata in Newfoundland, Nova Scotia, other areas of Quebec, Maine, and even the Baltic Provinces of Europe. Twenhofel achieved international recognition as an authority on the Ordovician to Silurian transition in northeastern North America, which he showed to be gradual.

While in the Midwest, William Twenhofel established a vigorous research program on local strata. He became embroiled in a controversy about the position of the Mazomanie glauconitic sands within the Upper Cambrian stratigraphy of the Upper Mississippi Valley. He argued for and finally proved the lateral equivalence of the unit to the Franconia Formation and set the stage for the development of the facies concept in sedimentary geology. He also did research on heavy minerals in sedimentary rocks, and a number of pioneering studies on lacustrine deposits in several of the lakes in Wisconsin. He studied such topics as marine conglomerates and unconformities, deep-sea sediments, and corals and coral reefs, among others. Several of these studies established the basis for future expansion that would form new and important directions in sedimentology. The ideas are now standard concepts that appear in introductory textbooks worldwide. It is this pioneering spirit that earned Twenhofel his reputation.

William H. Twenhofel was born in Covington, Kentucky, on April 16, 1875. He grew up on a farm in Covington and attended public school for his primary education but was forced to attend a private school for his secondary education. Being of modest means, Twenhofel had to work for six years as a teacher and a railway conductor to earn enough money to attend college. William Twenhofel married his childhood sweetheart, Virgie Mae Stevens, in 1899. They had three children. He attended the National Normal School in Lebanon, Ohio, where he received a bachelor of arts degree in 1904. Upon graduation, he accepted a position teaching mathematics at the East Texas Normal College in Commerce, Texas. By 1907 at age 32, he had saved enough money to attend Yale University, Connecticut. He earned a second bachelor of arts degree in 1908, a master of arts degree in 1910, and a Ph.D. in 1912, all of which were in geology. He completed his dissertation work under the advisement of paleontologist Charles Schuchert. Twenhofel joined the faculty at the University of Kansas in Lawrence in 1910 and became the state geologist of Kansas in 1915. In 1916, he accepted a position at the University of Wisconsin in Madison,

where he remained throughout the rest of his career. He retired to professor emeritus in 1945 but remained in active research for many years. William Twenhofel died on January 4, 1957.

William Twenhofel contributed greatly to the geologic profession in terms of literature. He is an author of more than 75 scientific publications including articles in international journals, governmental reports, and textbooks. He is probably best known for these textbooks, which were widely adopted and include *Invertebrate Paleontology* with Robert Schrock in 1935, *Principles of Sedimentation* in 1939, *Methods of Study of Sediments* in 1941, and *Principles of Invertebrate Paleontology* in 1953. Perhaps the most prestigious of the awards that Twenhofel received was the creation of the Twenhofel Medal as the highest award of the Society of Economic Paleontologists and Mineralogists.

Perhaps the main reason that Twenhofel was so effective in charting the direction of sedimentary geology was his effectiveness in service to the profession. He served as president of the Paleontological Society (United States) in 1931 as well as on numerous committees. However, his work for the National Research Council is legendary. He served as chair of the Committee on Sedimentation from 1923 to 1931 and was active on it from 1919 to 1949. He also served as director of the Division of Geology and Geophysics (1934–1937), the chair of the Committee on Paleoecology, and he helped organize the Committee on Stratigraphy. Twenhofel was a cofounding editor of the *Journal of Sedimentary Petrology*, and served as editor from 1933 to 1946.

V

Vail, Peter R.
(1930–   )
American
Stratigrapher

Although mean sea level is used as a point of reference for all elevation data, it does not remain the same. At times, the sea level rises significantly as it is doing now in a worldwide transgression and at other times it can fall several hundred feet in a worldwide regression. These major changes can be further compounded by local changes in the height of the continents. Peter Vail studied these changes and produced a spectacular series of sea-level curves to show the relative height of the oceans at any given time. One of the best places to chart such changes is the interior of North America during the Paleozoic and Mesozoic. Sea level was so high at the time that it flooded the interior of the continent, forming a huge, shallow epicontinental sea. Because it was so shallow, even a small change in sea level resulted in a huge shift in the position of the shoreline oceanward in a regression or inland in a transgression. By studying the positions of the shorelines over time, Vail produced a very sensitive curve. Because the stratigraphic record is so complete in this area over such a long period of time, Vail was able to model the cyclicity of sea level changes. Many of the large changes could be shown to be the result of major plate tec-

tonic changes primarily as the result of inflation and deflation of mid-ocean ridges, thus displacing more or less space in the ocean basins. Other very regular cycles, however, were discovered to be astronomical in nature, resulting from the regular shifts in the distance between Earth and the Sun called, Milankovitch Cycles. This change in distance produces a regular change in average temperature and thus climate. It not only varies sea level but also sediment character as a result.

Peter Vail also considered sea-level changes at other times by studying the stratigraphy of the continental shelves that are also relatively flat and thus quite sensitive to change. These regular changes in sea level produce distinct packages of sedimentary successions. Not only can these packages be seen in outcrop and in geophysical logs from petroleum exploration wells, but they also can even be seen in seismic reflection profiles that are used in oil exploration. Seismic reflection profiles are like sonograms of the Earth that show the character of the layering of sediments. Their study is called seismic stratigraphy, which can be quite intricate, given good data. Vail is likely the foremost authority in this field. Vail's paper "Seismic Stratigraphy and the Global Change of Sea Level, Part IV Global Cycles of Relative Changes of Sea Level" combines both of his areas of expertise. This grouping of sedimentary layers into repeating packages has been called sequence stratigraphy, of which Vail is one of the chief proponents. By such grouping of rocks, sometimes the big picture of the depositional systems become apparent that is otherwise lost in the typical viewing of each layer on a standard individual basis.

Peter Vail was born on January 13, 1930, in New York, New York, where he spent his youth. He attended Dartmouth College, New Hampshire, where he earned a bachelor of arts degree in 1952. He earned both his master of science and Ph.D. degrees from Northwestern University, Illinois, in 1956. Vail spent most of his career as a research geologist with Exxon (Esso at the time) beginning in 1956 with an affiliate company called Carter Oil Company in Tulsa, Oklahoma. It was at this time that he married his wife, Carolyn. They have three children. In 1965, Exxon consolidated its research activities into the Exxon Production Research Company in Houston, Texas, and Vail and his family relocated there. He remained with Exxon until 1986, when he joined the faculty at Rice University in Houston, Texas. He was named the W. Maurice Ewing Professor of oceanography, a position he holds today. Vail was a visiting scientist at the Woods Hole Oceanographic Institution, Massachusetts, in 1976 and a Gallagher Visiting Scientist at the University of Calgary, Canada, in 1980.

Because Peter Vail worked in industry for most of his career, by necessity, he has fewer professional publications than some of the other scientists in this book. Much of his work was proprietary for Exxon and released as internal reports. However, the internationally published articles on which he is an author are still quite numerous. Many are benchmarks in sea level changes that have spawned a whole new field of research with direct impact on climate change studies. Vail has received numerous honors and awards for his contributions to geology. He received the Virgil Kaufman Gold Medal from the Society of Exploration Geophysicists, the William Smith Medal from the Geological Society of London, and the Individual Achievement Award from

the Offshore Technology Conference. From the American Association of Petroleum Geologists (AAPG), he received the President's Award for the Best Published Paper and the Matson Award for the best conference presentation. Vail was also named a distinguished lecturer twice by AAPG and a William Smith Lecturer by the Geological Society of London.

Vail has performed significant service to the geological profession. He served on several important boards for the National Academy of Sciences, as well as the U.S. Department of Energy. He also served on numerous committees as well as performing editorial work for AAPG, the Society of Economic Paleontologists and Mineralogists, and the Geological Society of America.

⊠   **Valley, John W.**
(1948–   )
American
Metamorphic Petrologist, Geochemist

There are radioactive isotopes that decay with time and there are stable isotopes that do not. However, geologic, atmospheric and hydrospheric processes will concentrate certain stable isotopes. John Valley is an expert on stable isotope geochemistry and perhaps the foremost expert on stable isotopes in Precambrian rocks of high metamorphic grade. He studies stable isotopes (mostly oxygen, carbon, hydrogen, and sulfur), within individual minerals as well as whole rock systems, to help determine the processes involved in their formation. Isotopes are especially good monitors of fluid of thermal history and fluid interactions. They are useful for studying the genesis of igneous and metamorphic rocks at high temperatures or for paleoclimate and sedimentation at low temperatures. Valley's book, *Stable Isotopes in High Temperature Geologic Processes,* summarizes the igneous and metamorphic work.

Over the years Valley branched out. As a result, he has also been involved with a wide range of isotope studies under other conditions including: individual minerals, in sedimentary rocks during burial, in overthrust sheets in the Appalachians, in volcanic rocks from ocean islands and Yellowstone National Park, granites from the Sierra Nevada Mountains, and even Martian meteorites. He was even involved in analyzing isotopes from the fossilized teeth of herbivores to determine paleo diets. In each case, the stable isotopes provide a key piece of information to determine the process of formation or alteration of a preexisting rock. They are typically used as a tracer in many of these processes. The ratio of isotopes can be used as a fingerprint for the specific origin of fluids in many cases.

Even with all of this diversification, Valley's first interest is in the Precambrian rocks of the North American shield, and he periodically returns to projects there. Many projects are in the Grenville Province of Canada and the Adirondack Mountains of New York. He has been involved in experimental geochemistry, metamorphic petrology, the origin of anorthosite intrusive complexes, bulk and trace-element mineral chemistry, and geothermometry (temperatures of formation). This extensive work has established Valley as one of the foremost experts on the petrology and geochemistry of the North American shield, if not shield provinces in general. In one of his most recent studies, he documents evidence for the existence of continental crust and oceans on the Earth some 4.4 billion years ago from these complex rocks of the North American shield. This is a radical idea considering that the Earth is 4.6 billion years old and has been traditionally considered to still have been a relatively undifferentiated mass at 4.4 billion. Several examples of Valley's papers on these ancient rocks include, "Metamorphic Fluids in the Deep Crust: Evidence from the Adirondack Mountains, New York" and "Granulites: Melts and Fluids in the Deep Crust."

John Valley was born in Winchester, Massachusetts, on February 28, 1948. He enrolled in

John Valley conducts research on an ion microprobe in Edinburgh, Scotland *(Courtesy of J. Valley)*

Dartmouth College, New Hampshire, and earned a bachelor of arts degree in geology in 1970. He earned master of science and Ph.D. degrees in geology from the University of Michigan in 1977 and 1980, respectively. John Valley married Andrée Taylor in 1972; they have two children. He joined the faculty at Rice University in Houston, Texas, in 1980. In 1983, he accepted a position at the University of Wisconsin at Madison where he is currently a professor. He served as chair of the department from 1996 to 1999.

John Valley has had a very productive career. He is an author of some 145 articles in international journals and professional volumes. He also edited one professional volume and wrote another. Although he has collaborated with some of the top petrologists and geochemists in the profession, his ability to motivate his students to publish their work in top journals is even more impressive. The profession has recognized John Valley for his contributions in terms of honors and awards. He earned a William Hobbs Fellowship and a Horace H. Rackham Fellowship while still in graduate school. He received the ARCO prize in 1985, the Vilas Associate Award in 1999–2001 and the Kellett Award at the University of Wisconsin in 2001. He was also a Romnes Fellow at the University of Wisconsin in 1989–1994 and a Fulbright Scholar at the University of Edinburgh, Scotland, in 1989–1990.

John Valley has performed extensive service to the profession. He served as associate editor for *Journal of Geophysical Research* (1992), *American Journal of Science* (1996–present), and *Geological Society of America Bulletin* (1985–1991). He served as member and chair of numerous committees for the Mineralogical Society of America, Geological Society of America, American Geophysical Union, and the Geochemical Society. He served on the review panels for both the National Science Foundation and U.S. Department of Energy.

⊠ **Van der Voo, Rob**
(1940–)
Dutch
Paleomagnetist

If a magnet is heated above its Curie temperature, it becomes nonmagnetic. If cooled back down it will again be magnetized but its north and south poles will be oriented parallel to the Earth's magnetic field regardless of its orientation prior to heating. When an igneous rock cools through the Curie temperature for magnetite (578°C), the poles in the magnetite grains will align with the Earth's field. If magnetite grains are carried in suspension in water, they will spin like a compass needle and align with the Earth's field as they settle to the ocean floor. These records of the position of the Earth's magnetic field are paleomagnetics and are measured by a paleomagnetist or paleomagician as some geologists fondly call them. Rob Van der Voo is undoubtedly the world's foremost authority on paleomagnetics. He established himself in this position during the most important time for paleomagnetics: the plate tectonic revolution; and his papers "Paleomagnet-

ics, Continental Drift and Plate Tectonics" and "Paleomagnetism in Orogenic Belts" are classics.

By measuring the orientations of the magnetic field through a sequence of rock with known ages, Van der Voo charted changing field orientations with time, yielding apparent polar wandering paths. However, it was not the pole that was wandering but rather the continent that was wandering within the magnetic field as described in his paper, "A Method for the Separation of Polar Wander and Continental Drift." The technique is very sensitive to latitude positions but insensitive to changes in longitude. By correlating the paleomagnetism with paleoenvironmental analysis of the sedimentary rocks, paleoecological analysis of the fossils and any other pertinent information, Van der Voo, along with colleagues Scotese, Ziegler, and Bambach, was able to construct full animations of plate movements and interactions throughout the Paleozoic. This mammoth project was a giant step in plate tectonics, and results of their work now appear in every historical geology textbook. This reconstruction continues to be revised on a yearly basis as new data are received. Several studies on this work include *Paleozoic Base Maps* and *A Paleomagnetic Reevaluation of Pangea Reconstructions.*

Van der Voo is not only involved in such large-scale projects. He also addresses local problems using paleomagnetics and continues to add local information to the large database on plate movements with these regional projects. These projects chart deformation and bending or rotation of areas that either support geological studies or identify processes that would not otherwise be recognized without paleomagnetics. He also discovered how easily some types of rock become remagnetized through burial processes, which he continues to study. Finally, he is involved in establishing paleomagnetics as a method of geochronology.

Rob Van der Voo was born on August 4, 1940, in Zeist, the Netherlands. He attended University of Utrecht, the Netherlands, where he earned a bachelor of science degree in geology in

Rob Van der Voo at a forum at the University of Michigan *(Courtesy of R. Van der Voo)*

1961, a master of science degree in geology in 1965, a master of science degree in geophysics in 1969, and a doctorate in geology and geophysics in 1969. He began his career at the University of Michigan in Ann Arbor in 1970, and remains there today. He served as department chair in 1981 to 1988 and in 1991 to 1995 and director of the honors program in 1998 to the present. He was the Arthur F. Thurnau Professor of geology in 1994 to 1997. During his residence at University of Michigan, he was a visiting scholar at many programs, including Lamont-Doherty Observatory of Columbia University, New York (1976), University of Rennes, France (1977), University of Kuwait (1979), University of Texas at Arlington (1984), Greenland Geological Survey (1985), Instituto Jaume Almera, Barcelona, Spain (1990–1991), and Universities of Utrecht and Delft, the Netherlands (1997–1998). Rob Van der Voo married Tatiana M. C. Graafland in 1966 and they have two children. He is fluent in English, French, Dutch, Spanish, and German.

Van der Voo is an author of an impressive 225 articles in international journals and professional volumes. Many of these are the seminal studies on paleomagnetism and plate tectonic reconstructions. He also edited one professional vol-

ume and wrote one book. He has received many honors and awards in recognition of his contribution to the field. He was elected to both the Royal Academy of Sciences of the Netherlands (1979) and the Royal Norwegian Society of Sciences and Letters (1995). He received the G.P. Woollard Award from the Geological Society of America in 1992 and was named an A.V. Cox Lecturer by the American Geophysical Union in 1997. From University of Michigan he received the Henry Russel Award (1976), the Distinguished Faculty Achievement Award (1990), three Excellence in Education Awards (1991, 1992, 1993), and he was named a Distinguished Faculty Lecturer in 1998. In 2001, he received the Benjamin Franklin Medal in Earth Sciences from the Franklin Institute in Philadelphia.

Van der Voo's service to the profession is remarkable. The committees upon which he has served are too numerous to list here but between Geological Society of America and American Geophysical Union, they number in the 20s. He served as president of the Geomagnetism and Paleomagnetism Section of American Geophysical Union in 1988 to 1992. He has served as editor for *Geophysical Research Letters,* and *Earth and Planetary Science Letters* and associate editor of *Tectonics, Geology, Tectonophysics, Geological Society of America Bulletin,* and several others. He was part of several National Research Council and National Science Foundation committees and panels as well as a subcommittee for the National Academy of Sciences. He was on the evaluation committee for geoscience departments at eight universities.

⊠ **Veblen, David R.**
(1947–   )
American
Mineralogist

Transmission Electron Microscopy (TEM) is the technique that yields the highest magnification. It yields images rather than the direct observations of an optical microscope but it can magnify features to thousands of times their size, right down to the atoms. TEM with lower energy (typically 90,000 volts) yields less penetrating power and is used for observational biological applications. Analytical TEM operates on a higher power (typically 120,000 volts) for higher penetration and quantitative and observational use in material science and geology. David Veblen has established himself as one of the foremost experts of TEM to geology. The applications clearly must involve minerals or glass simply because of the scale. TEM can image the relations among the atoms within these materials whether it be the ordering, observations of reactions at the atomic scale, or defects in the structures.

David Veblen is a mineralogist who first established his reputation looking at a confusing group of minerals called pyroboles and biopyroboles. These minerals are a strange combination of amphiboles and pyroxene that are interlayered on the atomic level. It took TEM to determine the structure of these complex minerals. Two of his books and volumes on these minerals are *Amphiboles: Petrology and Experimental Phase Relations* and *Amphiboles and Other Hydrous Pyroboles*. Once established Veblen branched out and applied TEM along with X-ray techniques and crystal chemistry to many different minerals and materials. The materials he worked with include volcanic glass (obsidian) in which there is no atomic structure (ordering of atoms) because it is a supercooled liquid. This work has implications for high level nuclear waste because it is typically encased in glass before being buried. Even more exciting was his work on the non-geological substances, superconductors. With numerous researchers, many from the Carnegie Institution of Washington, D.C., Veblen helped determine the structure of synthetic superconducting materials. In fact they helped to establish multiple new high temperature superconductors. This work is reported in the paper "Crystallography, Chemistry

and Structural Disorder in the New High-Tc Bi-Ca-Sr-Cu-O Superconductor."

The list of minerals that David Veblen has worked on alone and with colleagues is long. He branched out from pyroboles to pyroxene and amphibole. Other areas of study include asbestos, both amphibole and serpentine, micas and especially reactions between biotite and chlorite, and clay minerals. The work on clay minerals is not only to unravel their complex structures but also to help define weathering processes. Veblen's paper "High-Resolution Transmission Electron Microscopy Applied to Clay Minerals," is a summary of that work. The reactions from one mineral to another can be traced on the atomic level in this process that has strong implications for environmental geology. The release or uptake of certain chemical species is governed by these reactions. Many prominent mineralogists and petrologists have performed research with Veblen on minerals and mineral reactions, too numerous to list here. In addition to all of this mineral research, Veblen also finds time to refine and develop new techniques for mineralogical TEM both in terms of analytical procedures and software routines for reducing data. David Veblen defines the cutting edge of TEM research in geology.

David R. Veblen was born on April 27, 1947, in Minneapolis, Minnesota. As a toddler, he collected minerals, rocks, and fossils, and by the time he reached the age of five, he had decided to pursue a career in mineralogy and geology. He attended Harvard University and earned all three of his degrees in geology there. He earned a bachelor of arts degree in 1969 with highest honors, magna cum laude and Phi Beta Kappa. He earned a master of arts and a Ph.D. in 1974 and 1976, respectively. He spent his next three years as a postdoctoral research fellow at Arizona State University. He then accepted a position on the faculty at Arizona State University. Veblen joined the faculty at the Johns Hopkins University in 1981 as a joint appointment in earth and planetary sciences and materials science and engineering and has remained ever since. He was named a Morton K. Blaustein Professor of Earth and Planetary Sciences in 1998. He was a visiting professor at California Institute of Technology in 1990 and a Tage Erlander Guest Professor in Sweden. David Veblen has two children.

He is an author of some 130 articles in international journals and professional volumes. Many of these studies set the standard for the profession. He is editor of two volumes. He has received approximately $6.7 million in grant funding, mostly from the National Science Foundation. In recognition of his contributions to the profession, Veblen received the Mineralogical Society of America Award in 1983, among others.

Veblen has performed extensive service to the profession. He served as vice president (1995–1996) and president (1996–1997) of the Mineralogical Society of America, among numerous committees. He also served as councilor for the Clay Minerals Society. He was the associate editor for *The American Mineralogist* (1982–1985) and served on the editorial board for *Phase Transitions*.

W

Walcott, Charles D.
(1850–1927)
American
Paleontologist

The story of Charles D. Walcott is straight out of a Horatio Alger novel. This self-educated high school dropout from modest beginnings worked his way up to become one of the best known and most influential geologists ever in the United States. His accomplishments went way beyond geology to the founding and administration of some of the most prestigious scholarly and governmental institutions in the United States today. In terms of paleotology, he was responsible for determining that trilobites were arthropods through careful study of the limbs of fossils. He wrote a major volume on Paleozoic fossils and resolved the stratigraphic problems of the position of the Taconic system. This work led him to confirm trilobite zone sequences in the Cambrian, and as a result, he summarized the stratigraphy of the Cambrian System of North America.

The research that Walcott conducted during his famous tenure at the U.S. Geological Survey continued his efforts on Cambrian life with the publication of research papers on trilobites and jellyfish from China. He also researched data for Monograph 51, *Cambrian Brachiopoda,* with the volume of plates equaling the size of the volume of text. Later, while with the Smithsonian Institution, he made his most famous discovery, the legendary Burgess Shale in the Canadian Rocky Mountains. This spectacularly rich and well-preserved fossil location was made famous in STEPHEN JAY GOULD's modern popular book, *Wonderful Life,* and SIMON CONWAY MORRIS'S book, *Cauldron of Life.* Walcott's areas of study were concentrated in Alberta and British Columbia, Canada, during his later research years.

Charles D. Walcott was born on March 31, 1850, in New York Mills, New York. His family was in the business of cotton milling. He received his formal education at the Utica Free Academy until the age of 18. Due to the lack of formal science training at the academy and encouragement at home, Walcott's interest in science was not actively pursued. During this time, Colonel Jewett, a retired New York State Museum curator, had moved to Utica and began to interest young Walcott in fossils. At age 12, Walcott was working summers in Trenton Falls, New York, on a farm during the Civil War. Trenton Falls is a haven for Ordovician fossils. It cannot be confirmed whether Walcott graduated from high school because the records were lost, but it is suspected that he did not. With no future in sight, he went to work first in a hardware store and then on a farm owned by a local farmer named William Rust.

Rust also had a keen interest in collecting fossils. During this time they amassed such an abundant collection of unique and well preserved fossils, they earned close to $150,000 by 1995 standards.

Walcott sold his collection to the most famous naturalist of the time, Professor Louis Agassiz. As a condition of the sale, Walcott was required to ship the fossils to Agassiz's laboratory in the Museum of Comparative Zoology at Harvard University, Massachusetts, during the summer of 1873. During this time, Agassiz expressed the importance of studying trilobite appendages. Even though Walcott had never attended college, he often consulted with Professor Agassiz when he needed guidance during his fossil research. Walcott heeded Agassiz's advice regarding trilobite appendages and began cutting rocks with fossils into thin sections (rocks ground thin enough to see through) in order to study trilobites even closer (using a microscope). At the time, trilobite legs had never been found or studied. Through Walcott's persistence and making hundreds of thin sections, he proved that trilobites had jointed appendages and were therefore arthropods.

In late 1876, Walcott took a position as special assistant to James Hall, the state paleontologist of New York. He spent countless hours studying Hall's large collection of fossils and his library. During this time, Walcott also lobbied for Hall in the state legislature. In July 1879, with a letter of support from Hall and Hall's former assistant, R. P. Whitfield, Walcott was hired as one of the original members of the United States Geological Survey as an assistant geologist. His concentrations were focused on biostratigraphy (determining the age of sedimentary rocks by studying fossils). This was a change from his usual research efforts in paleobiology.

In 1894, Walcott was appointed the third director of the U.S. Geological Survey, succeeding John Wesley Powell, and kept this position for the next 13 years. He was responsible for expanding the research efforts into water resources, more to-

Charles Walcott (left) with John Wesley Powell (center) and Archibald Geike in 1897 *(Courtesy of the U.S. Geological Survey, J.S. Diller Collection)*

pographic mapping, and studying national forests. During this time, Walcott was also responsible for establishing the Carnegie Institution of Washington and the contained Geophysical Laboratory. After his tenure with the U.S. Geological Survey, he became the fourth secretary of the Smithsonian Institution in 1907. As Walcott conducted his research and oversaw the Smithsonian Institution, he also served first as vice president of the National Academy of Sciences for 10 years and then president from 1916–1922. In recognition of this service and his vast contributions to the field, a medal for paleontology has been named there in his honor. In 1915, Walcott founded the National Advisory Committee for Aeronautics and was ultimately responsible for the construction of the Freer Art Gallery of the Smithsonian Institution. Charles D. Walcott died on February 9, 1927, in Washington, D.C.

⊠ **Walter, Lynn M.**
(1953–)
American
Aqueous Geochemist

Some people think that research that benefits environmental conditions is diametrically opposed to that which benefits the petroleum industry, and yet the research that Lynn Walter performs seems to span the gap. Lynn Walter is one of the foremost experts on aqueous geochemistry of sediments and soils under surface to shallow subsurface conditions. She achieves this research by performing detailed analyses on samples taken in the field as well as performing experimental work under laboratory conditions. She has worked carefully on precipitation and dissolution kinetics of carbonates. These carbonates can form cements in hydrocarbon reservoir rocks, reducing oil flow rates, and thus this research is of interest to petroleum companies. However, because carbonates also readily interact with surface waters and air, the research also has implications for environmental processes. Walter's paper, "Dissolution

Lynn Walter in her office at the University of Michigan *(Courtesy of L. Walter)*

and Recrystallization in Modern Shelf Carbonates: Evidence from Pore Water and Solid Phase Chemistry," is an example of her main body of work.

Most of Lynn Walter's research involves the interactions of soils and rocks with the pore fluids that they contain. This work involves detailed geochemistry and isotope geochemistry of the fluids, coupled with detailed observations on the soil and rock using an analytical electron microscope. Coupling these two data sets yields powerful predictive capabilities for diagenetic (burial and lithification of sediments) processes and groundwater chemistry. This work can be done on a regional basis to evaluate the oil and gas potential of a specific rock unit or of a basin. An example of this work is the paper, "Fluid Migration, Hydrogeochemical Evolution and Hydrocarbon Occurrence: Eugene Island Block, Gulf of Mexico Basin." However, each sample is analyzed in painstaking detail to make these prognoses. Because soils and modern sediments interact with the contained fauna and flora, there is also a component of biogeochemistry to this work. For example, the paper, "Carbon Exchange Dynamics and Mineral Weathering in a Temperate Forested Watershed (Northern Michigan): Links Between Forested Ecosystems and Groundwaters," illustrates this research.

Lynn Walter was born on April 18, 1953, in Chicago, Illinois. She attended Washington University in Saint Louis, Missouri, where she earned a bachelor of arts degree in geology in 1975. She did graduate work at Louisiana State University in Baton Rouge and earned a master of science degree in geology in 1978. She studied the hydrogeochemistry of Saint Croix for her thesis. She earned her doctoral degree at the University of Miami, Florida, in marine geology in 1983. Her dissertation was on phosphate interaction with carbonate sediments. She received a postdoctoral fellowship at the University of Miami in 1982–83 before becoming a research assistant professor in 1983. Her postdoctoral research was an experimental study of the growth rate of carbonate cement. Lynn Walter joined the faculty at her alma mater at Washington University, Saint Louis, in 1984. She accepted a position at the University of Michigan in Ann Arbor in 1989, where she remains today as a full professor.

Lynn Walter has been very productive. She has published more than 50 papers in professional journals and volumes. Her first paper was in the prestigious journal *Science,* and all of the others are in the top international journals, including one in the equally prestigious journal *Nature* and several in the high-profile journal *Geology.* She has been extremely successful with research funding, obtaining approximately $3.6 million from the National Science Foundation, Gas Research Institute, oil companies, foundations, and the Environmental Protection Agency.

Lynn Walter has received many awards and honors. She has been Phi Beta Kappa since 1975 at Washington University. She was awarded both Chevron and Pennzoil scholarships at Louisiana State University. She was awarded the Koczy Fellowship and the F.G. Walton Smith Prize at University of Miami. In 1987, she received the Presidential Young Investigator Award from the National Science Foundation, and she received the Distinguished Service Award from the Geological Society of America in 1999.

Lynn's professional service is also exemplary. She served as editor for the *Geological Society of America Bulletin* from 1995 to 1999. She served as associate editor for many top professional journals, including *Journal of Sedimentary Petrology* from 1989 to 1992, *Geological Society of America Bulletin* from 1990 to 1995, *Geology* from 1991 to 1996, and *Geochimica et Cosmochimica Acta* from 1999 to the present. She served as a member of several important panels, including two for the National Research Council, one for the National Science Foundation, one for the Environmental Protection Agency, and one for the American Geophysical Union.

Watson, E. Bruce
(1950–)
American
Experimental Geochemist

How are Earth's deepest properties and processes known if they cannot be seen? The answer is to establish a high-temperature, high-pressure experimental research facility to simulate those conditions. One such facility from which have originated some of the best research, cutting-edge ideas, and elegant solutions is that of Bruce Watson at Rensselaer Polytechnic Institute, New York. Watson's research can be described as "materials science of the Earth" because he studies the physicochemical processes of Earth materials under extreme conditions. His laboratory consists of solid-media, piston-cylinder apparatuses that can generate conditions up to 4 GigaPascals and 2,000°C, as well as internally and externally heated gas–medium pressure vessels that generate conditions up to 300 MegaPascals and 1,300°C. With these pieces of equipment, Watson and his group seek to understand the processes that distribute and redistribute chemical elements and isotopes in the solid deep Earth at scales ranging from micrometers to kilometers at depths up to 150 km. The results help to form a clearer picture of deep-Earth systems and the evolution of the mantle and lower crust.

The specific processes that Bruce Watson researches can be divided into three categories. The first is the movement (diffusion) of elements in melts and fluids and the permeability of rocks to those melts and fluids at high pressures and temperatures. The second is the partitioning or preferential concentration of certain trace elements (very low concentrations) among minerals, melts and fluids under lower crustal and upper mantle conditions. Finally, the behavior of minor minerals that concentrate trace elements are studied.

Bruce Watson was born on October 16, 1950, in Nashua, New Hampshire. He attended

Bruce Watson in his high-pressure research laboratory at Rensselaer Polytechnic Institute *(Courtesy of B. Watson)*

Williams College, Massachusetts, in 1968 and 1969, but transferred to the University of New Hampshire, and earned his bachelor of arts degree in geology in 1972. He then entered Massachusetts Institute of Technology as a graduate student and earned his Ph.D. in geochemistry in 1976. He was awarded a postdoctoral fellowship to the Carnegie Institution of Washington, D.C., for 1976 and 1977 before accepting a faculty position at Rensselaer Polytechnic Institute in Troy, New York, in 1977. He was chairman of the department from 1990 to 1995 and since 1995, he has been an institute professor of science. During his time at Rensselaer Polytechnic Institute he has been a visiting scientist at Macquarie University in Australia in 1981 and at the Max-Planck Institut für Chemie in Mainz, Germany, in 1984, as well as a participating guest at the Lawrence Livermore National Laboratory, California, in 1999.

Watson has published some 105 articles in international journals and professional volumes. Many of these papers set new benchmarks in the understanding of lower crustal and mantle geochemistry. He received recognition for his research through numerous honors and awards from the profession. He became a fellow of the American

Academy of Arts and Sciences in 1996 and a member of the National Academy of Sciences in 1997. He received the Early Career Award from Rensselaer Polytechnic Institute in 1982 and the F. W. Clarke Medal of the Geochemical Society in 1983. He was awarded the Presidential Young Investigator Award from the National Science Foundation from 1984 to 1989. He was designated an R. A. Daly Lecturer by the American Geophysical Union in 1999 and was awarded the Arthur L. Day Medal by the Geological Society of America in 1998.

Bruce Watson has performed outstanding service to the profession throughout his career. He was the president of the Mineralogical Society of America in 1998 after having served on numerous committees in prior years. He was a councilor for the Geochemical Society in 1991 to 1994 and served on several other committees as well. He also served on several committees for the American Geophysical Union. He served as editor for *Chemical Geology* from 1991 to 1995 and for *Neues Jahrbuch für Mineralogie* from 1988 to 1996. He was associate editor for *Geochimica et Cosmochimica Acta* from 1985 to 1988 and served on the editorial board from 1997 to 1999. He was also on evaluation committees for McGill University of Canada (1991), the Carnegie Institution of Washington, D.C. (1992 and 2000), Brown University, Rhode Island (1993), Harvard University, Massachusetts (1994 to present), Rice University, Texas (2000). He has also served on numerous panels for the National Science Foundation, U.S. Department of Energy, and the National Research Council.

⊠   **Weeks, Alice M. D.**
(1909–1988)
American
Mineralogist

Alice M. D. Weeks is one of the top pioneering women of geology. She achieved positions of responsibility and respect for her work in geology at a time when women were scarce in the profession. The road was not an easy one, but her tenacity carried her to success in the end. She worked in many positions below her ability, such as a draftsperson, as well as an instructor and teacher in many capacities to earn her wings in geology. In the end, she broke through this social barrier and established herself as one of the experts on uranium mineralogy in the years when uranium exploration was one of the most important fields in geology, thanks to the cold war. Most of this research involved the defining of many new uranium minerals and their occurrence as well as compounds with other radioactive elements. This work includes the processes involved in concentrating the ore, as well. She considered many of these processes both under high temperature hydrothermal conditions as well as at near surface conditions related to clay mineralogy. Much of this research was conducted in the southwestern United States (Utah, Colorado, Arizona, New Mexico, and Texas), among others. Several important papers resulting from this research include "Mineralogy and Oxidation of the Colorado Plateau Uranium Ores" and "Coconninoite a New Uranium Mineral from Utah and Arizona."

Even after she achieved her well-earned position of authority, Alice Weeks was forced to dress as a man to gain access to many of the uranium mines to obtain samples. There were superstitions against allowing women into mines in those times. It is a wonder that Weeks achieved an illustrious career in the face of such adversity.

Alice Mary Dowse and her twin sister, Eunice, were born on August 26, 1909, in Sherborn, Massachusetts. After being home schooled in her early years, Alice received diplomas from Sawin Academy and Dowse High School in Sherborn in 1926. She attended Tufts University, Massachusetts, and earned a bachelor of science degree in mathematics and science, cum laude, in 1930. Upon graduation, she taught at the Lancaster School for girls in Massachusetts for $2\frac{1}{2}$ years be-

fore returning to Tufts University to attend several geology courses. Alice Dowse did her graduate studies at Harvard University, Massachusetts, and earned a master of science degree in 1934, but was financially unable to continue toward her doctorate. It is reported that she was not permitted to attend certain classes because she was female and was forced to sit in the hall outside of the classroom to take notes. She accepted a research fellowship at Bryn Mawr College, Pennsylvania, in 1934 for one year and remained a second year as a laboratory instructor. She returned to Harvard University in 1936 to work toward her doctorate. She also began teaching at Wellesley College, Massachusetts, first as an instructor and later as a member of the faculty. Between the time constraints and rationing during World War II, it took until 1949 before she was finally awarded her doctoral degree. She mapped two $7\frac{1}{2}$-minute quadrangles in Massachusetts under the supervision of MARLAND P. BILLINGS.

In May of 1950, Alice Dowse married Dr. Albert Weeks, a petroleum geologist. In 1949, she

Alice Weeks conducts research using a reflecting petrographic microscope in 1958 *(Courtesy of the U.S. Geological Survey, E. F. Patterson Collection)*

took a leave from Wellesley College to work for the U.S. Geological Survey. It became a career position in 1951 when Weeks became a project leader in uranium mineralogy through the Trace Elements Lab. Most of this work was done in the area of the Colorado Plateau. In 1962, she left the U.S. Geological Survey to build a geology program at Temple University in Philadelphia, Pennsylvania. When she retired to professor emeritus in 1976, the department had seven full-time faculty and 14 full-time graduate students to her credit. Alice Weeks died of complications related to Alzheimer's disease on August 29, 1988.

Alice Weeks led a very productive career. She is an author of numerous articles and reports in international journals, professional volumes, and governmental reports. Her research on uranium mineralogy is published in seminal papers in several top-quality journals. In recognition of her contributions, the uranium mineral "weeksite" was named in her honor. Weeks performed significant service to the profession. She was a charter member of the Women Geoscientists Committee of the American Geological Institute. She was a Fellow of the American Association for the Advancement of Science and to all other societies of which she belonged. She served on numerous committees for the Geological Society of America and the Mineralogical Society of America.

⊠ **Wegener, Alfred**
(1880–1930)
Germany
Meteorologist (Plate Tectonics)

Although trained as an astronomer and employed as a meteorologist, Alfred Wegener is recognized as the "father of plate tectonics." But he proposed his theory so far in advance of its acceptance that he was viewed essentially as a heretic. He made his first presentations on this idea in 1912 and published them in 1915 in a book entitled *The Origin of Continents and Oceans*. Be-

cause of World War I, the book went largely unnoticed outside of Germany until its third printing in 1922 when it was translated into English, French, Russian, Spanish, and Swedish. Wegener's theory rejected the popular idea that land bridges had once connected the continents but had sunk into the sea as the Earth cooled. Instead, he likened the continents to icebergs floating in the ocean, drawing from his Arctic experience. He argued that the continents are made of less dense granitic rock, whereas oceanic rocks are dense volcanic rocks. He developed the still accepted theory of isostasy, which is the balance of the height of crust based upon density and thickness, like wood, ice, or other materials of varying density floating in a swimming pool. He cited the glacial rebound (rising) of land since the last ice age and removal of the mile-thick ice sheet in the northern hemisphere. Mountain ranges were to have formed like wrinkles on a shriveling apple at that time but Wegener proposed that they formed as the result of collisions of existing continents as they drifted around the Earth. He even proposed that all continents had once formed a supercontinent that he named Pangea. This proposal was based not only on the shapes and inferred paths but also on fossils and paleoclimatic evidence. Enigmatic glacial deposits clustered at the South Pole when Pangea was reconstructed, among others.

In 1926, he was invited to an international symposium in New York to discuss his theory. Phrases like "Utter, damned rot!" and "Anyone who valued his reputation for scientific sanity would never dare to support such a theory" and other such criticisms were abundant at the meeting. Stoically, Wegener listened to his critics and murmured, "Nevertheless, it moves!" just as Galileo did as he was forced to recant his support of Copernicus's theory of the Earth moving around the Sun. Wegener, however, admitted that he had not come up with a satisfactory mechanism to drive the massive plates around the Earth. That would remain a mystery until the late 1950s and 1960s when the rest of the plate tectonic paradigm was derived.

Alfred Wegener was born on November 1, 1880, in Berlin, Germany, where he grew up. He studied natural sciences at the University of Berlin, where he earned all of his degrees including a Ph.D. in astronomy in 1904. In 1905, he obtained a position with the Royal Prussian Aeronautical Observatory near Berlin, where he studied the upper atmosphere using weather balloons and kites. He also flew hot-air balloons and set a world endurance record for staying aloft with his brother Kurt Wegener for 52 hours in 1906. Because of his balloon experience he was invited to participate in a 1906 Danish expedition to Greenland's unmapped northeast coast. He performed research on the polar atmosphere while there. When he returned to Germany, his success on the Arctic expedition was rewarded with a faculty position at the small University of Marburg, Germany. He led a second expedition to Greenland in 1912 and narrowly escaped death when a glacier his team was climbing suddenly calved. They were the first research team ever to overwinter on the ice cap. In 1924, Wegener joined the faculty at the University of Graz in Austria as a professor of meteorology and geophysics. Wegener returned to Greenland in 1930 to lead a team of 21 scientists on a systematic study of the great ice cap and its climate. The ambitious study wound up 38 days behind schedule because the harbor was iced in. On July 15, a small party headed inland to establish the mid-ice camp at Eismitte on July 30. Because of bad weather, the team got stranded. A rescue team that included Wegener was sent on September 21 to save the first team. The four that made the rescue braved temperatures of -58°F but the group at Eismitte were fine. On November 1, Wegener and a young Greenlander set out for the coast to establish the second camp. They were never heard from again. The next April, a search party was sent out. On May 12, 1931, Wegener's body was found buried in his sleeping bag. It appears that he died in his tent, likely of a heart at-tack from the extreme exertion in driving through the snow. The theory could not be verified because Wegener's young companion was never found. The remaining team built an ice-block mausoleum marked with a 20-foot iron cross. It has since disappeared into the snow to become part of the great glacier.

⊠ Wenk, Hans-Rudolf
(1941–)
Swiss
Mineralogist (Textural Analysis)

The Earth has a very strong magnetic field compared with the other terrestrial planets. The reason given has always been that the interaction of the solid and liquid core produces a self-exciting dynamo created by the spinning of the Earth. The details of this interaction, however, have never really been explained. One piece of evidence that adds to this investigation is the stark difference in seismic wave velocity depending upon the direction that they pass through the solid core. They travel much faster parallel to the poles than through the equator. New research by Hans-Rudolf Wenk and his associates appears to be quickly leading to a solution to this fundamental question. Wenk has done experimental work on iron at high pressures and temperatures, which forms hexagonal crystals. These crystals appear to have aligned parallel to the poles through processes of dynamic recrystallization. An example of this work is Wenk's paper "Plastic Deformation of Iron in the Earth's Core."

This work on the core is a natural progression of the research of Wenk on convection in the Earth's mantle. The upper mantle undergoes thermally induced convective circulation. However, the upper mantle is primarily solid with only small pockets of liquid at the plate margins. The circulation is therefore accomplished dominantly through dynamic recrystallization processes or crystal plasticity. The movement of structural defects on the atomic level throughout the mineral crystal lattice causes shape changes and the alignment of crystals to produce a preferred orientation. Seismic waves travel at different speeds depending upon which direction they travel through the crystal. Seismic waves in the upper mantle travel 10 percent faster perpendicular to mid-ocean ridges than parallel to them as a result of this alignment. Seismologists can map the fabric of the upper mantle using seismic velocities otherwise known as teleseismic imaging.

The scale difference between these Earth scale processes and the minute sizes of the samples that Hans-Rudolf Wenk typically analyzes makes it almost incredible that they could have any bearing on each other. Wenk studies the crystallographic alignment of minerals and other materials and the atomic scale processes that produce them. His true expertise is therefore better called textural analysis and he is arguably the foremost expert. He has studied these alignment processes in minerals such as calcite, dolomite, mica, staurolite, and olivine, in rocks such as quartzite, mylonite, eclogite, and opal in chert, and even in mollusk shells, ice, bones, and calcified tendons, and wires, nickel plating, and hexagonal metals. It requires the most sophisticated of analytical equipment to attempt such research. Wenk utilizes a variety of equipment in this analysis including pulsed neutron sources, high-resolution transmission electron microscopy (HRTEM), three-dimensional transmission electron microscopy (3DTEM), synchotron X-ray diffraction, X-ray goniometry, neutron diffraction, and scanning electron microscopy (SEM), among others. Modeling of the processes involved in these alignments involves sophisticated processes and typically supercomputers. Wenk has produced a software package for this application called *BEARTEX*. He has also written several books on this work including *Texture and Anisotropy. Preferred Orientation in Polycrystals and their Effect on Material Properties* and *An Introduction to Modern Textural Analysis*.

Hans-Rudolf Wenk was born on October 25, 1941, in Zurich, Switzerland, where he spent his youth. He attended the University of Basel, Switzerland, where he earned a bachelor of arts degree in geology in 1963. He completed his graduate studies at the University of Zurich, Switzerland, where he earned his Ph.D. in crystallography in 1965. In 1966–1967, Wenk was a research geophysicist at the University of California at Los Angeles under DAVID T. GRIGGS. In 1967, he joined the faculty at the University of California at Berkeley, where he remains today. Hans-Rudolf Wenk married Julia Wehhausen in 1970. He has been a visiting researcher or professor 17 times at such schools as the Universities of Frankfurt, Hamburg, and Kiel in Germany; the Universities of Grenoble, Lyon, and Metz, in France; Nanjing University, China; University of Hiroshima, Japan; and University of Perugia, Italy, among others. Wenk is a seasoned mountaineer and technical mountain climber in his spare time.

Hans-Rudolf Wenk is amid a very productive career having been an author of more than 300 papers in international journals, professional volumes, and governmental reports. Many of these papers set new benchmarks in the study of textural analysis of rocks as well as Earth mantle and core processes. He is also an author or editor of four books and volumes and of an American Geophysical Union-sponsored videotape on anisotropic mantle convection entitled *Texturing of Rocks in the Earth's Mantle. A Convection Model Based on Polycrystal Plasticity.* His book *Electron Microscopy in Mineralogy* is quite popular. In recognition of his contributions to geology, Hans-Rudolf Wenk has received several honors and awards. He has received two Alexander von Humboldt Senior Research Awards and a Humboldt Research Fellowship from the Humboldt Society, and the Berndt Mathias Scholarship from Los Alamos National Laboratory.

Most of the service that Wenk has performed has been with the American Geophysical Union and societies involved in material science. He has also served on committees for the Mineralogical Society of America.

⊠ Whittington, Harry B.
(1916–   )
British
Invertebrate Paleontologist

Harry Whittington has been called the "dean of trilobites" and the "vice chancellor of the lower Paleozoic" in recognition of the profound contributions to geology that he has made in both of these areas. Beginning in Wales, Whittington painstakingly studied the morphology and relations of trilobite fossils. This work was quickly expanded to a worldwide basis, especially in Europe, North America, and China. Theses studies on trilobites spanned taxonomy, stratigraphic uses and distribution, limb structure, silicified trilobites, functional morphology, and evolution, to name a few. He masterminded the now famous 1959 volume entitled *Treatise on Invertebrate Paleontology,* which includes trilobites and related forms. In 1966, he assembled a master synthesis showing the global distribution of Ordovician trilobite faunas in terms of the former positions of continents and oceans at that time. This work was quickly used to help constrain plate tectonic history and processes and began a new plate tectonic reconstruction method that blossomed in the 1970s.

Beginning in the early 1960s, Whittington began to study the trilobites that CHARLES D. WALCOTT discovered and collected from the Burgess Shale in British Columbia, Canada. He was quite perplexed because these "trilobites" did not contain the usual features, and he decided that they were really not trilobites. In 1966 and 1967, Whittington joined the Geological Survey of Canada in a field expedition to reexamine the Burgess Shale over 7,000 feet up the slopes of the Rocky Mountains. He brought two of his now fa-

mous graduate students, SIMON CONWAY MORRIS and Derek Briggs. This research resulted in the discovery of dozens of new species unrelated to those of the early Paleozoic or any other fauna. The account of this research is recorded in the book *Wonderful Life* by STEPHEN JAY GOULD, which popularized the story of the Burgess Shale. Beginning in 1971, Whittington and his students wrote the real scientific contributions that resulted from this research. This work indicates a true explosion of life during the Cambrian with many complex species that are still not fully understood followed by a contraction of the groups as competition culled the less well adapted. These findings added greatly to our understanding of Paleozoic evolution and evolutionary processes in general.

Harry Whittington was born on March 24, 1916, in Handsworth in Yorkshire, England. He attended Handsworth Grammar School and later Birmingham University, England, where he earned bachelor of science and Ph.D. degrees in geology in 1937 and 1940, respectively. During his final two years in graduate school (1938–1940) he was at the U.S. National Museum and a Commonwealth Fund Fellow at Yale University, Connecticut. Harry Whittington married Dorothy Arnold in 1940. That year he accepted a position as lecturer in geology at Judson College in Rangoon, Burma. In 1943, he became a professor of geography at Ginling College in Chengtu, western China. In 1945, he returned to his alma mater at Birmingham University to become a lecturer in geology. Whittington left England again to join the faculty at Harvard University, Massachusetts, beginning as a visiting lecturer, but quickly moving through the ranks. He was also the curator of invertebrate paleontology at the Agassiz Museum of Comparative Zoology at Harvard. In 1966, he moved back to England to join the faculty at Cambridge University, where he was named the Woodwardian professor of geology. Whittington retired to professor emeritus in 1983 at which point he was named an Uppingham Scholar until 1991.

Harry Whittington led a very productive career. He was not only an author of numerous articles in international journals and professional volumes, he also wrote several famous monographs on trilobites. Many of his studies are seminal reading on trilobites, the Burgess Shale, and the early Paleozoic. He is the author of two semipopular and widely read books entitled *The Burgess Shale* and *Trilobites*. For his research contributions to paleontology and geology, Whittington has received numerous honors and awards. He is a Fellow of the Royal Society of London and received an honorary degree from Harvard University. He received the Paleontological Society Medal (U.S.), both the Lyell Medal and the Wollaston Medal from the Geological Society of London, the Mary Clark Thompson Medal from the U.S. National Academy of Sciences, the Lapworth Medal from the Paleontological Association (U.K.), and the Geological Society of Canada Medal.

Whittington has performed significant service to the profession. He has served in numerous positions for the Geological Society of London, the Paleontological Association (U.K.), and the Geological Society of America. He is also a trustee of the British Museum (Natural History).

⊠ Williams, Harold
(1934–   )
Canadian
Regional Tectonics

Harold Williams is one of the premier field mappers in the history of geology. His career overlapped the plate tectonic revolution and he keenly watched its development. After the basic paradigm had been established for the current plate configuration and interactions, the next group of geologists began applying those processes to observations of ancient rocks. Harold Williams was prominent in the group that was attempting to reconstruct the Appalachians. He interpreted and reinterpreted his vast geologic mapping in this context and established himself as the foremost expert on the tectonics of Newfoundland and indeed, the entire Canadian Appalachians. He then performed the unimaginable at the time. He produced a tectonic map of the entire Appalachian Orogen both in Canada and the United States entitled *Tectonic-Lithofacies Map of the Appalachian Orogen*. This project involved the compilation of existing maps and reinterpretation of them into a tectonic context, which was a feat in itself. Because he was respected and well liked, he was able to obtain the assistance of numerous other regional geologists orogen wide. Considering the territorial nature of regional geologists, this feat borders on the miraculous. The result was an internally consistent map with the general consent of the geologic community both Canadian and American. Within one or two years of publication, this map was hanging on the wall in most geology departments throughout the Appalachians as well as many other departments throughout the United States, Canada, and western Europe. He later produced geophysical maps that cover the same area (*Magnetic Anomaly Map of the Appalachian Orogen* and *Bouguer Gravity Anomaly Map of the Appalachian Orogen*).

The main reason that Williams became such a leader in regional tectonics is the rich tectonic geology of Newfoundland. It is doubtful that there is another area in the entire Appalachian-Caledonian chain with more or better preserved plate tectonic elements. They record several plate collision events. Beautiful subduction zone complexes are marked by the Dunnage, Cold Spring, Teakettle, and Carmanville Melanges. There are well-preserved fragments of ancient oceanic crust in the Bay of Islands ophiolite complex and Fleur de Lys Supergroup. There are large sheets of rock that were slid in atop existing rock during plate collisions including the Humber Arm Allochthon, Hare Bay Allochthon, and Coney Head Complex.

Harold Williams (third from left in open jacket), flanked by John Dewey (left) and James Skehan, S.J. (right), on a field conference in 1994 *(Courtesy of J. Skehan, S.J.)*

There is even a back-arc-basin complex (Noggin Cove Formation) and several plate suture zones. These are just of few of the many elements that Williams has studied. Papers on this work include "Acadian Orogeny in Newfoundland" and "Appalachian Suspect Terranes," among many others.

Harold (Hank) Williams was born on March 14, 1934, in Saint John's, Newfoundland, Canada. He attended Memorial University of Newfoundland and earned both a diploma in engineering and a bachelor of science degree in geology in 1956. He earned a master of science degree in geology in 1958 on a Dominion Command scholarship. Williams earned a Ph.D. from University of Toronto, Canada, in 1961. He joined the faculty at Memorial University of Newfoundland that year and remained there for the rest of his career. In 1984, he was named university research professor, one of the first two at Memorial

University. He was also the Alexander Murray professor from 1990 to 1995. Williams retired to professor emeritus in 1997. Harold Williams is an avid folk musician and is known for his gregarious nature.

Harold Williams has had a very productive career. He is an author of numerous articles in international journals and professional volumes and he is an editor of six professional volumes. He is also an author of some 15 maps. Several of these are among the most-cited works on regional geology ever (highest number of citations of any Canadian geologist in 1984). Many of these works set new benchmarks in the understanding of the Appalachian orogen. For his research contributions to geology, Williams has received numerous honors and awards. From the Geological Association of Canada, he received both the Past President's Medal (1976) and the

Logan Medal (1988), as well as being named a Distinguished Fellow in 1996. He was the first recipient of the Douglas Medal from the Canadian Society of Petroleum Geologists in 1981. He received the Miller Medal from the Royal Society of Canada in 1987. From Memorial University, he received the Governor General's Medal (1956), the Dominion Command Scholarship (1956 and 1957), and the Issak Walton Killam Memorial Scholarship (1976 to 1979), in addition to those already listed. He was also named the James Chair Professor at Saint Francis Xavier University, Nova Scotia, Canada, in 1989. Williams also performed significant service to the profession to the Geological Association of Canada where he served as president, in addition to numerous committees, and to the Geological Society of America, where he was an associate editor for the *Geological Society of America Bulletin*, among others.

⊠ Wilson, J. (John) Tuzo
(1908–1993)
Canadian
Geophysicist, Plate Tectonics

J. Tuzo Wilson was one of the true powerhouses of the earth sciences and a giant of plate tectonics. His statement, "I enjoy, and always have enjoyed, disturbing scientists," served as a kind of a motto for him as he splashed his way through the profession. Surprisingly, he was an outspoken opponent of plate tectonics for a good part of his career. Through the late 1940s and 1950s, Wilson argued vehemently for the mountain building theory of Sir Harold Jeffreys of a contracting Earth. Even then he was considered somewhat of a maverick and a brazen promoter of ideas that made many uncomfortable. Most of his research at this time was on the Canadian Shield where he coupled basic geologic relations with early geochronology to interpret the growth of continents on a worldwide basis. However, even in his out-

spoken support, privately he admitted that much of the theory was inadequate.

It was a true reflection of his mettle when he was able to switch directions and embrace the continental drift concept still in its infancy after being one of its strongest critics. At 50 years old, he became one of the strongest supporters and contributors to the development of the theory. His first contribution was to interpret the Hawaiian Islands. By looking at the current volcanic activity, the ages of the islands, and the extension into the Hawaiian and Emperor seamounts, he interpreted them to represent a stationary plume of magma from the mantle over which the Pacific plate moves. The train of islands tracks past movements of the Pacific plate. His paper "A Possible Origin of the Hawaiian Islands" summarizes this work. His second contribution was to interpret the huge fracture systems that offset mid-ocean ridges as a new class of plate boundary called transform boundaries. These huge strike-slip faults occur all along mid-ocean ridges throughout the Earth as compensation features to the spreading that takes place on the ridges as described in the paper "A New Class of Faults and their Bearing on Continental Drift." LYNN R. SYKES proved Wilson's theory, by analyzing earthquakes from these features, that they are indeed strike-slip (of lateral motion) and currently active all over. His third most famous contribution was to propose that the Atlantic Ocean closed and then reopened virtually along the same line (i.e.: "Did the Atlantic Close and Then Re-Open?"). With an enormous amount of additional research, it was shown that indeed an early ocean basin called Tethys was closed during the building of the supercontinent Pangea during the Paleozoic. The Atlantic then opened nearly along the old suture zone of this closure. This idea of zones of weakness in the Earth's crust that would be repeatedly reactivated has subsequently been shown to be a very common phenomenon. These are three fundamental pieces of plate tectonics that have withstood the test of time, and they are only a sampling of the outstanding body of work produced by Wilson during his career.

J. Tuzo Wilson was born on October 24, 1908, in Ottawa, Canada, where he spent his youth. His mother was a famous mountaineer for whom Mount Tuzo in western Canada was named. At age 17, Wilson became a field assistant of the famous Mount Everest mountaineer Noell Odell, who showed him the wonders of field geology. Wilson enrolled at the University of Toronto, Canada, where he earned a bachelor of science with majors in both physics and geology in 1930. He considered himself to be Canada's first ever graduate in geophysics. He earned a scholarship for graduate studies at Cambridge University, England, under Sir Harold Jeffreys, but there was no real program in geophysics and he wound up earning another bachelor of arts degree in geology in 1932. He returned to Canada to work at the Geological Survey of Canada, but the director urged him to complete his education instead. He enrolled at Princeton University, New Jersey, with classmates HARRY H. HESS and W. MAURICE EWING and graduated in 1936 with a Ph.D. in geology/geophysics. He returned to the Geological Survey of Canada until the outbreak of World War II, when he joined the Canadian army as an engineer and spent three years overseas. When he returned to Canada as a colonel, he remained in the army for an additional four years. As director of operational research, he organized and carried out Exercise Musk Ox, the first ever motorized expedition—some 3,400 miles—to cross the Canadian Arctic.

J. Tuzo Wilson married Isabel Dickson in 1938. He joined the faculty at his alma mater at the University of Toronto in 1946. In 1968, Wilson left his position at the main campus at University of Toronto to serve as principal of the new Erindale College of the University of Toronto. He was forced to retire from his academic position in 1974 to become the director of the Ontario Science Center, the largest of its kind thanks to his efforts. Wilson died on April 15, 1993, of a heart attack.

J. Tuzo Wilson was a dynamo, relentlessly carrying all of his projects to success. His scientific publications were no exception, numbering well over 100 in international journals and professional volumes. Several of these are benchmarks in the plate tectonic paradigm and appear in prestigious journals like *Nature.* In recognition of these research contributions, Wilson received numerous honors and awards. He was a fellow of the Royal Society of England and the Royal Society of Canada. He received the Penrose Medal from the Geological Society of America, the Walter H. Bucher Medal from the American Geophysical Union, the John J. Carty Medal from the U.S. National Academy of Sciences, the Vetlesen Prize from the Vetlesen Foundation, the Wollaston Medal from the Geological Society of London, the Huntsman Award from the Bedford Institute of Oceanography, and the Maurice Ewing Medal from the Society of Exploration Geophysicists.

Wise, Donald U.
(1931–   )
American
Structural Geologist

Most geologists have one or two areas of specialization at which they excel. This is not true for Don Wise. He dabbles in many different aspects of geology, generally related to structural geology, and yet he never fails to make an impact in each. His apparent motto is "Variety is the spice of geology." His regional geology studies have concentrated on the Pennsylvania Piedmont of the Appalachian orogen and the Beartooth Mountains in Wyoming and Montana in the middle Rocky Mountains. He studies small details on the scale of an outcrop, including cleavages, fractures, and folds for both of these areas and publishes these results.

Wise is well known for his work on planetary geology. His first foray into this field was a wild hypothesis that the Moon may have been derived from the Earth by splitting away during the formation of the Earth's core, as summarized in his paper, "Origin of the Moon from the Earth, Some New Mechanisms and Comparisons." At least it was wild at the time. Now it appears as a possible mode of formation in many introductory textbooks. He studied cratering on the Moon as well as its planetary architecture. Wise then moved on to study Mars. He and colleague Gerhart Neukum devised a method of extending lunar cratering densities to obtain the time scale for Mars in the paper "Mars: A Standard Crater Curve and Possible New Time Scale." Based upon these dates and photo interpretation of images obtained from Martian orbiters, he proposed a model for the gross tectonics of Mars. This model involves an early convection cell that ingested the crust from the northern lowland third of the planet to produce the great Tharsis Bulge and its giant volcanoes. It was also controversial.

Wise may be best known for his work on fractures and lineaments. He developed methods for the statistical evaluation of fractures and lineaments on the outcrop, on maps, on aerial photographs and on satellite images. His paper "Topographic Lineament Swarms: Clues to their Origin from Domain Analysis of Italy" includes much of this work. He even developed inversion techniques to determine the stress field that produced fracture sets. This work was done on several areas in New England as well as Kentucky, Wyoming, Italy, and other planets. Wise also worked regional problems in Italy and New Zealand. More recently he has taken on creationism and even written a paper called "Creationism's Geologic Time Scale."

Don Wise was born on April 21, 1931, in Reading, Pennsylvania. He spent his youth in the Pennsylvania Dutch country. He attended Franklin and Marshall College in Lancaster, Pennsylvania, where he graduated Phi Beta Kappa with a bachelor of science degree in geology in 1953. He earned a master of science degree from the California Institute of Technology, Pasadena, in

geology in 1955. Wise then moved back to the East Coast to continue his graduate studies at Princeton University, New Jersey, where he earned a Ph.D. in geology in 1957. Upon graduation, he joined the faculty at his alma mater, Franklin and Marshall College. Don Wise was married in 1965; he and his wife have two children. In 1968, he became the chief scientist and deputy director of the lunar exploration office of NASA in Washington, D.C., where he served through the first lunar landing. In 1969, Wise joined the faculty at the University of Massachusetts at Amherst, where he remained until his retirement in 1993. He served as department head from 1984 to 1988. Wise was a visiting scientist at the Max Planck Institute in Heidelberg, Germany, in 1975, at the University of Rome, Italy, in 1976, and at Canterbury University in Christchurch, New Zealand, in 1988–1989. In addition to being professor emeritus at the University of Massachusetts, Wise is also currently a research associate at Franklin and Marshall College.

Donald Wise has led a productive career. He is an author of more than 50 articles in international journals, professional volumes, and governmental reports. Several of these papers are seminal studies on multiple folding terminology, fracture systems, the regional geology of Wyoming-Montana, Mesozoic basins of New England, regional tectonics of the Pennsylvania Piedmont, and planetary geology of the Moon and Mars. In recognition of his contributions to geology, Wise was awarded the Career Contribution Award from the structure and tectonics division of the Geological Society of America in 2001.

Wise has also performed significant service to the profession. He was the founding chair of the structure and tectonics division of the Geological Society of America, as well as the chair of the planetary geology division. He was a consultant for the U.S. Nuclear Regulatory Commission, and various geotechnical, oil, and power companies.

Withjack, Martha O.
(1951–   )
American
Structural Geologist, Petroleum Geologist

There are several types of subsurface "traps" in which oil and gas can accumulate. The old adage in the oil and gas industry is that one "drills structure" which means that traps based in structural geology give better chance for success. It is therefore no surprise that structural geologists are in demand in the petroleum industry. Martha Withjack is one of the premier structural geologists to grace the petroleum industry, although she recently switched to academia. Using geophysical data, both taken within drilled wells called well logs as well as seismic reflection profiles, which are a kind of sonogram of the shapes of the rock layers deep underground, Withjack studies the structural features within sedimentary basins. She models both the sedimentation patterns, in a method called "seismic stratigraphy," as well as the faulting and folding that has been imposed upon these sedimentary rocks. Typically, these two aspects are tightly associated because the sedimentation patterns respond to the active faulting and are distributed accordingly.

Martha Withjack is an expert on extensional tectonics and associated basin development. She has investigated the geometry of normal faults and the relationship of associated structures with a variety of sedimentary sequences. This research involved both modeling of existing basins as well as experimental structural models, typically using clay and sand layers in a moveable vise. She looked at the development of different types of folds in this process (forced and rollover) as well as the fault patterns. An example is her paper "Experimental Models of Extensional Forced Folds." She looked at patterns with the two sides of a basin pulling directly away from each other, as well as when they slide sideways (strike-slip movement) while pulling apart, a process called "oblique rifting," which produces a far different structural pattern shown in a paper entitled "Deformation Produced by Oblique Rifting." Asymmetric extension also involves differences on each side of the extending crust but vertically rather than horizontally. She also investigated an odd but common phenomenon in which faults may start out as normal faults but abruptly switch into reverse faults, or vice versa, in a process called "tectonic inversion" as discussed in her paper "Estimating Inversion—Results from Clay-Model Studies." This abrupt switch causes the sedimentation patterns and folds to also change abruptly and form a characteristic geometrical relationship. With the resources of major oil companies, her field areas of study are worldwide and include

Martha Withjack takes a rest on a field trip in the Appalachians *(Courtesy of M. Withjack)*

eastern India, the Gulf of California, the Gulf of Aden, the Gulf of Suez, offshore Norway, the Chukchi Sea, Alaska, offshore Vietnam, the Caspian Sea, and offshore Newfoundland, among others.

Martha Withjack was born on January 10, 1951, in Orange, New Jersey. She attended Rutgers University in New Brunswick, New Jersey, where she earned a bachelor of arts degree, Phi Beta Kappa in mathematics, in 1973. She did her graduate studies at Brown University, Rhode Island, where she earned a master of science degree in 1975 and a Ph.D. in 1977, both in geology. Upon graduation Withjack accepted a position as research geologist with Cities Service Company in Tulsa, Oklahoma, but moved to ARCO oil and gas company in Plano, Texas, in 1983. In 1988, she accepted a position as senior research geologist with Mobil Technology Corporation in Dallas, Texas. In 2000, Withjack joined the faculty at her alma mater, Rutgers University in New Brunswick, where she remains.

Martha Withjack is leading a very productive career, but because of her extensive industry experience it is a bit different from those with purely academic experience. She is an author of 25 articles in international journals and professional volumes, but also of 28 major internal technical reports for petroleum companies. Many of these are seminal papers on extensional tectonics and appear in top journals. Withjack has received several honors and awards in recognition of her research contributions to geology. She received two Best Paper Awards from the American Association of Petroleum Geologists (AAPG), including the George C. Matson Award in 1999 and the Cam Sproule Memorial Award in 1986. She was also chosen as a distinguished lecturer by both the Petroleum Exploration Society of Australia and AAPG.

Withjack has performed service to the profession. She served on several committees for both the Geological Society of America and AAPG. She also served in several editorial positions including associate editor of both the *Geological Society of America Bulletin* and the *American Association of Petroleum Geologists Bulletin*. She has taught numerous short courses both in industry and through professional societies.

Wones, David R.
(1932–1984)
American
Mineralogist, Petrologist

It requires solid laboratory experimentation on rocks and minerals to provide the physical constraints of formation (pressure-temperature, etc.) on what is observed in their natural setting in the field. Experimental geochemists and petrologists typically spend most of their careers in the laboratory with rare excursions into the field to observe a phenomenon or to touch base with the reality of the chemical systems that they are modeling. In contrast, field geologists might use analytical equipment in the laboratory to aid in their field interpretations but they typically do not have the patience or perhaps the interest to conduct experiments to model their observations. David Wones was an exception to this apparent exclusivity. His experimental research was mostly on micas and assemblages of minerals in which mica is a major component. This outstanding research established him as the leading authority on micas. He even had a newly discovered mica named after him, wonesite, in recognition of his contributions.

On the other hand, Wones was an outstanding field petrologist. Most of his research was on granite plutons but he dabbled with other rocks as well. He conducted extensive field research on the Sierra Nevadas in California, as well as the granite plutons of New England, and especially Maine. In fact, as with micas, he was the leading authority on the plutons of New England and among the top few in the world on the processes of granite emplacement and crystallization.

David Wones on a field trip into the Boulder Batholith, Montana *(Courtesy of Arthur Sylvester)*

Even with all of his research prowess, one of David Wones's greatest traits was his ability to mentor and to inspire camaraderie. He took a highly fractured department at Virginia Tech and inspired an astonishing period of cooperation and mutual respect almost immediately. He inspired his students and colleagues alike with his unbridled passion for geology of all types while maintaining his genuine concern and respect for them. His tragic death at an early age left a void in the lives of many who knew him.

David Wones was born on July 13, 1932, in San Francisco, California. His father was a colonel in the U.S. Army and as a result he lived in various places as the assignments changed. He graduated from Thomas Jefferson High School in San Antonio, Texas, in 1950. He attended Massachusetts Institute of Technology (MIT) and earned a bachelor of science degree in geology in 1954. He also did his graduate studies at MIT but earned a Vannevar Bush Fellowship and did his dissertation research at the Geophysical Laboratory at the Carnegie Institution of Washington, D.C., from 1957 to 1959. He earned his doctorate in 1960. He joined the U.S. Geological Survey in 1959 as an experimental petrologist, but he also did field research in California and New Mexico. In 1967, Wones joined the faculty at MIT. He returned to the U.S. Geological Survey in 1971 as chief of the Branch of Experimental Geochemistry and Mineralogy. He then accepted a position at Virginia Polytechnic Institute and State University in 1977 and served as department chair from 1980 to 1984. David Wones married Constance Gilman in 1958 and they had four children. David Wones was killed in 1984 in an automobile accident while going to pick up a lecturer from the airport for a departmental seminar.

David Wones led a highly productive career, publishing numerous articles in international journals and collected volumes. Many of these are seminal works on micas as well as granite petrology and emplacement. David Wones performed extensive service to the profession. He was president of Geological Society of America in 1978–1979 as well as of the Mineralogical Society of America. He played a major role in the International Geological Correlation Project of the International Geological Congress, including the running of a major convention.

⊠ Wyllie, Peter J.
(1930–   )
British
Experimental Petrologist

Considering that the Earth's crust and mantle are composed of solid rock, the very existence of magma and lava is anomalous. Yet there are active volcanoes all over the Earth and magma being continuously emplaced into the crust. It is this first order problem of the Earth that Peter Wyllie first addressed and the impetus for his 1967 volume *Ultramafic and Related Rocks.*

Peter Wyllie began to conduct innovative experimental determinations of mineral and melt chemical phase relations at high pressures and temperatures with his mentor O. FRANK TUTTLE. The experiments reproduced conditions deep within the Earth and were designed to determine the processes in the upper mantle and crust where melting occurs. This experimental work was done using both complex real rock systems and related simple synthetic systems. His initial focus was on two problems, the origin of granitic magma, both plutons and volcanics and the origin of carbonatites. Carbonatites are rare calcite-rich plutonic rocks. Wyllie's discovery of conditions for the precipitation of calcite from melts at moderate temperatures confirm their origin.

In his later experiments, Peter Wyllie pursued progressively more complex silicate-carbonate systems. One project involves the metamorphism and melting of ocean crust as it is progressively heated in a subduction zone. The melting of the wet ocean crust has applications for the origin of water-charged andesites that form explosive volcanoes in island arcs as well as introduction of water and carbon dioxide in the mantle. This research has applications to the origin of kimberlites (origin of diamonds) and carbonatites. More recently, Wyllie has been studying the origin of granites, tonalities and trondheimites from the Archean age gray gneiss of the continental interiors. It involves a backwards approach to determine the composition of the source rock by the composition of the melt that came out of it at elevated temperature and pressure. This work gives us a new understanding of the processes in the early crust.

Wyllie has often incorporated these applications into broader reviews involving plate tectonics and global processes. His two textbooks captured the spirit and the history of the plate tectonics revolution. The first, *The Dynamic Earth,* was written for graduate students and the second, *The Way the Earth Works,* was written for non-science majors.

Peter Wyllie was born on February 8, 1930, in London, England. He joined the British Royal Air Force in 1948 where he was an aircraftsman first class (radiotelephony operator). He received the Best Recruit Award in basic training for the Royal Air Force at Padgate in 1948 and he was the heavyweight boxing champion for the Royal Air Force in Scotland in 1949. He attended the University of Saint Andrews, Scotland, after his discharge in 1949 and earned a bachelor of science degree in physics and geology in 1952 with the Miller Prize for outstanding achievement. Wyllie was a geologist with the British North Greenland Expedition from 1952 to 1954 before returning to the University of Saint Andrews. He earned a second bachelor of science degree with first class honors in geology in 1955 and a Ph.D. in geology in 1958. In 1956, Peter Wyllie married Romy Blair and they would have three children. The same year, he accepted a position as a research assistant at Pennsylvania State University at University Park before becoming an assistant professor in 1958. Wyllie was a lecturer at Leeds University in England in 1959 to 1961 before

Peter Wyllie lecturing on a high-pressure experiment. He is holding a cold-seal "test-tube" high-pressure vessel *(Courtesy of P. Wyllie)*

returning to Pennsylvania State University in 1961. In 1965, he joined the faculty at the University of Chicago, Illinois. He served as associate dean in 1972–1973 and department chair from 1979 to 1982. He was also the Homer J. Livingston professor of geology from 1978 to 1983. In 1983, Wyllie moved to the California Institute of Technology in Pasadena, where he remains today. There he served as chair from 1983 to 1987 and academic officer from 1994 to 1999 when he retired and became professor emeritus. He was a Louis Murray Visiting Fellow at the University of Cape Town, South Africa, in 1987, and he was appointed an honorary professor at the China University of Geosciences at Beijing in 1996.

Peter Wyllie has led an extremely productive career. He is an author of some 310 articles in international journals and professional volumes. He is also the editor or author of four books and volumes. Many of these publications are definitive studies of experimental petrology and mantle processes. His research has been well recognized by the profession as is evident in his numerous honors and awards. He is a member of the U.S. National Academy of Sciences, the Russian National Academy of Sciences, the Indian National Academy of Sciences, the Indian Science Academy, and the Chinese Academy of Sciences. He is also a Fellow of the Royal Society of London. He received an honorary doctorate from the University of Saint Andrew. He received the Polar Medal from Queen Elizabeth II, the Mineralogical Society of America Award, the Quantrell Teaching Award from the University of Chicago, the Wollaston Medal from the Geological Society of London, the Leopold von Buch Medal of the German Geological Society, the Roebling Medal from the American Mineralogical Society, and the Abraham-Gottlob-Werner Medaille from the German Mineralogical Society.

Wyllie has performed significant service to the profession. He was president and vice president of the Mineralogical Society of America (1977–1978 and 1976–1977, respectively), the International Mineralogical Association (1986–1990 and 1978–1986), and the International Union of Geodesy and Geophysics (1995–1999 and 1991–1995, respectively). He has served on numerous committees and panels for these societies as well as the National Science Foundation, National Research Council, International Council of Scientific Unions, American Geophysical Union, among others. He was the chief editor for the *Journal of Geology* from 1967 to 1983 and the Springer-Verlag monograph series *Rocks and Minerals* (22 volumes) from 1967–1999, as well as serving on the editorial boards of 12 other journals.

Y

Yoder, Hatten S., Jr.
(1921–   )
American
Petrologist, Geochemist

There are a few Earth scientists whose accomplishments are so numerous that they are difficult to fit in the allocated space of this book and Hatten Yoder is one. He is among the most influential experimental mineralogists-petrologists in the history of the science. His early research involved the design and construction of an innovative high-pressure experimental apparatus that was unmatched in the world. He conducted experiments on metamorphic rocks and minerals that established whole new stability fields of metamorphism and quantified them. Indeed, through his experimental studies, he was able to quantify for the first time, the pressures and temperatures of various metamorphic isograds and grades. This work was far ahead of its time. Yoder's book *Geochemical Transport and Kinetics* and his paper "Thermodynamic Problems in Petrology" summarize much of this work.

Yoder's other major area of interest is the origin of basalts. He produced the first quantitative study of the crystallization of basaltic magma by applying his experimental results. He also conducted experimental research on the origin of basaltic magma and developed elegant models that now appear in every textbook on physical geology not to mention igneous petrology. A book entitled *Generation of Basaltic Magma* is a result of this research. He developed a new model for the generation of contemporaneous bimodal volcanism in extensional plate tectonic settings. Even physical properties like viscosity of lavas were determined in Yoder's lab. He edited the volume *The Evolution of Igneous Rocks: Fiftieth Anniversary Perspectives* (on NORMAN L. BOWEN's book) which itself was an instant classic.

There are many other areas of geology that Hatten Yoder has influenced, whether it is the development of the early atmosphere, experimental constraints on minerals leading to a mineral being named after him (yoderite), the formation of ore deposits or advising the U.S. Congress on natural resources and the environment. Each of these is a story in itself. However, one of Yoder's main areas of interest is the history of geology. He has taken on the task to preserve the record of the historical development of Earth sciences through society work and publication. Many of the publications are biographies of people who worked with Yoder. Other publications preserve the history of classic geologists but there are also discussions on intellectual development of modern geology.

Hatten Yoder Jr. was born on March 20, 1921, in Cleveland, Ohio. He attended the University of Chicago, Illinois, where he earned an as-

Hatten Yoder at work in his laboratory in the Geophysical Laboratory at the Carnegie Institution of Washington, D.C. *(Courtesy of H. Yoder Jr.)*

sociate of arts degree in 1940, a bachelor of science in geology in 1941, and a certificate of proficiency in meteorology in 1942. He joined the U.S. Navy in 1942 and saw active duty during World War II until 1946, including the MOKO expedition to Siberia as a meteorological officer. He retired as a decorated lieutenant commander from the U.S. Naval Reserves after 16 years of service. He attended graduate school at the Massachusetts Institute of Technology and earned a Ph.D. in 1948. He accepted a position as a petrologist at the Geophysical Laboratory of the Carnegie Institution of Washington, D.C., in 1948 and remained there for the rest of his career.

He served as director of the Geophysical Lab from 1971 until his retirement in 1986. He remains a director emeritus today. During his career, he was a visiting professor at California Institute of Technology, University of Texas, University of Colorado, and the University of Cape Town. Yoder married Elizabeth Bruffey in 1959 and they had two children. Mrs. Yoder passed away in 2001.

Hatten Yoder has had a very productive career. He is an author of some 103 articles in international journals, professional volumes, and governmental reports, as well as book reviews, forewords, and encyclopedia entries. Many of these articles are benchmarks in mineralogy, petrology, and geochemistry. He is also the author of some 25 biographies of prominent geologists, which appear in books and journals. His research as well as his interest in the history of geology have earned Yoder several important honors and awards from the profession. He is a member of the National Academy of Sciences and a fellow of the American Academy of Arts and Sciences. He was awarded honorary degrees from the University of Paris VI in 1981 and the Colorado School of Mines in 1995. He received both the Mineralogical Society of America Award and the Columbia University Bicentennial Medal in 1954. He received the Arthur L. Day Medal and the History of Geology Award from the Geological Society of America, the Arthur L. Day Prize and Lectureship from the National Academy of Sciences, the A.G. Werner Medal from the German Mineralogical Society, the Wollaston Medal from the Geological Society of London, and the Roebling Medal from the Mineralogical Society of America, among others.

He was also named International Scientist of the Year for 2001.

Yoder has also performed service to the geologic profession too extensive to completely describe here. He served as president (1971–1972), vice president (1970–1971) and numerous panels and committees for the Mineralogical Society of America. He served as section president (1961–1964) and numerous positions for the American Geophysical Union. He served on numerous U.S. National Committees (geology, geochemistry, history of geology), and on committees and panels for the National Research Council, Geological Society of America, and Geochemical Society (organizing and founding member). He served on review committees for numerous universities including Harvard University, Massachusetts Institute of Technology, and Institut de Physique du Globe de Paris, France, among others. Yoder served in editorial positions for *American Journal of Science, Journal of Petrology* and *Geochimica et Cosmochimica Acta,* among others.

Z

Zen, E-An
(1928–   )
Chinese
Field Geologist, Petrologist

Many researchers have the reputation for being thorough in their work, but few are more thorough than E-An Zen. He is highly focused and tenacious in his projects. His geological career has concentrated on three areas of interest, mineralogy and petrology of marine sediments, granite petrology, and geological advocacy. Zen's interest in marine sediments started with his dissertation on the Taconic sequence of sedimentary rocks in western Vermont and easternmost New York (see "Taconic Stratigraphic Names: Definitions and Synonymies" and "Time and Space Relationships of the Taconic Allochthon and Autochthon," for example). This work could have been simply a mapping exercise as was common at the time. It may have then led to a career of regional Appalachian geology. However, Zen was not satisfied with the status quo. He studied modern marine sediments that were recovered from the Peru-Chile trench to better understand those comparable sedimentary rocks in the Taconics as described in "Mineralogy and Petrology of Marine Bottom Sediment Samples off the Coast of Peru and Chile." His geochemical work on these rocks also led him to an interest in the thermodynamics of clay minerals.

His position with the U.S. Geological Survey led Zen to work in the Pioneer Mountains of southwestern Montana and introduced him to his second main area of interest, granite magma, and granite batholiths. He studied granitic plutons from there all the way to southern Alaska, devising methods to determine depths of emplacement through mineral chemistry. These data in turn allowed him to evaluate the uplift and erosion rates for the entire western United States. This interest also led Zen to help organize and run parts of two Hutton international conferences on granite emplacement. His papers "Using Granite to Image the Thermal State of the Source Terrane" and "Plumbing the Depths of Batholiths" are two summaries of this work. He is still working to determine thermal budgets for granite emplacement.

His third area of interest is public advocacy for the science of geology, which he began later in his career. Once again he attacked the project with the same zest that he had for his research projects. Primarily using the forum of *GSA Today*, Zen attempted to educate the educators about global sustainability and recipes for how to live in harmony with our planet. In collaboration with ALLISON R. PALMER, Zen wrote a series of articles entitled "Engaging 'my neighbor' in the Issue of Sustainability," Parts II, V, IX, X, and XII. Once again his commitment and doggedness propelled him to the forefront of the movement and he was

E-An Zen shows weathering features on a rock exposure in Great Falls National Park in northern Virginia *(Courtesy of E. Zen)*

alogy of modern marine sediments recovered from the Peru-Chile trench. Zen was a visiting assistant professor at the University of North Carolina at Chapel Hill in 1958–1959 before becoming a geologist and research geologist with the U.S. Geological Survey from 1959 to 1990. During that time he held a visiting professorship at California Institute of Technology, 1962–1963, a Crosby visiting professorship at Massachusetts Institute of Technology in 1972, a Harry Hess visiting fellowship at Princeton University in 1981, and a visiting fellowship at the Australian National University in 1991. In 1990, Zen became a scientist emeritus at the U.S. Geological Survey and an adjunct professor at the University of Maryland, positions he still holds today.

E-An Zen has published 115 articles to date in international journals, professional volumes, and U.S. Geological Survey reports and maps. He also published some 23 articles on geology and society. Zen was also an editor of an important volume entitled *Studies of Appalachian Geology, Northern and Maritime.* He has received numerous honors and awards for his research. He is a member of the National Academy of Science (1976–) and a fellow of both the American Academy of Arts and Sciences and the American Association for the Advancement of Science. He was awarded the Arthur Day Medal from the Geological Society of America in 1986, the Roebling Medal from the Mineralogical Society of America in 1991, and the Major John Sacheverell Coke Medal from the Geological Society of London in 1992. He also received the 1995 Thomas Jefferson Medal from the Virginia Museum of Natural History.

Zen has performed outstanding service to the geological profession. He served on evaluating committees for California Institute of Technology, Harvard University, and Princeton University. He served on the National Research Council, the Scholarly Studies Committee for the Smithsonian Institution, the U.S. Committee on Geodynamics and on Geochemistry. He served as a member and

named to several very important committees at the National Academy of Science and National Research Council.

E-An Zen was born on May 31, 1928, in Peking, China. He immigrated to the United States as a teenager soon after World War II. He attended Cornell University, New York, and earned a bachelor of arts degree in geology in 1951. He entered Harvard University as a graduate student the same year and earned a master of arts and a Ph.D. in 1951 and 1955, respectively. His dissertation was on the petrology and stratigraphy of the Taconic Allochthon in western Vermont. He was a research fellow and associate at the Woods Hole Oceanographic Institution from 1955 through 1958, where he studied the miner-

officer for committees for the Geological Society of America and Mineralogical Society of America too numerous to mention. In addition, he served on the Committee on Human Rights and the Committee on Transition to Sustainability at the National Academy of the Sciences and the National Committee on Science Education Standards and Assessment at the National Research Council in 1992.

⊠ Zoback, Mary Lou
(1952–   )
American
Seismologist

When an earthquake occurs on the San Andreas Fault or any one of the numerous related faults in California, the scientific spokesperson for the event is Mary Lou Zoback. All information distributed to the press is funneled through her office and she is typically the person on the radio or television interviews that surround the earthquake. Because of her high-profile position, she is periodically seen on documentaries about earthquakes in California and in general. In her position of chief scientist for the U.S. Geological Survey Hazards Team, she has a staff of 200 people and a $26-million budget to monitor all earthquake activity and research the causes and effects.

In addition to this highly public side, Mary Lou Zoback is also a talented geophysicist who studies the interrelations of plate-scale stresses of the Earth's crust and current seismicity. She studies both plate margins, like the San Andreas Fault, and intraplate regions like the Basin and Range Province of the southwestern United States and the New Madrid seismic zone. She developed methods to determine current distribution of tectonic forces using geologic, seismic, and in situ stress measurements taken from boreholes. She demonstrated that large regions of the Earth's crust are being subjected to a relatively constant and uniformly oriented stress field to great depths. This stress field results from large-scale plate tectonic processes. Papers on this subject include "State of Stress and Intraplate Earthquakes in the Central and Eastern United States" and "Tectonic Stress Field of the Continental U.S." She extended this research in collaboration with 40 scientists from 30 countries to cover the entire Earth. In a global stress mapping project, they demonstrated that broad areas within the tectonic plates are subjected to uniform stresses that can be predicted from the geometry of the plate and the forces that drive the plates (this was discussed in the paper "Global Patterns of Intraplate Stress: A Status Report on the World Stress Map Project of the International Lithosphere Program"). These findings mean that intraplate earthquakes like New Madrid (1812), among others, result from these large-scale regional stresses rather than local ones.

Mary Lou Chetlain was born on July 5, 1952, in Sanford, Florida, where she grew up. She entered college at Florida Institute of Technology in Oceanology and remained for two years before transferring to Stanford University, California, where she graduated with a bachelor of science degree in geophysics in 1974. She remained at Stanford for her graduate studies and earned master of science and Ph.D. degrees in geophysics in 1975 and 1978, respectively. She was awarded a National Research Council postdoctoral fellowship for 1978–1979 with the U.S. Geological Survey in Menlo Park, California, to study heat flow. In 1979, she became a research geophysicist for the U.S. Geological Survey in Menlo Park and later project chief. Zoback became the chief scientist for the Earthquake Hazard Team for the western United States (U.S. Geological Survey) in 1999 and remains in that position as of 2002. Mary Lou Zoback works on many of her research projects with her husband, geophysicist Mark Zoback, whom she married in 1973. The Zobacks have two children and reside in Palo Alto, California.

Mary Lou Zoback has authored numerous articles in international journals and professional volumes. She was also editor of two high-profile volumes. Her research has been well recognized by the profession in terms of honors and awards. She has been a member of the National Academy of Sciences since 1995. She was awarded a G.K. Gilbert Fellowship by the U.S. Geological Survey (1990–1991). She received the James B. MacElwane Award from the American Geophysical Union in 1997.

Mary Lou Zoback's service to the profession is unparalleled for someone so early in his or her career. She was a member of the President's Medal of Science selection committee (1999–2001). She is on the membership selection committee for the National Academy of Sciences. She was the chair of the World Stress Map Program (2000–2001). She worked on numerous committees for the National Research Council, including science standards, radioactive waste management, geodynamics, and geosciences, environment and resources. She is a member of the NASA Steering Committee for solid Earth sciences. For the Geological Society of America, she was councilor (1986–1988), chair of the Geophysics Division (1987–1988), president of the Cordilleran Section (1989–1990), vice president (1998–1999) and president (1999–2000). She was president of the Tectonophysics Section of the American Geophysical Union in 1996–1998. She has served on the editorial board for *Geology* (1982–1984) and the *Journal of Geodynamics* (1992–1999), among other editorial positions and duties. She serves on other national and international committee and panel positions but they are too numerous to list here.

Zuber, Maria T.
(1958–   )
American
Planetary Scientist, Geophysicist

The rise of Maria Zuber to her position as one of the foremost authorities on planetary geology has been almost as spectacular as the rockets used in the many NASA missions in which she has participated. She has been involved in at least three missions to Mars, including the Mars Global Surveyor Science Team, as well as missions to Mercury and to nearby asteroids. She even discovered an asteroid before she was 30 years old. Using data from topographic and corresponding geophysical surveys collected during these NASA missions, Zuber uses theoretical models and numerical analysis to evaluate the evolution of planets and their various components. She has studied the evolution of the crust and mantle on Mars, Venus, and the Moon, in addition to that on the Earth. These studies evaluate the thickness of crust and circulation in the underlying mantle using surface topography coupled with airborne gravity and magnetic surveys. These analyses involve sophisticated advanced mathematical studies using the polynomial fitting of the shapes. She evaluated the physical properties of Mercury's core using data from the NASA Messenger mission, of which she was a team member. More recently, she has been studying asteroids and their interiors. Using data from Magellan, she evaluated the tectonics, structure, and volcanism on Venus. She even studied clouds and snow depth on Mars in addition to wrinkle ridges and other structural features.

Closer to home, Zuber has studied mantle convection and the development of mid-ocean ridges on Earth. She has also done extensive work on the tectonics of Australia. Her lunar studies evaluated the volcanism and its relationship to the development of the lunar crust. To show that these numerical models of continental development can have more familiar analogs, she wrote a paper entitled "Folding of a Jelly Sandwich" in an illustration of rheological analyses.

Maria Zuber was born on June 27, 1958, in Norristown, Pennsylvania, where she grew up. She attended the University of Pennsylvania in Philadelphia, where she earned a bachelor of arts degree in astrophysics and geology in 1980. She completed her graduate studies at Brown University, Rhode Island, and earned a master of science degree in 1983 and a Ph.D. in 1986, both in geophysics. Her dissertation adviser was Mark Parmentier. Zuber earned a National Research Council Fellowship in 1985–1986 with NASA at the Goddard Space Center, Maryland. She remained there on staff as a geophysicist in 1987. She joined the faculty at Johns Hopkins University in 1991 as an associate research professor of geophysics after holding the position of visiting professor the previous year. In 1993, Zuber was named to the prestigious first-ever Second Decade Society endowed associate professorship by the Johns Hopkins Alumni Association. Just as she was promoted to full professor in 1995, Maria Zuber moved to the Massachusetts Institute of Technology in Cambridge, where she remains today. In 1998, she was named to the E.A. Griswold Professorship, another endowed position that she still holds. She also rejoined NASA in 1994 as a senior research scientist concurrent with her faculty positions. In this role, she is associated with the Goddard Space Center's Laboratory for Terrestrial Physics. From 1996 to 1999, she was a part-time visiting scientist at the Woods Hole Oceanographic Institution, Massachusetts. Maria Zuber has a husband, Jack, and two children.

Maria Zuber is amid a very productive career with authorship on some 93 scientific articles in international journals, professional volumes, and governmental reports. Many of these papers are seminal works on the development of the Earth's crust, as well as the evolution of planets. In recognition of her many contributions to the Earth sciences both in teaching and in research, several prestigious honors and awards have been bestowed upon her. She received the Thomas O. Paine Memorial Award from the Planetary Society. She received Outstanding Performance Awards every year from 1988 through 1992 and the Group Achievement Awards in 1991, 1993, 1994, 1998, and 2000, all from NASA. From the Johns Hopkins University, she was given the Oraculum Award for Excellence in Teaching and the David S. Olton Award for contributions to undergraduate research. She was also a distinguished leader in science lecturer for the National Academy of Sciences and the inaugural Carl Sagan lecturer for the American Geophysical Union.

Zuber has also performed significant service to the profession. She was president of the Planetary Sciences Section of the American Geophysical Union, in addition to serving as member and chair of numerous committees. She also served on numerous panels and committees at NASA and the National Academy of Sciences. She was also the editor for *Planetary Geosciences,* associate editor for the *Journal of Geophysical Research* and on the board of reviewing editors for *Science,* among others.

ENTRIES BY FIELD

ARCHAEOLOGICAL GEOLOGY
Folk, Robert L.
Herz, Norman

CLIMATE CHANGE
Alley, Richard B.
Broecker, Wallace
Fairbridge, Rhodes W.
Holland, Heinrich D.
Imbrie, John
Raymo, Maureen E.
Shackleton, Sir Nicholas J.

EARTH SCIENCE ADVOCACY
Grew, Priscilla C.
Palmer, Allison R. (Pete)
Revelle, Roger
Zen, E-An

ECONOMIC GEOLOGY
Bodnar, Robert J.
Herz, Norman
Kerr, Paul F.
Kerrich, Robert
Logan, Sir William Edmond
Skinner, Brian J.
Smith, Joseph V.

GEOCHEMISTRY
Albee, Arden L.
Allègre, Claude

Berner, Robert
Bodnar, Robert J.
Bowen, Norman L.
Bowring, Samuel A.
Brantley, Susan L.
Craig, Harmon
Day, Arthur L.
DePaolo, Donald J.
Ernst, W. Gary
Eugster, Hans P.
Fyfe, William S.
Garrels, Robert M.
Gilbert, M. Charles
Goldsmith, Richard
Harrison, T. Mark
Hayes, John M.
Helgeson, Harold C.
Hochella, Michael F., Jr.
Holland, Heinrich D.
Holmes, Arthur
Kerrich, Robert
Lindsley, Donald H.
Mahood, Gail A.
Montanez, Isabel Patricia
Mukasa, Samuel B.
O'Nions, Sir R. Keith
Patterson, Claire (Pat) C.
Raymo, Maureen E.
Ringwood, Alfred E. (Ted)
Roedder, Edwin W.
Skinner, Brian J.

Spear, Frank S.
Stolper, Edward M.
Suess, Hans E.
Taylor, Hugh D., Jr.
Thompson, James B., Jr.
Tilton, George R.
Turekian, Karl K.
Tuttle, O. Frank
Valley, John W.
Walter, Lynn M.
Watson, E. Bruce
Yoder, Hatten S., Jr.

GEOMORPHOLOGY OR QUATERNARY GEOLOGY
Ashley, Gail
Fairbridge, Rhodes W.
Gilbert, G. Karl
Holmes, Arthur
Keller, Edward A.
Morisawa, Marie
Porter, Stephen C.

GEOPHYSICS
Anderson, Don L.
Atwater, Tanya
Birch, A. Francis
Bromery, Randolph W. (Bill)
Bullard, Sir Edward C.
Cox, Allan V.
Day, Arthur L.

Drake, Charles L.
Ewing, W. Maurice
Griggs, David T.
Gutenberg, Beno
Holmes, Arthur
Hubbert, M. King
Karig, Daniel E.
Kent, Dennis V.
Lehmann, Inge
Liebermann, Robert C.
Matthews, Drummond H.
McKenzie, Dan P.
McNally, Karen C.
McNutt, Marcia
Molnar, Peter
Oliver, Jack E.
Press, Frank
Richter, Charles F.
Romanowicz, Barbara
Rosendahl, Bruce R.
Sykes, Lynn R.
Talwani, Manik
Turcotte, Donald L.
Van der Voo, Rob
Wilson, (John) Tuzo
Zoback, Mary Lou

HYDROGEOLOGY

Bethke, Craig M.
Bredehoeft, John D.
Cherry, John A.

MINERALOGY

Bloss, F. Donald
Bragg, Sir William Lawrence
Dana, James D.
Goldsmith, Julian R.
Hochella, Michael F., Jr.
Jahns, Richard H.
Kerr, Paul F.
Liebermann, Robert C.
Navrotsky, Alexandra
Smith, Joseph V.
Veblen, David R.

Weeks, Alice, M.D.
Wenk, Hans-Rudolf
Wones, David R.

OCEANOGRAPHY (MARINE GEOLOGY)

Broecker, Wallace S.
Bullard, Sir Edward C.
Craig, Harmon
Dietz, Robert R.
Ewing, W. Maurice
Hayes, John
Hess, Harry
Karig, Daniel E.
Menard, H. William
Morse, John W.
Revelle, Roger
Rizzoli, Paola Malanotte
Turekian, Karl K.

PALEONTOLOGY

Berry, William B.N.
Clark, Thomas H.
Cloud, Preston E., Jr.
Conway Morris, Simon
Dawson, Sir (John) William
Dunbar, Carl O.
Gould, Stephen Jay
Imbrie, John
Landing, Ed
Miller, Kenneth
Olsen, Paul E.
Ostrom, John H.
Palmer, Allison R. (Pete)
Raup, David M.
Stanley, Steven M.
Teichert, Curt
Walcott, Charles D.
Whittington, Harry B.

PETROLOGY

Albee, Arden L.
Bascom, Florence
Bowen, Norman L.

Brown, Michael
Buddington, Arthur F.
Carmichael, Ian S.
Cashman, Katherine V.
Crawford, Maria Luisa
Gilbert, M. Charles
Grew, Priscilla
Holmes, Arthur
Jahns, Richard H.
Kuno, Hisashi
Lindsley, Donald H.
Mahood, Gail A.
McSween, Harry Y., Jr.
Pitcher, Wallace S.
Selverstone, Jane
Spear, Frank J.
Stolper, Edward M.
Tuttle, O. Frank
Valley, John W.
Wones, David R.
Wyllie, Peter J.
Yoder, Hatten S., Jr.

PETROLEUM GEOLOGY

Bally, Albert W.
Friedman, Gerald
Klein, George D.
Withjack, Martha O.

PLANETARY GEOLOGY

Head, James W., III
McSween, Harry Y., Jr.
Melosh, H.J.
Sagan, Carl E.
Shoemaker, Eugene M.
Zuber, Maria T.

REGIONAL OR FIELD GEOLOGY

Bascom, Florence
Clark, Thomas
Crawford, Maria Louisa
Dewey, John
Glover, Lynn, III
Hatcher, Robert, Jr.

Rast, Nicholas
Rodgers, John
Selverstone, Jane
Stose, Anna I. Jonas
Sylvester, Arthur G.
Thompson, James B., Jr.
Williams, Harold
Zen, E-An

SEDIMENTOLOGY OR STRATIGRAPHY

Alvarez, Walter
Ashley, Gail Mowry
Berner, Robert
Bouma, Arnold H.
Burke, Kevin
Chan, Marjorie A.
Dickinson, William R.
Dott, Robert H., Jr.
Folk, Robert L.
Friedman, Gerald M.
Glover, Lynn, III
Hoffman, Paul
Hsu, Kenneth J.
Jordan, Teresa E.
Kay, Marshall
Klein, George D.
Landing, Ed
McBride, Earle F.
Miller, Kenneth G.
Montanez, Isabel Patricia

Pettijohn, Francis J.
Sloss, Laurence L.
Twenhofel, William H.
Vail, Peter R.

STRUCTURAL GEOLOGY

Bally, Albert W.
Billings, Marland P.
Burchfiel, B. Clark
Cloos, Ernst
Handin, John W.
Hatcher, Robert D., Jr.
Hsu, Kenneth J.
Hubbert, M. King
Marshak, Stephen
Means, Winthrop D.
Moores, Eldridge M.
Muehlberger, William R.
Nance, R. Damian
Price, Raymond A.
Ramberg, Hans
Ramsay, John G.
Sibson, Richard H.
Simpson, Carol
Stock, Joann M.
Suppe, John E.
Sylvester, Arthur G.
Tullis, Julia A. (Jan)
Wenk, Hans-Rudolf
Wise, Donald U.
Withjack, Martha O.

TECTONICS

Alvarez, Walter
Atwater, Tanya
Burchfiel, B. Clark
Burke, Kevin C.A.
Cox, Allan V.
Dewey, John F.
Dickinson, William R.
Drake, Charles L.
Ernst, W. Gary
Hess, Harry H.
Hoffman, Paul
Hsu, Kenneth J.
Matthews, Drummond
McKenzie, Daniel
Menard, H. William
Molnar, Peter
Moores, Eldridge M.
Muehlberger, William R.
Nance, R. Damian
Oliver, Jack E.
Ramberg, Hans
Rast, Nicholas
Rodgers, John
Rosendahl, Bruce R.
Sengor, A.M. Celal
Stock, Joann M.
Wegener, Alfred
Wilson, (John) Tuzo

ENTRIES BY COUNTRY OF BIRTH

AUSTRALIA
Bragg, Sir (William) Lawrence
Fairbridge, Rhodes W.
Ringwood, Alfred E. (Ted)
Skinner, Brian

AUSTRIA
Suess, Hans E.

CANADA
Bowen, Norman L.
Cherry, John A.
Dawson, Sir (John) William
Harrison, T. Mark
Hoffman, Paul
Kay, Marshall
Logan, Sir William Edmond
Price, Raymond A.
Williams, Harold
Wilson, (John) Tuzo

CHINA
Hsu, Kenneth J.
Zen, E-An

CZECHOSLOVAKIA
Kent, Dennis

DENMARK
Lehmann, Inge

FRANCE
Allègre, Claude
Romanowicz, Barbara

GERMANY
Cloos, Ernst
Friedman, Gerald M.
Gutenberg, Beno
Holland, Heinrich D.
Teichert, Curt
Wegener, Alfred

GREAT BRITAIN
Brown, Michael
Bullard, Sir Edward C.
Burke, Kevin C.A.
Carmichael, Ian S.
Clark, Thomas H.
Conway Morris, Simon
Dewey, John F.
Holmes, Arthur
Kerrich, Robert
Matthews, Drummond H.
McKenzie, Dan P.
Nance, R. Damian
O'Nions, Sir R. Keith
Pitcher, Wallace S.
Ramsay, John G.
Shackleton, Sir Nicholas J.
Simpson, Carol
Smith, Joseph V.

Whittington, Harry B.
Wyllie, Peter J.

INDIA
Talwani, Manik

IRAN
Rast, Nicholas

ITALY
Rizzoli, Paola Malanotte

JAPAN
Hochella, Michael F., Jr.
Kuno, Hisashi

NETHERLANDS
Bally, Albert W.
Bouma, Arnold H.
Klein, George D.
Van der Voo, Robert

NEW ZEALAND
Fyfe, William S.
Sibson, Richard H.

NORWAY
Ramberg, Hans

SWITZERLAND
Eugster, Hans P.

Montanez, Isabel Patricia
Wenk, Hans-Rudolf

TURKEY
Sengor, A.M. Çelal

UNITED STATES
Albee, Arden L.
Alley, Richard B.
Alvarez, Walter
Anderson, Don L.
Ashley, Gail Mowry
Atwater, Tanya
Bascom, Florence
Berner, Robert
Berry, William B.N.
Bethke, Craig M.
Billings, Marland P.
Birch, A. Francis
Bloss, F. Donald
Bodnar, Robert J.
Bowring, Samuel A.
Brantley, Susan L.
Bredehoeft, John D.
Broecker, Wallace S.
Bromery, Randolph W. (Bill)
Buddington, Arthur F.
Burchfiel, B. Clark
Cashman, Katherine V.
Chan, Marjorie A.
Cloud, Preston E., Jr.
Cox, Allan V.
Craig, Harmon
Crawford, Maria Luisa
Dana, James D.
Day, Arthur L.
DePaolo, Donald J.
Dickinson, William R.
Dietz, Robert S.
Dott, Robert H., Jr.
Drake, Charles L.
Dunbar, Carl O.
Ernst, W. Gary
Ewing, W. Maurice
Folk, Robert L.

Garrels, Robert M.
Gilbert, G. Karl
Gilbert, M. Charles
Glover, Lynn, III
Goldsmith, Julian R.
Gould, Stephen Jay
Grew, Priscilla C.
Griggs, David T.
Handin, John W.
Hatcher, Robert D., Jr.
Hayes, John M.
Head, James W., III
Helgeson, Harold C.
Herz, Norman
Hess, Harry H.
Hubbert, M. King
Imbrie, John
Jahns, Richard H.
Jordan, Teresa E.
Karig, Daniel E.
Keller, Edward A.
Kerr, Paul F.
Landing, Ed
Liebermann, Robert C.
Lindsley, Donald H.
Mahood, Gail A.
Marshak, Stephen
McBride, Earle F.
McNally, Karen C.
McNutt, Marcia
McSween, Harry Y., Jr.
Means, Winthrop D.
Melosh, H.J.
Menard, H. William
Miller, Kenneth G.
Molnar, Peter
Moores, Eldridge M.
Morisawa, Marie
Morse, John W.
Muehlberger, William R.
Mukasa, Samuel B.
Navrotsky, Alexandra
Oliver, Jack E.
Olsen, Paul E.
Ostrom, John H.

Palmer, Allison R. (Pete)
Patterson, Claire (Pat) C.
Pettijohn, Francis J.
Porter, Stephen C.
Press, Frank
Raup, David M.
Raymo, Maureen E.
Revelle, Roger
Richter, Charles F.
Rodgers, John
Roedder, Edwin W.
Rosendahl, Bruce R.
Sagan, Carl E.
Selverstone, Jane
Shoemaker, Eugene M.
Sloss, Laurence L.
Spear, Frank S.
Stanley, Steven M.
Stock, Joann M.
Stolper, Edward M.
Stose, Anna I. Jonas
Suppe, John E.
Sykes, Lynn R.
Sylvester, Arthur G.
Taylor, Hugh P., Jr.
Thompson, James B., Jr.
Tilton, George R.
Tullis, Julia A. (Jan)
Turcotte, Donald L.
Turekian, Karl K.
Tuttle, O. Frank
Twenhofel, William H.
Vail, Peter R.
Valley, John W.
Veblen, David R.
Walcott, Charles D.
Walter, Lynn M.
Watson, E. Bruce
Weeks, Alice, M.D.
Wise, Donald U.
Withjack, Martha O.
Wones, David R.
Yoder, Hatten S., Jr.
Zoback, Mary Lou
Zuber, Maria T.

ENTRIES BY COUNTRY OF MAJOR SCIENTIFIC ACTIVITY

AUSTRALIA
Fairbridge, Rhodes W.
Liebermann, Robert C.
Ringwood, Alfred E. (Ted)
Skinner, Brian J.

AUSTRIA
Suess, Hans E.

CANADA
Cherry, John A.
Clark, Thomas H.
Dawson, Sir (John) William
Fyfe, William S.
Hoffman, Paul
Kerrich, Robert
Logan, Sir William Edmond
Nance, R. Damian
Price, Raymond A.
Rast, Nicholas
Williams, Harold
Wilson, (John) Tuzo

DENMARK
Lehmann, Inge

FRANCE
Allègre, Claude
Romanowicz, Barbara

GERMANY
Cloos, Ernst
Gutenberg, Beno
Holland, Heinrich D.
Teichert, Curt
Wegener, Alfred

GREAT BRITAIN
Bragg, William Lawrence, Sir
Brown, Michael
Bullard, Sir Edward C.
Burke, Kevin C.A.
Carmichael, Ian S.
Conway Morris, Simon
Dewey, John
Fyfe, William
Holmes, Arthur
Matthews, Drummond H.
McKenzie, Dan P.
Nance, R. Damian
O'Nions, Sir R. Keith
Pitcher, Wallace S.
Ramsay, John G.
Rast, Nicholas
Shackleton, Sir Nicholas J.
Smith, Joseph V.
Whittington, Harry B.
Wyllie, Peter J.

ITALY
Rizzoli, Paola Malanotte

JAPAN
Kuno, Hisashi

NETHERLANDS
Bally, Albert
Bouma, Arnold

NEW ZEALAND
Fyfe, William S.
Sibson, Richard H.

NORWAY
Ramberg, Hans

SWITZERLAND
Eugster, Hans
Hsu, Kenneth
Ramsay, John G.
Wenk, Hans-Rudolf

TURKEY
Sengor, A.M. Çelal

UNITED STATES
Albee, Arden L.
Alley, Richard B.

Alvarez, Walter
Anderson, Don L.
Ashley, Gail Mowry
Atwater, Tanya
Bally, Albert W.
Bascom, Florence
Berner, Robert
Berry, William B.N.
Bethke, Craig M.
Billings, Marland P.
Birch, A. Francis
Bloss, F. Donald
Bodnar, Robert J.
Bouma, Arnold H.
Bowen, Norman L.
Bowring, Samuel A.
Brantley, Susan L.
Bredehoeft, John D.
Broecker, Wallace S.
Bromery, Randolph W. (Bill)
Brown, Michael
Buddington, Arthur F.
Burchfiel, B. Clark
Carmichael, Ian S.
Cashman, Katherine V.
Chan, Marjorie A.
Cloos, Ernst
Cloud, Preston E., Jr.
Cox, Allan V.
Craig, Harmon
Crawford, Maria Luisa
Dana, James D.
Day, Arthur L.
DePaolo, Donald J.
Dewey, John F.
Dickinson, William R.
Dietz, Robert S.
Dott, Robert H., Jr.
Drake, Charles L.
Dunbar, Carl O.
Ernst, W. Gary
Eugster, Hans P.

Ewing, W. Maurice
Fairbridge, Rhodes W.
Folk, Robert L.
Friedman, Gerald M.
Garrels, Robert M.
Gilbert, G. Karl
Gilbert, M. Charles
Glover, Lynn, III
Goldsmith, Julian R.
Gould, Stephen Jay
Grew, Priscilla C.
Griggs, David T.
Gutenberg, Beno
Handin, John W.
Harrison, T. Mark
Hatcher, Robert D., Jr.
Hayes, John M.
Head, James W., III
Helgeson, Harold C.
Herz, Norman
Hess, Harry H.
Hochella, Michael F., Jr.
Holland, Heinrich D.
Hsu, Kenneth J.
Hubbert, M. King
Imbrie, John
Jahns, Richard H.
Jordan, Teresa E.
Karig, Daniel E.
Kay, Marshall
Keller, Edward A.
Kent, Dennis V.
Kerr, Paul F.
Klein, George D.
Landing, Ed
Liebermann, Robert C.
Lindsley, Donald H.
Mahood, Gail A.
Marshak, Stephen
McBride, Earle F.
McNally, Karen C.
McNutt, Marcia

McSween, Harry Y., Jr.
Means, Winthrop D.
Melosh, H.J.
Menard, H. William
Miller, Kenneth G.
Molnar, Peter
Montanez, Isabel Patricia
Moores, Eldridge M.
Morisawa, Marie
Morse, John W.
Muehlberger, William R.
Mukasa, Samuel B.
Nance, R. Damian
Navrotsky, Alexandra
Oliver, Jack E.
Olsen, Paul E.
Ostrom, John H.
Palmer, Allison R. (Pete)
Patterson, Claire (Pat) C.
Pettijohn, Francis J.
Porter, Stephen C.
Press, Frank
Ramberg, Hans
Rast, Nicholas
Raup, David M.
Raymo, Maureen E.
Revelle, Roger
Richter, Charles F.
Rizzoli, Paola Malanotte
Rodgers, John
Roedder, Edwin W.
Romanowicz, Barbara
Rosendahl, Bruce R.
Sagan, Carl E.
Selverstone, Jane
Shoemaker, Eugene M.
Simpson, Carol
Sloss, Laurence L.
Smith, Joseph V.
Spear, Frank S.
Stanley, Steven M.
Stock, Joann M.

ENTRIES BY YEAR OF BIRTH

1750–1800
Logan, Sir William Edmond

1801–1850
Dana, James D.
Dawson, Sir (John) William
Gilbert, G. Karl
Walcott, Charles D.

1851–1880
Bascom, Florence
Day, Arthur L.
Twenhofel, William H.
Wegener, Alfred

1881–1890
Bowen, Norman L.
Bragg, Sir William Lawrence
Buddington, Arthur F.
Gutenberg, Beno
Holmes, Arthur
Lehmann, Inge
Stose, Anna I. Jonas

1891–1900
Clark, Thomas H.
Cloos, Ernst
Dunbar, Carl O.
Kerr, Paul F.
Richter, Charles F.

1901–1910
Billings, Marland P.
Birch, A. Francis
Bullard, Sir Edward C.
Ewing, W. Maurice
Hess, Harry H.
Hubbert, M. King
Kay, Marshall
Kuno, Hisashi
Pettijohn, Francis J.
Revelle, Roger
Suess, Hans E.
Teichert, Curt
Weeks, Alice M.D.
Wilson, (John) Tuzo

1911–1920
Bloss, F. Donald
Cloud, Preston E., Jr.
Dietz, Robert S.
Fairbridge, Rhodes W.
Garrels, Robert M.
Goldsmith, Julian R.
Griggs, David T.
Handin, John W.
Jahns, Richard H.
Menard, H. William
Morisawa, Marie
Pitcher, Wallace S.
Ramberg, Hans

Rodgers, John
Roedder, Edwin W.
Sloss, Laurence L.
Tuttle, O. Frank
Whittington, Harry B.

1921–1930
Albee, Arden L.
Bally, Albert W.
Bromery, Randolph W. (Bill)
Burke, Kevin C.A.
Carmichael, Ian S.
Cox, Allan V.
Craig, Harmon
Dott, Robert H., Jr.
Drake, Charles L.
Eugster, Hans P.
Folk, Robert L.
Friedman, Gerald M.
Fyfe, William S.
Glover, Lynn, III
Herz, Norman
Holland, Heinrich D.
Hsu, Kenneth J.
Imbrie, John
Muehlberger, William R.
Oliver, Jack E.
Ostrom, John H.
Palmer, Allison R. (Pete)
Patterson, Claire (Pat) C.

Press, Frank
Rast, Nicholas
Ringwood, Alfred E. (Ted)
Shoemaker, Eugene M.
Skinner, Brian J.
Smith, Joseph V.
Thompson, James B., Jr.
Tilton, George R.
Turekian, Karl K.
Vail, Peter R.
Wyllie, Peter J.
Yoder, Hatten S., Jr.
Zen, E-An

1931–1940
Allègre, Claude
Alvarez, Walter
Anderson, Don L.
Berner, Robert
Berry, William B.N.
Bouma, Arnold H.
Bredehoeft, John D.
Broecker, Wallace S.
Burchfiel, B. Clark
Crawford, Maria Luisa
Dewey, John F.
Dickinson, William R.
Ernst, W. Gary
Gilbert, M. Charles
Grew, Priscilla C.
Hatcher, Robert D., Jr.
Hayes, John M.
Helgeson, Harold C.
Karig, Daniel E.
Klein, George D.
Lindsley, Donald H.
Matthews, Drummond H.
McBride, Earle F.
McNally, Karen C.
Means, Winthrop D.
Moores, Eldridge M.

Porter, Stephen C.
Price, Raymond A.
Ramsay, John G.
Raup, David M.
Sagan, Carl E.
Shackleton, Sir Nicholas J.
Sykes, Lynn R.
Sylvester, Arthur G.
Talwani, Manik
Taylor, Hugh P., Jr.
Turcotte, Donald L.
Van der Voo, Robert
Williams, Harold
Wise, Donald U.
Wones, David R.

1941–1950
Ashley, Gail Mowry
Atwater, Tanya
Bodnar, Robert J.
Brown, Michael
Cherry, John A.
Gould, Stephen Jay
Head, James W., III
Hoffman, Paul
Keller, Edward A.
Kent, Dennis V.
Kerrich, Robert
Landing, Ed
Liebermann, Robert C.
McKenzie, Dan P.
McSween, Harry Y., Jr.
Melosh, H.J.
Molnar, Peter
Morse, John W.
Navrotsky, Alexandra
O'Nions, Sir R. Keith
Rizzoli, Paola Malanotte
Romanowicz, Barbara
Rosendahl, Bruce R.
Sibson, Richard H.

Simpson, Carol
Spear, Frank S.
Stanley, Steven M.
Suppe, John E.
Tullis, Julia A. (Jan)
Valley, John W.
Veblen, David R.
Watson, E. Bruce
Wenk, Hans-Rudolf

1951–1960
Alley, Richard B.
Bethke, Craig M.
Bowring, Samuel A.
Brantley, Susan L.
Cashman, Katherine V.
Chan, Marjorie A.
Conway Morris, Simon
DePaolo, Donald J.
Harrison, T. Mark
Hochella, Michael F., Jr.
Jordan, Teresa E.
Mahood, Gail A.
Marshak, Stephen
McNutt, Marcia
Miller, Kenneth G.
Montanez, Isabel Patricia
Mukasa, Samuel B.
Nance, R. Damian
Olsen, Paul E.
Raymo, Maureen E.
Selverstone, Jane
Sengor, A.M. Çelal
Stock, Joann M.
Stolper, Edward M.
Walter, Lynn M.
Withjack, Martha O.
Zoback, Mary Lou
Zuber, Maria T.

CHRONOLOGY

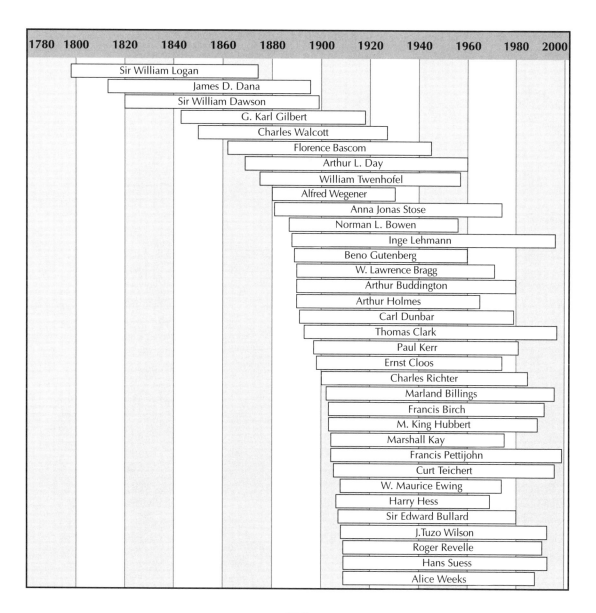

1780	1800	1820	1840	1860	1880	1900	1920	1940	1960	1980	2000

Sir William Logan

James D. Dana

Sir William Dawson

G. Karl Gilbert

Charles Walcott

Florence Bascom

Arthur L. Day

William Twenhofel

Alfred Wegener

Anna Jonas Stose

Norman L. Bowen

Inge Lehmann

Beno Gutenberg

W. Lawrence Bragg

Arthur Buddington

Arthur Holmes

Carl Dunbar

Thomas Clark

Paul Kerr

Ernst Cloos

Charles Richter

Marland Billings

Francis Birch

M. King Hubbert

Marshall Kay

Francis Pettijohn

Curt Teichert

W. Maurice Ewing

Harry Hess

Sir Edward Bullard

J. Tuzo Wilson

Roger Revelle

Hans Suess

Alice Weeks

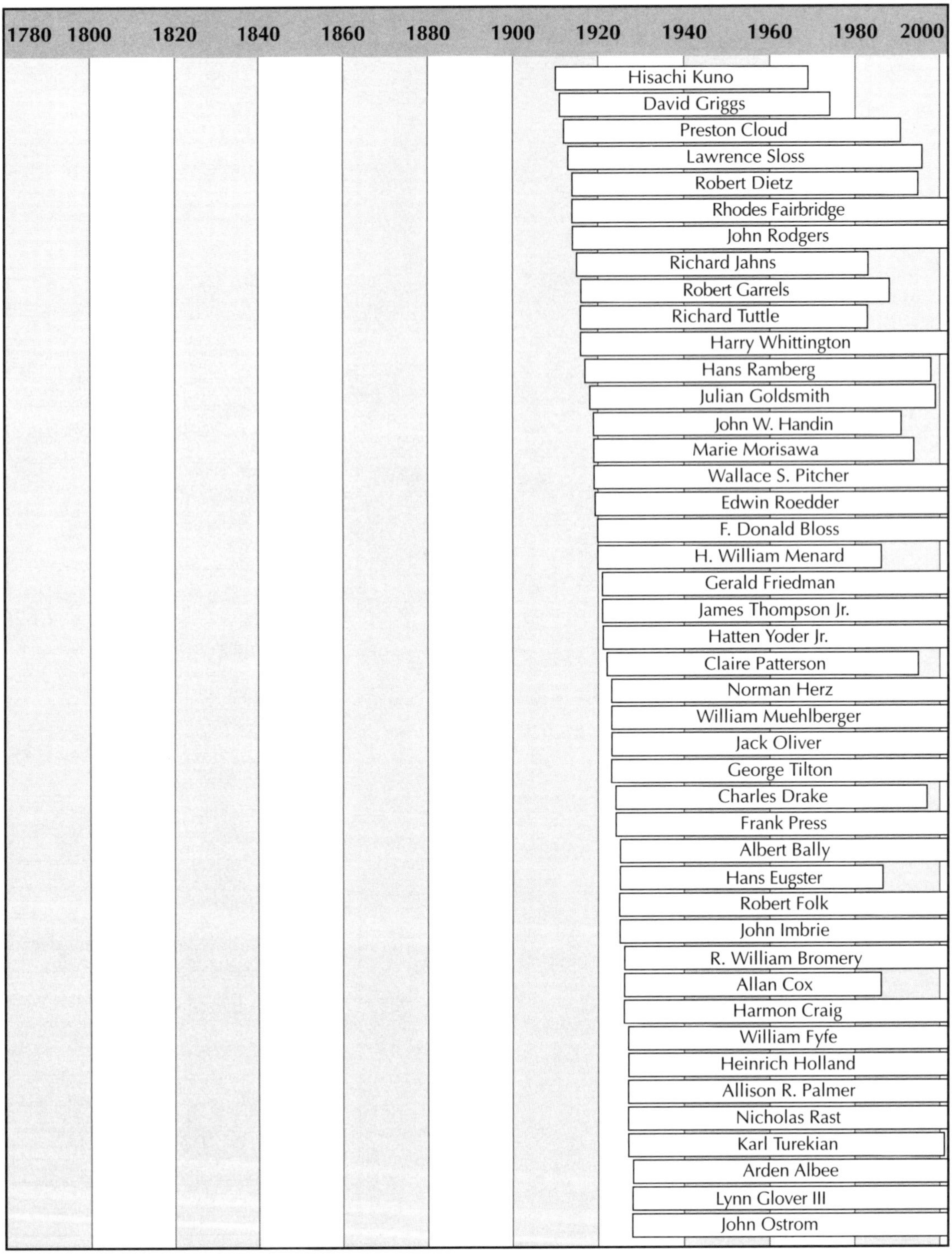

1780	1800	1820	1840	1860	1880	1900	1920	1940	1960	1980	2000

Hisachi Kuno
David Griggs
Preston Cloud
Lawrence Sloss
Robert Dietz
Rhodes Fairbridge
John Rodgers
Richard Jahns
Robert Garrels
Richard Tuttle
Harry Whittington
Hans Ramberg
Julian Goldsmith
John W. Handin
Marie Morisawa
Wallace S. Pitcher
Edwin Roedder
F. Donald Bloss
H. William Menard
Gerald Friedman
James Thompson Jr.
Hatten Yoder Jr.
Claire Patterson
Norman Herz
William Muehlberger
Jack Oliver
George Tilton
Charles Drake
Frank Press
Albert Bally
Hans Eugster
Robert Folk
John Imbrie
R. William Bromery
Allan Cox
Harmon Craig
William Fyfe
Heinrich Holland
Allison R. Palmer
Nicholas Rast
Karl Turekian
Arden Albee
Lynn Glover III
John Ostrom

1780	1800	1820	1840	1860	1880	1900	1920	1940	1960	1980	2000

Eugene Shoemaker

Brian Skinner

Joseph Smith

E-An Zen

Kevin Burke

Robert Dott Jr.

Kenneth Hsu

Ian Carmichael

A. Edward Ringwood

Peter Vail

Peter Wyllie

William Berry

Wallace Broecker

William Dickinson

Gary Ernst

Harold Helgeson

Drummond Matthews

John Ramsay

Donald Wise

Arnold Bouma

Earle McBride

Hugh Taylor

Donald Turcotte

David Wones

Don Anderson

John Bredehoeft

Winthrop Means

Raymond Price

David Raup

Manik Talwani

B. Clark Burchfiel

Donald Lindsley

Stephen Porter

Carl Sagan

Harold Williams

Robert Berner

M. Charles Gilbert

Claude Allegre

John Dewey

Daniel Karig

Sir Nicholas Shackleton

Lynn Sykes

Eldridge Moores

Arthur Sylvester

1780	1800	1820	1840	1860	1880	1900	1920	1940	1960	1980	2000
									Maria Louisa Crawford		
									George Klein		
									Walter Alvarez		
									Priscilla Grew		
									Robert Hatcher Jr.		
									John Hayes		
									Karen C. McNally		
									Robert van der Voo		
									Gail Ashley		
									John Cherry		
									Stephen Jay Gould		
									James Head III		
									Paul Hoffman		
									Stephen Stanley		
									Hans-Rudolf Wenk		
									Tanya Atwater		
									Edward Keller		
									Robert Lieberman		
									Daniel McKenzie		
									John Suppe		
									Peter Molnar		
									Alexandra Navrotsky		
									Jan Tullis		
									Sir Keith O'Nions		
									Harry McSween Jr.		
									Richard Sibson		
									Dennis Kent		
									John Morse		
									Paola Malanotte Rizzoli		
									Bruce Rosendahl		
									Michael Brown		
									H. Jay Melosh		
									Carol Simpson		
									David Veblen		
									Robert Kerrich		
									John Valley		
									Robert J. Bodnar		
									Ed Landing		
									Frank Spear		
									Barbara Romanowicz		
									Bruce Watson		
									Simon Conway Morris		
									Donald DePaolo		
									Gail Mahood		

1780	1800	1820	1840	1860	1880	1900	1920	1940	1960	1980	2000
									R. Damian Nance		
									Martha Withjack		
									T. Mark Harrison		
									Marcia McNutt		
									Edward Stolper		
									Mary Lou Zoback		
									Samuel Bowring		
									Maureen Chan		
									Michael Hochella Jr.		
									Teresa Jordan		
									Paul Olsen		
									Lynn Walter		
									Katherine Cashman		
									Stephen Marshak		
									Samuel Mukasa		
									Çelal Sengor		
									Kenneth Miller		
									Jane Selverstone		
									Richard Alley		
									Craig Bethke		
									Susan Brantley		
									Maria Zuber		
									Maureen Raymo		
									Joann Stock		
									Isabel Montanez		

BIBLIOGRAPHY

American Geophysical Union, Biographies, http://www.agu.org.

Boston University, Department of Earth Sciences website, (Raymo, Simpson) http://www.bu.edu/es.

Brown University, Department of Geological Sciences website, (Head and Tullis) http://www.geo.brown.edu.

California Institute of Technology, Department of Geological and Planetary Sciences website, (Albee, Anderson, Stolper, Taylor, Wyllie) http://www.gps.caltech.edu.

Cornell University, Department of Geological Sciences website, (Jordan and Turcotte) http://www.geology.cornell.edu.

Database Publishing Group. *American Men and Women of Science.* New Providence, N.J.: R.R. Bowker, 1983.

Debus, Allen G., ed. *World Who's Who in Science.* Chicago, Ill.: Marquis Who's Who, 1968.

Elliott, Clark A. *Biographical Dictionary of American Science, the 17th through the 19th Centuries.* Westport, Conn.: Greenwood Press, 1979.

Famous Earth Scientists, http://www.geology.about.com/cs/biographies.

Fenton, Carroll L., and Mildred A. Fenton. *Giants of Geology.* Garden City, N.Y.: Doubleday and Co., 1952.

Geological Society (of London). *Award Citations* and *Memorials,* http://www.geolsoc.org.uk.

Gillispie, Charles C., ed. *Dictionary of Scientific Biography.* New York: Charles Scribner's Sons, 1970.

Ginsburg, Robert, ed. "Rock Stars," *GSA Today,* Boulder, Colo.: 1998. http://gsahist.org/gsat/gsat.html.

Harvard University, Department of Geological Sciences website, (Hayes, Hoffman, Holland), http://www.harvard.edu.

Holmes, Fredric L., ed. *Dictionary of Scientific Biography, Supplement II.* New York: Charles Scribner's Sons, 1990.

Hust, Adele. *Who's Who in Frontier Science and Technology.* Chicago, Ill.: Marquis Who's Who, 1984–85.

Johns Hopkins University, Department of Earth and Planetary Sciences website, (Stanley, Veblen), http://www.jhu.edu.

Journal of the History of Earth Sciences Society, vol. 1–18, 1983–1999.

Kessler, James H., J.S. Kidd, Renee A. Kidd, and Katherine A. Morin. *Distinguished African American Scientists of the 20th Century.* Phoenix, Ariz.: Oryx Press, 1996.

Lamont-Doherty Earth Observatory website (Broecker, Olsen), http://www.ldeo.columbia.edu.

Louisiana State University, Department of Geological Sciences website, (Bouma) http://www.geol.lsu.edu.

Massachusetts Institute of Technology, Department of Earth, Atmospheric and Planetary Sciences website, (Bowring, Burchfiel, Molnar, Rizzoli, Zuber) http://www.mit.edu.

Memorials. Boulder, Colo.: Geological Society of America, 1969.

Memorial University, Department of Geological Sciences website, (Williams) http://www.esd.mun.ca.

Mineralogical Society of America, Award Citations, Washington, D.C.: *American Mineralogist* (vol. 1–86), 1915–2001.

Mineralogical Society of America, Memorials, Washington, D.C.: *American Mineralogist* (vol. 1–86), 1915–2001.

National Academy of Sciences, *National Academy Biographical Memoirs* (vol. 1–71). Washington, D.C. 1925–2001. http://www4.nationalacademies.org/nas/nashome.nsf/webLink/members.

Ogilvie, Marilyn B. *Women in Science, Antiquity through the 19th Century.* Cambridge, Mass.: Massachusetts Institute of Technology Press, 1988.

Ohio University, Department of Geological Sciences website, (Nance) http://www-as.phy.ohiou.edu/departments/geology.

Pennsylvania State University, Department of Geosciences website, (Alley, Brantley) http://www.geosc.psu.edu.

Princeton University, Department of Geosciences website, (Suppe) http://www.geoweb@princeton.edu.

Rensselaer Polytechnic Institute, Department of Earth and Environmental Sciences website, (Spear, Watson) http://www.rpi.edu/dept/geo.

Rice University, Department of Geology and Geophysics website, (Talwani) http://www.rice.edu.

Rutgers University, Department of Geological Sciences website, (Ashley, Kent, Miller, Withjack), http://www-rci.rutgers.edu/~geolweb/

Society of Exploration Geophysics, Virtual Museum: Biographies, http://www.seg.org/museum/VM/

Stanford University, Department of Geological and Environmental Sciences website, (Ernst, Mahood) http://www.pangea.stanford.edu.

State University of New York at Stony Brook, Department of Geosciences website, (Liebermann, Lindsley) http://www.sunysb.edu.

Texas A & M University, Department of Geological Sciences website, (Morse) http://geoweb.tamu.edu.

University of Arizona, Department of Geosciences website, (Melosh) http://www.geo.az.edu.

University of California at Berkeley, Department of Geology and Geophysics website, (Berry, Carmichael, Depaolo, Helgeson, Wenk), http://www.seismo.berkeley.edu.

University of California at Davis, Department of Geology website, (Montanez, Moores, Navrotsky), http://www.geology.ucdavis.edu.

University of California at Los Angeles, Department of Earth and Space Sciences website, (Harrison) http://www.igpp.ucla.edu.

University of California at San Diego, Scripps Institution of Oceanography website, (Craig, Revelle, Suess), http://www.sio.ucsd.edu.

University of California at Santa Barbara, Department of Geological Sciences website, (Atwater, Keller, Sylvester), http://www.geol.ucsb.edu.

University of California at Santa Cruz, Department of Earth Sciences website, (McNally) http://www.earthsci.ucsc.edu.

University of Chicago, Department of Geophysical Sciences website, (Smith) http://geosci.uchicago.edu.

University of Georgia, Department of Geology website, (Herz) http://www.gly.uga.edu.

University of Houston, Department of Geosciences website, (Burke) http://www.uh.edu.

University of Illinois at Urbana-Champaign, Department of Geology website, (Bethke, Klein, Marshak), http://www.geology.uiuc.edu.

University of Maryland, Department of Geology website, (Brown) http://www.geol.umd.edu.

University of Michigan, Department of Geological Sciences website, (Mukasa, Van der Voo, Walter) http://www.geo.lsa.umich.edu.

University of Miami, Division of Marine Geology and Geophysics website, (Rosendahl) http://www.rsmas.miami.edu.

University of Nebraska at Lincoln, Department of Geosciences website, (Grew) http://www.unl.edu/geology/geohome.html.

University of New Mexico, Department of Earth and Planetary Sciences website, (Selverstone) http://epswww.unm.edu.

University of Oklahoma, School of Geology and Geophysics website, (Gilbert, M.C.) http://geology.ou.edu.

University of Oregon, Department of Geological Sciences website, (Cashman) http://darkwing.uoregon.edu.

University of Tennessee, Department of Geological Sciences website, (Hatcher, McSween), http://geoweb.gg.utk.edu.

University of Utah, Department of Geology and Geophysics website, (Chan) http://www.mines.utah.edu.

University of Washington, Department of Geological Sciences website, (Porter) http://www.geology.washington.edu.

University of Wisconsin at Madison, Department of Geology and Geophysics website, (Valley) http://www.geology.uwisc.edu.

Veglahn, Nancy J. *Women Scientists*. New York: Facts On File, Inc., 1991.

Virginia Polytechnic Institute and State University, Department of Geological Sciences website, (Bodnar, Hochella), http://www.geol.vt.edu.

Yale University, Department of Geology and Geophysics website, (Berner, Rodgers, Skinner, Turekian) http://geology.yale.edu.

Yount, Lisa. *A to Z of Women in Science and Math*. New York: Facts On File, Inc., 1999.

Note: Page numbers in **boldface** indicate main topics. Page numbers in *italic* refer to illustrations.